Bioorganic Chemistry in Healthcare and Technology

NATO ASI Series

Advanced Science Institutes Series

A series presenting the results of activities sponsored by the NATO Science Committee, which aims at the dissemination of advanced scientific and technological knowledge, with a view to strengthening links between scientific communities.

The series is published by an international board of publishers in conjunction with the NATO Scientific Affairs Division

A	**Life Sciences**	Plenum Publishing Corporation
B	**Physics**	New York and London
C	**Mathematical and Physical Sciences**	Kluwer Academic Publishers
D	**Behavioral and Social Sciences**	Dordrecht, Boston, and London
E	**Applied Sciences**	
F	**Computer and Systems Sciences**	Springer-Verlag
G	**Ecological Sciences**	Berlin, Heidelberg, New York, London,
H	**Cell Biology**	Paris, Tokyo, Hong Kong, and Barcelona
I	**Global Environmental Change**	

Recent Volumes in this Series

Volume 202—Molecular Aspects of Monooxygenases and Bioactivation of Toxic Compounds
edited by Emel Arınç, John B. Schenkman, and Ernest Hodgson

Volume 203—From Pigments to Perception: Advances in Understanding Visual Processes
edited by Arne Valberg and Barry B. Lee

Volume 204—Role of Melatonin and Pineal Peptides in Neuroimmunomodulation
edited by Franco Fraschini and Russel J. Reiter

Volume 205—Developmental Patterning of the Vertebrate Limb
edited by J. Richard Hinchliffe, Juan M. Hurle, and Dennis Summerbell

Volume 206—Alcoholism: A Molecular Perspective
edited by T. Norman Palmer

Volume 207—Bioorganic Chemistry in Healthcare and Technology
edited by Upendra K. Pandit and Frank C. Alderweireldt

Volume 208—Vascular Endothelium: Physiological Basis of Clinical Problems
edited by John D. Catravas, Allan D. Callow, C. Norman Gillis, and Una S. Ryan

Series A: Life Sciences

Bioorganic Chemistry in Healthcare and Technology

Edited by

Upendra K. Pandit

University of Amsterdam
Amsterdam, The Netherlands

and

Frank C. Alderweireldt

University of Antwerp (RUCA)
Antwerp, Belgium

Plenum Press
New York and London
Published in cooperation with NATO Scientific Affairs Division

Proceedings of a NATO Advanced Research Workshop on
Bioorganic Chemistry in Healthcare and Technology,
held September 18–21, 1990,
in Houthalen-Helchteren, Belgium

Library of Congress Cataloging in Publication Data

NATO Advanced Research Workshop on Bioorganic Chemistry in Healthcare and Technology (1990: Houthalen-Helchteren, Belgium)
Bioorganic chemistry in healthcare and technology / edited by Upendra K. Pandit and Frank C. Alderweireldt.
p. cm.—(NATO ASI series. Series A, Life sciences; v. 207)
"Proceedings of a NATO Advanced Research Workshop on Bioorganic Chemistry in Healthcare and Technology, held September 18–21, 1990, in Houthalen-Helchteren, Belgium"—T.p. verso.
Includes bibliographical references and index.
ISBN 0-306-44007-5
1. Bioorganic chemistry—Congresses. 2. Clinical biochemistry—Congresses. 3. Biochemical engineering—Congresses. I. Pandit, Upendra K. (Upendra Kumar), 1930- . II. Alderweireldt, Frank C. III. Title. IV. Series.
[DNLM: 1. Biochemistry—congresses. 2. Catalysis—congresses. 3. Chemistry, Organic-congresses. 4. Drug Design—congresses. 5. Enzymes—metabolism—congresses. QU 4 N279b 1990]
QP550.N38 1990
574.19'2—dc20
DNLM/DLC 91-24099
for Library of Congress CIP

ISBN 0-306-44007-5

A Division of Plenum Publishing Corporation
233 Spring Street, New York, N.Y. 10013

Printed in the United States of America

PREFACE

In current thinking, Bioorganic Chemistry may be defined as the area of chemistry which lies in the border region between organic chemistry and biology and which describes and analyzes biological phenomena in terms of detailed molecular structures and molecular mechanisms. This molecular-level view of biological processes is not only essential to their fuller understanding but also serves as the platform for the application of the principles of such processes to areas of health-care and technology.

The objective of the ASI workshop on " Bioorganic Chemistry in Healthcare and Technology", held in the Hengelhoef Congress Centre in Houthalen-Helchteren, Belgium, from September 18-21, 1990, was to bring together most of the international experts in the field to discuss the current developments and new trends in bioorganic chemistry, especially in relation to the selected theme.

The book presents nineteen invited plenary and session lectures and eighteen posters. These cover areas of (i) molecular design of therapeutic and agronomical agents based upon mechanistic rationale or drug-receptor interactions, (ii) production of substances of commercial value via combined organic chemical and bio-chemical methodologies, (iii) fundamental studies on the molecular mechanisms of enzymes and (iv) the evolution of conceptually new molecular systems which are programmed to execute specific recognition and/or catalytic functions. An abstracted version of the plenary discussion held at the end of the workshop is also included. We feel confident that the subject matter of this book will be of interest to a broad group of chemists engaged in academic or industrial research.

The workshop and the Proceedings volume were made possible by a NATO grant from its Scientific Affairs Division and the financial support of the undermentioned industries.
Amylum (Aalst, Belgium), DSM (Geleen, The Netherlands), Glaxo S.p.A (Verona, Italy), International Biosynthetics (Rijswijk, The Netherlands), Janssen Pharmaceutica (Beerse, Belgium), Kontaktgruppe für Forschungsfragen: Ciba-Geigy A.G., F. Hoffmann-La Roche and Co., A.G., Lonza A.G., Sandoz A.G. (Basel, Switzerland), N.O.V.O. (Bagsvaerd, Denmark).

We gratefully acknowledge the valuable help of members of the organizing committee: Prof. J.P. Kutney (Vancouver, Canada), Prof. A. Marquet (Paris, France) and Prof. G. Wulff (Düsseldorf, Germany). Thanks are also due to Ms.E.H. Hoogeveen, Mr. J. Schrooten and Mr. J. Verreydt for their competent administrative and technical assistance.

U.K. Pandit, Amsterdam

F.C. Alderweireldt, Antwerp

CONTENTS

BIOGENERATION OF AROMAS: GAMMA AND DELTA LACTONES FROM C-6 TO C-12

Claudio Fuganti and Piero Grasselli

Dipartimento di Chimica del Politecnico
Centro CNR per la Chimica delle Sostanze Organiche Naturali
20133 Milano, Italy

Massimo Barbeni

San Giorgio Flavors, SPA, Via Fossata 114
10147 Torino, Italy

INTRODUCTION

The overall sensation perceived when food is consumed, defined by the word flavour, is due to the interaction of taste, odour, and textural feeling. Flavour can result from compounds that are divided in two broad classes: those responsible for taste and those responsible for odour, the latter often designated as aroma substances. Many chemically defined substancees have been identified in natural and processed food, [1] whose significant function is flavouring rather than nutrition. Several compounds are being currently used in the flavour industry, but some materials play a key role in specific formulations. This is the case of the C-6 -- C-12 gamma and delta lactones, important flavour components in fruits such as peach, apricot, strawberry, mango, coconut, milk products, and fermented foods.[2]

Margarine[3] producers started using lactones as flavouring agents in their products[3] following the observation that butter triglycerides contain a higher amount of hydroxy fatty acids than other bovine fats e.g., tallow; which is one of the reasons that the latter fats do not smell of butter. Fresh cream contains hardly any free lactones, but when the product is stored, traces of hydroxy fatty acids are split off by hydrolysis. Beta and delta homologues are present, but the delta hydroxy fatty acids result predominant. They are easily converted into lactones by ring closure. These materials are optically active and of (R) absolute configuration.[4]

Microbial reduction of gamma and delta keto acids to the corresponding (R) gamma and delta hydroxy derivatives was explored, in the expectation to develop by these means a practical synthesis of nature-identical, chiral lactones. Several microorganisms were found which were able to reduce the above substrates to (R) gamma and delta hydroxy acids, besides a few which yielded the (S) enantiomers.[5]

Bioorganic Chemistry in Healthcare and Technology, Edited by U.K. Pandit and F.C. Alderweireldt, Plenum Press, New York, 1991

Saccharomyces cerevisiae (baker's yeast) was chosen for the manufacture of (R) delta lactones from the corresponding delta keto acids. It has been reported that the industrial process involves runs on 30 m^3 aqueous suspension of 3.000 kg wet yeast, to which 60 Kg keto acid are added, to give, in 18-24 h, 45 Kg of delta dodecanolide of (R) absolute configuration, after extraction of the acidified solution and distillation.[6] By this procedure, which seems[7] one of the most significant applications of baker's yeast to the production of chiral compounds through transformation of non conventional substrates, tons of nature identical lactones have been produced over the years.

In the meantime analytical studies have indicated that gamma and delta lactones isolated from different natural sources showed much lower optical rotations with respect to the materials isolated from butter fat. Furthermore, delta dodecanolide from cocos meat showed (S) absolute configuration, whereas the accompanying C-10 and C-8 lower homologues contained an excess of the (R) enantiomers.[6] These observations, making weaker the link between chirality and naturality that had induced the production of (R) delta lactones as nature identical flavouring materials, contributed, at least in part, to the decision of stopping the manufacture of chiral lactones by bioreduction and to use the racemic materials instead. However, independently from that, new facts were emerging, which, together with a new definition of natural flavour, led eventually, to new methods of production of gamma and delta lactones.

NATURAL FLAVOURS

Until this century many natural flavour materials were obtained from animals and higher plants. However, supplies of many of these products have decreased as a result of a series of factors, including wild life protection and industrial growth. Thus, food supply started containing both naturally occurring and synthetic chemical compounds. Consumers and regulators became increasingly involved in and concerned about the source and the composition of food flavourings. Although no chemical distinction exists between a flavour compound synthesized by nature and the same compound prepared in the laboratory, many consumers perceive anything natural as healthful and anything synthetic as harmful. In general, legislators have recognized that many forms of enzymatic and thermal processing are needed to develop the flavours perceived to be natural by consumers in traditional food. Thus, recent rules [8] include amongst 'natural' any flavour material obtained from natural precursors through 'enzymolysis'. Since the last term is understood to cover not only hydrolysis but all the enzymatic processes, considerable efforts are being dedicated to the preparation of substantial quantities of flavouring agents through the action of isolated or whole-cell enzymic system(s) on advanced intermediates.[9] In this context the problem of the biogeneration of C-6 -- C-12 gamma and delta lactones was raised and we outline here some of the solutions that have been proposed uptil now.

BIOGENERATION OF LACTONES

Gamma and delta lactones are widely distributed in nature and their generation in fruits usually occurrs at the time of ripening, when catabolic processes are prevalent. However, the mechanism which activates their production is unknown. Simple structural considerations allow to envisage[10] for this class of compounds a derivation from fatty acids. However, the circumstance referred to above, namely, that both optical purity and absolute configuration can vary for identical lactones isolated from different sources supports the idea of the presence of different[11] biosynthetic pathways, involving either anabolic or degradative processes. Comparison of the structural formulas of the three C-18 unsaturated fatty acids (most abundant in plant glycerides) i.e. oleic **(1)**, linoleic **(2)** and linoleic **(3)** with those of the gamma and delta hydroxy acids **(4)-(9)**, - open forms of some relevant C-6 -- C-12 lactones - indicates C-2 degradation by beta oxidation of suitably hydroxylated derivatives of fatty acids as the metabolic link between the two classes of compounds (Scheme 1). The classical beta oxidation enzymes cannot directly metabolize some of the intermediates in the beta oxidation sequence of unsaturated fatty acids. However, the existence of isomerases which can convert the above 'anomalous' intermediates into the 'normal' beta oxidation intermediates has been demonstrated in many living organisms, including yeasts.[12]

(*R*) GAMMA DECANOLIDE FROM RICINOLEIC ACID

Indeed, (*R*) gamma decanolide **(14)** is at present manufactured via biodegradation of ricinoleic acid **(10)** by *Candida lipolytica*[13] and other microorganisms.[14] The process originates from the observation[15] that the former microorganism can perform beta oxidation of ricinoleic acid **(10)**, due to the presence of an isomerase at the level of C-12 intermediates (Scheme), with accumulation of the (4*R*) form of 4-hydroxy decanoic acid **(6)**. Besides the microorganisms which are able to perform the above biodegradation, few[16] were identified which produce, endogenously, components of the whole set of C-6 -- C-12 gamma and delta lactones, albeit in low quantities. Ricinoleic acid **(10)**, the major fatty acid component in the seed oil of castor beans, is formed from oleic acid by homoallylic hydroxylation.[17] Interestingly enough, the castor bean seed system elso hydroxylates other monoenoic acids, but the hydroxyl group is always inserted in the same position relative to the existing double bond. This capacity is present in plants of the genus *Lesquerella*[18] whose glycerides contain the hydroxy fatty acids **(25)** and **(26)**, containing the gamma hydroxy alkene structural unit present in ricinoleic acid **(10)**. Indeed, products **(25)** and **(26)** undergo beta oxidation in *Cladosporium suaveolens*,[14] affording (*R*) gamma octanolide **(24)** and (*R*) gamma dec-7-enolide **(27)**, respectively. Conversely, in fungal ergot oil, ricinoleic acid appears to be formed by a specific anaerobic hydration of linoleate. In the case of **(14)**, **(24)** and **(27)** the chirality present in the precursors is incorporated at position 4 of the derived lactones. Since, the most accessible hydroxy fatty acids useful for the bioproduction of lactones are racemic, it seemed useful in the light of the present discussion on chirality and naturality of lactones[11c] to study the mode of biodegradation of racemic ricinoleic acid and its analogs in which the gamma hydroxy alkene structural unit is shifted along the fatty acid framework.

To this end, the racemic C-14 -- C-19 hydroxy fatty acids **(15)-(20)**, all of which contain the structural unit (*Z*) $CH=CHCH_2CH(OH)R$, were prepared and fed to growing cultures of *C.suaveolens.* As expected, in agreement with C-2 degradation by beta oxidation, the C-14, C-17 and C-19 hydroxy acids **(15)**, **(18)** and **(20)** afforded C-8 and C-11 delta lactones, whereas the C-15, C-16 and C-18 acids **(16)**, **(17)** and **(19)** gave rise to C-7, C-8 and C-10 gamma lactones[19]. However, the absolute configuration of the prevalent enantiomer in the two series differd; the (*S*) lactones **(21)** and **(22)** and the (*R*) gamma analogues **(23)**, **(24)(14)** being obtained. Moreover, the optical purity (Table) is higher within the first set and decreases in each series on shortening the n-alkyl side chain. Furthermore, the *ee* values in entries 1 and 2 suggest the operation of kinetic resolution within the biodegradation.

The most relevant features of the above experiments are: i) the inversion of configuration, associated with the lactone ring size, as

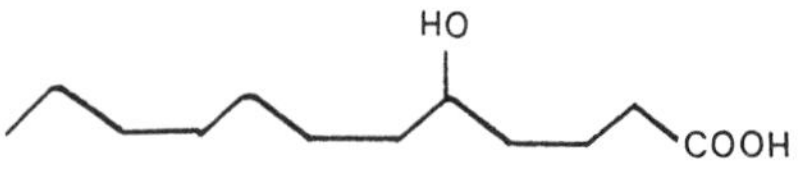

(4)

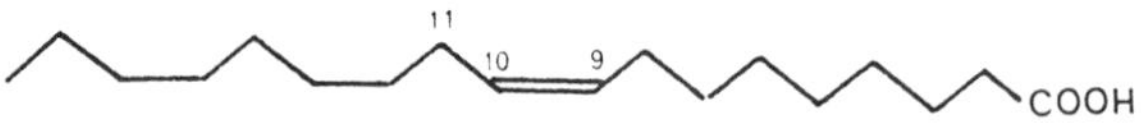

(1)

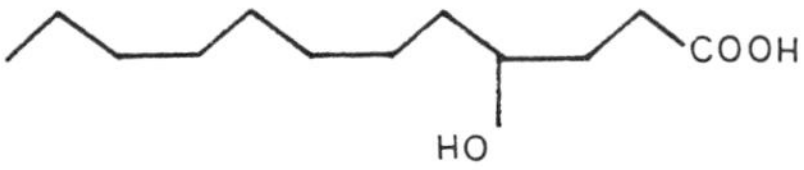

(5)

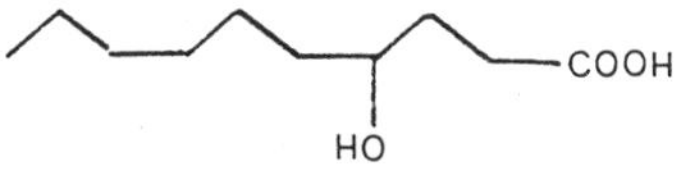

(6)

(2)

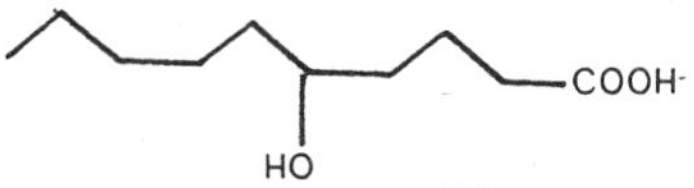

(7)

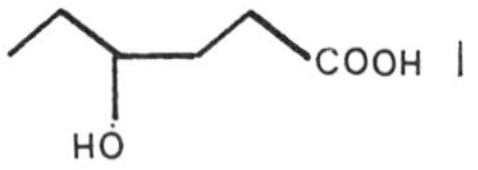

(8)

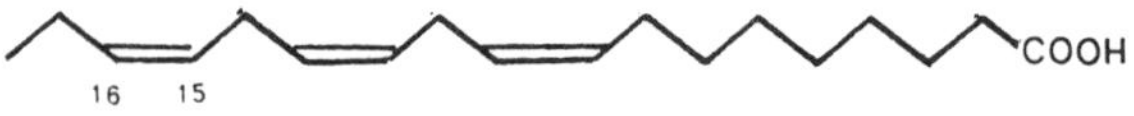

(3)

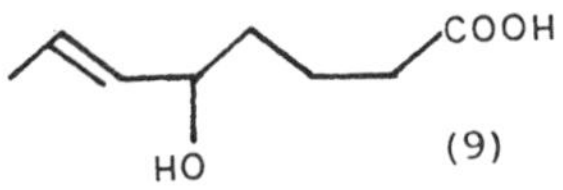

(9)

Scheme 1

TABLE 1

Enantiomeric Excess Values of Lactones **(21)-(24)** and **(14)**, Obtained from Racemic **(15)-(20)** in Growing Cultures of C.suaveolens

entry	precursor	lactone	R	ee values 24 h	48 h	120 h	198 h
1	**(15)**	(S)-**(21)**	n-C_3H_7	0.58	0.50	0.38	
2	**(16)**	(R)-**(23)**	n-C_3H_7		0.30	0.26	
3	**(17)**	(R)-**(24)**	n-C_4H_9			0.22	
4	**(18)**	(S)-**(22)**	n-C_6H_{13}		0.88		
5	**(19**	(R)-**(14)**	n-C_6H_{13}		0.54		
6	**(20)**	(S)-**(22)**	n-C_6H_{13}				0.8

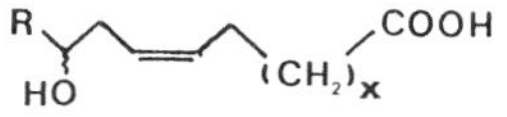

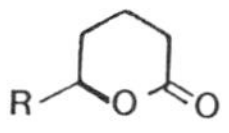

(15) R= n-C_3H_7; X= 5
(16) R= n-C_3H_7; X= 6
(17) R= n-C_4H_9; X= 6
(18) R= n-C_6H_{13}; X= 5
(19) R= n-C_6H_{13}; X= 6
(20) R= n-C_6H_{13}; X= 7

(21) R= n-C_3H_7
(22) R= n-C_6H_{13}

(23) R= n-C_3H_7
(24) R= n-C_4H_9
(14) R= n-C_6H_{13}

the consequence of the shift of the gamma hydroxy alkene moiety along the fatty acid chain; ii) the influence of the length of the R, n-alkyl side chain on the optical purity of the adducts, and iii) the remarkable susceptibility of the degradative enzyme (s) to the stereochemistry of the hydroxyl-bearing carbon atom located at a remote position, in the molecule, to the site of the initial beta oxidation.

In *C.suaveolens* the maximum yield of gamma and delta lactones from the hydroxy acids **(15)-(20)** is reached within 48-60 h after feeding. However, it was observed in these experiments that the gamma and delta hydroxy acids present in the fermentation medium are rapidly degraded soon after the peak concentration has been reached. In some instances, the concentration of the hydroxy acids is decreased by 50% in 4-6 h. Since this fact represents a serious practical drawback we decided to study the steric course of the degradation, in order to gain information on the phenomenon. We fed to growing cultures of *C.suaveolens* the racemic gamma and delta hydroxy decanoic acids **(6)** and **(7)**, as sodium salts (500 mg/100 ml, 2% Nutrient Merck), and determined the absolute configuration and the optical purity of the lactones, derived from the survived hydroxy acids. Under these conditions, *ca* 50% of the fed material is consumed in 48 h. However, the gamma and the delta decanolides obtained in these experiments were of apposite absolute configurations. From racemic **(6)** and **(7)**, *(R)* gamma decanolide **(14)** and *(S)* delta decanolide **(30)** were obtained, respectively. As expected for a kinetic resolution, the *ee* values of the survived materials increased as the degradation proceeded, changing from 0.33 to 0.62 for **(14)** and from 0.25 to 0.42 for **(30)** at 50% and 75% consumption. These results seem interesting, expecially when compared with those obtained in the degradation of racemic ricinoleic acid and its isomers in the same microorganism.[19] Indeed, the *(R)* enantiomer of ricinoleic acid is degraded more rapidly, as indicated by the formation from **(19)** of *(R)* gamma decanolide **(14)**, whose optical purity decreases as its generation proceeds. Conversely, on feeding 4-hydroxydecanoic acid **(6)**, a faster metabolism of the *(S)* enantiomer occurs, leaving unaltered the *(R)* enantiomer **(28)**, whose optical purity increases as the degradation proceeds. The reverse is true in stereochemical terms for the generation and metabolism of 5-hydroxydecanoic acid **(7)**, during which the (5*S*) enantiomer **(29)**, is left unaltered.

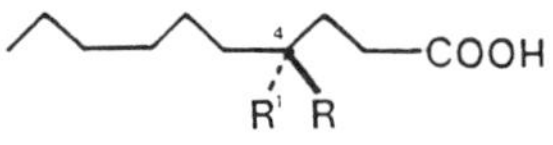

(6) R,R^1 = H, OH
(28) R= H; R^1 = OH

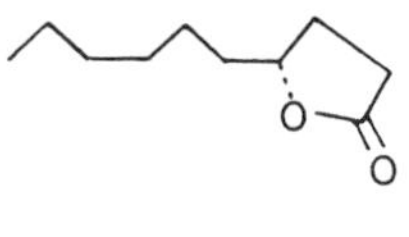

(14)

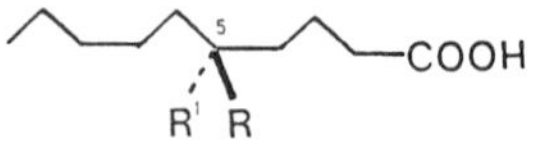

(7) R,R^1 = H, OH
(29) R= OH; R^1 = H

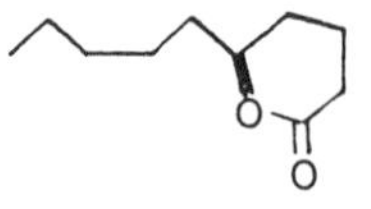

(30)

The above stereochemical course requires some comments. The breakdown of 4-hydroxydecanoic acid could follow a pathway similar to the one described in Scheme 3, proposed[21] for the conversion, of ricinoleate into acetyl CoA, in germinating castor bean seeds. It might well be that in both the proposed routes (beta oxidation to alpha-hydroxy octanoic acid or conversion to ketodecanoic acid, followed by beta oxidation[22]) there is a preference for the (4S) enantiomer.

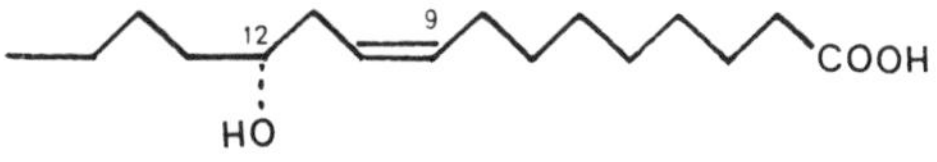

(25)

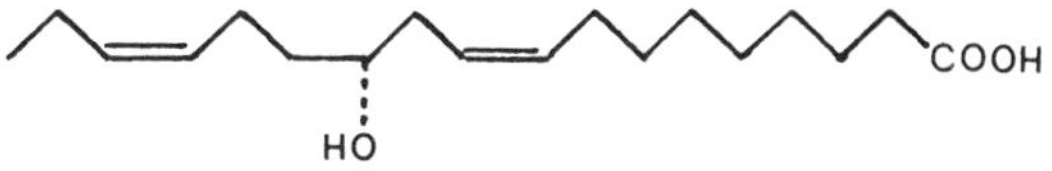

(26)

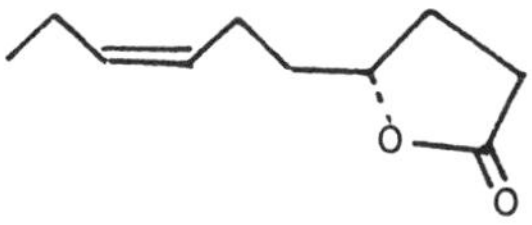

(27)

Conversely, 5-hydroxydecanoic acid seems structurally suited for a direct conversion through beta oxidation and involving 3-hydroxyoctanoyl CoA as an early intermediate, into 5 moles of acetyl CoA. The reported observation that the (R) enantiomer of 5-hydroxydecanoic acid is preferentially degraded contrasts with a straightforward degradation by beta oxidation, because it seems to require the (3S) configuration for the intermediate 3-hydroxyoctanoyl CoA.[23]

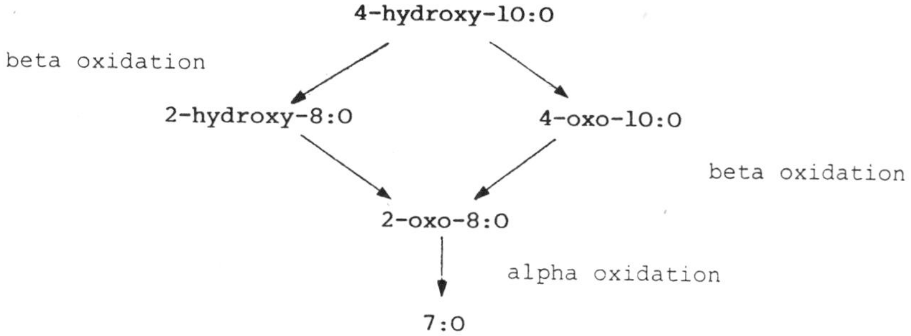

Scheme 2

(S) DELTA DECANOLIDE FROM LINOLEIC ACID 13-HYDROPEROXIDE

Lipoxygenase from many plant sources catalyzes the formation of (13S) 13-hydroperoxy-(9Z,11E)-octadienoic acid **(31)** and (13S) 13-hydroperoxy-(9Z,11E,15Z)-octadecatrienoic acid **(33)** from linoleic and linolenic acids **(2)** and **(3)**, respectively. This is the most common pathway of the enzymic enantiorelective introduction of an oxygen function in fatty acids which contain a 1-cis, 4-cis-pentadienyl system. The hydroperoxydes **(31)** and **(33)** play a key rôle in determining the organoleptic quality of natural and processed food. In a number of fruits and vegetables a hydroperoxide lyase occurs which fragments the 18-carbon chain leading to volatile C-6[24] and C-9 aldehydes. Alternatively, the hydroperoxides **(31)** and **(33)** can be transformed chemically or enzymatically into the carbinols **(32)** and **(34)**, or into the ketodienoic acids, which, in turn, are reduceable to **(32)** and **(34)** and/or their enantiomers, depending upon the nature of the reducing agents. Product **(32)** occurs in Mannina emarginata, its enantiomer in Coriaria nepalensis ((-)-coriolic acid), whereas the racemic modification is present in Absinthium oils.[25] Structural considerations suggested a biosynthetic relationship between delta decanolide and **(31)** and/or the products in its cascade. Thus, **(31)** obtained by the action of molecular oxygen onto linoleic acid in the presence of soya bean lipoxygenase (EC. 1.13.11.12) was incubated (25 mg/100 ml) with growing cultures of C. suaveolens. After 24 h delta decanolide was obtained in 15 % yield. Much higher incorporation (40%) was observed on feeding product **(32)**, prepared from **(31)** by sodium cysteinate reduction. The samples of the C-10 lactone isolated in the two experiments possess the (S) absolute configuration depicted in **(30)**. The ee values were ca 0.8, identical to those of the precursors determined through NMR and HPLC methods.[26]

Interestingly enough, the racemic modification of **(32)**, prepared according to known procedures[27], afforded in C.suaveolens (S) delta decanolide **(30)**. Its ee value was 0.8-0.85 and seemed independent a the incubation time. The formation of the (S) enantiomeric form of delta decanolide from the racemic modification of the precursor is interesting in the context the present debate on the relationship between naturality and chirality[11c] of flavouring substances, especially when one considers that delta decanolide has been found in nature in the (R) absolute configuration.[4] However, the above result fits into the rule drawn from the previous experiments,[19] on the mode of biodegradation of hydroxy fatty acids; viz. the formation of (S) delta lactones from hydroxy derivatives in which the oxygenated carbon occurs in odd position in the chain and of (R) gamma lactones from precursors in which this is located in even position, irrespective of the position and the number of the double bonds.

Configuration of this trend arises upon feeding **(35)**, the C-19 analog of **(32)**, obtained by sodium cysteinate reduction of the hydroperoxide prepared by lipoxygenation of the C-19 analog of linoleic acid. In this case, upon feeding the (S) precursor **(35)**, gamma nonanolide was obtained, which, on comparison with an authentic sample[28] was shown to contain 65-70% of the (R) enantiomer **(38)**hus, biodegradation of (S) **35)**to gamma nonanolide formally involves, at some stage of the sequence, inversion of configuration at the hydroxyl bearing carbon atom. Confirmation of the stereochemistry of the degradation arose from feeding the racemic material **(36)**, which was prepared from **(35)** by oxidatione to ketodienoic acid, followed by reduction. In this case, (R) gamma nonanolide **(38)** was obtained, with 0.7 ee. Any mechanistic hypothesis on the conversion of (S) **(35)** into (R) **(38)** requires determination of the fate of the hydrogen atom at position 14 of the precursors. Accordingly, product **(37)**, bearing a deuterium at the hydroxyl-bearing carbon atom, was prepared. The results of the feeding experiments with the latter material are reported in the context of stereochemical studies on the above bioconversions.

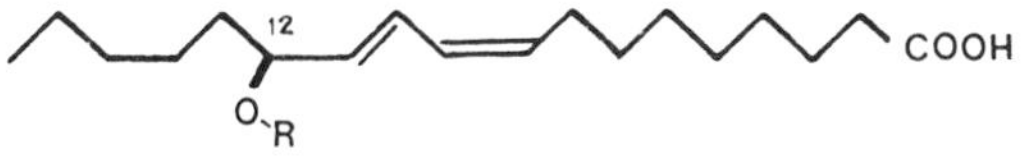

(31) R= OH

(32) R= H

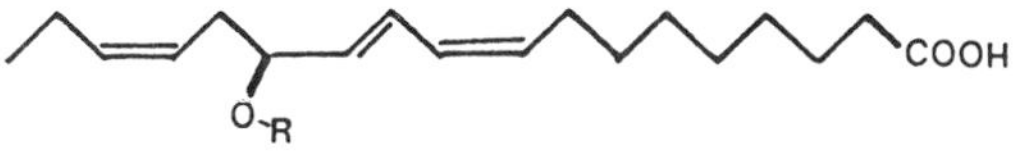

(33) R= OH

(34) R= H

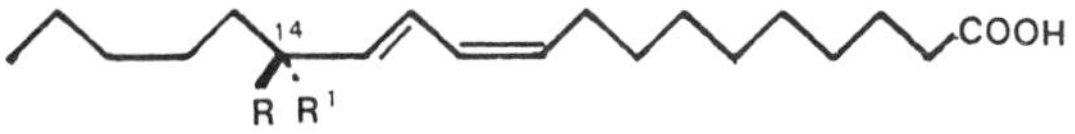

(35) R= OH; R^1 = H

(36) R,R^1 = H, OH

(37) R,R^1 = ^2H, OH

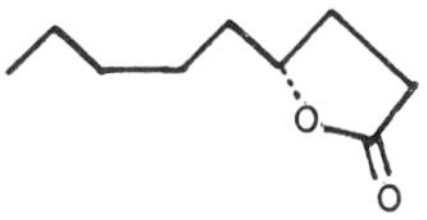

(38)

(R) GAMMA AND (S) DELTA LACTONES FROM THE PHOTOOXIDATION PRODUCTS OF C-18 UNSATURATED FATTY ACIDS

Photo-and autooxidations are the most common ways of introducing oxygen functionalities into the unsaturated fatty acid framework.[29] The materials obtained by these methods are racemic. The former process can be easily performed on a large scale, using a natural photoactivator; yielding, on sodium cysteinate reduction of the intermediate hydroperoxides, hydroxy fatty acids, which formally meet the requisites of naturality. The ratio of the positional isomers does not depend upon the mode of photooxidation. Oleic acid **(1)** giving a 1:1 mixture of **(39)** and **(40)**, whereas linoleic acid **(2)** affords **(43)-(46)** in ca a 2:1:1:1:2 ratio. On feeding **(39)** and **(40)**, to C.suaveolens the formation of gamma dodecanolide was observed; besides a small amount of a product which was assigned, on the basis of MS studies, the structure indicated in **(42)**. The sample of gamma dodecanolide was shown, by chiral analysis, to contain over 70% of the (R) enantiomer **(41)**. The biodegradation of the mixture **(43)-(46)** by the above mentioned microorgnism afford gamma and delta lactones **(30)** and **(47)-(49)**. However, the ratio amongst these substance changes significantly in time, due to differences in kinetics of fomation and degradation of the intermediate gamma and delta hydroxy acids.

The absolute configuration of the chain-unsatured gamma lactones **(47)** and **(48)** was determined through conversion, by hydrogenation, into (R) gamma decanolide **(14)** and gamma dodecanolide **(41)**, respectively. The ee values of the above mentioned materials, obtained at different incubation times, decreased from 0.8 at 24 h to 0.4-0.35 at 120 h.

Much more complex was the mixture of lactones arising in C.suaveolens from the hydroxy acids prepared from linolenic acid **(3)** by the photooxidation/reduction sequence. A whole series of C-6 -- C-12 lactones, containing unsaturated side chains in most cases was obtained.

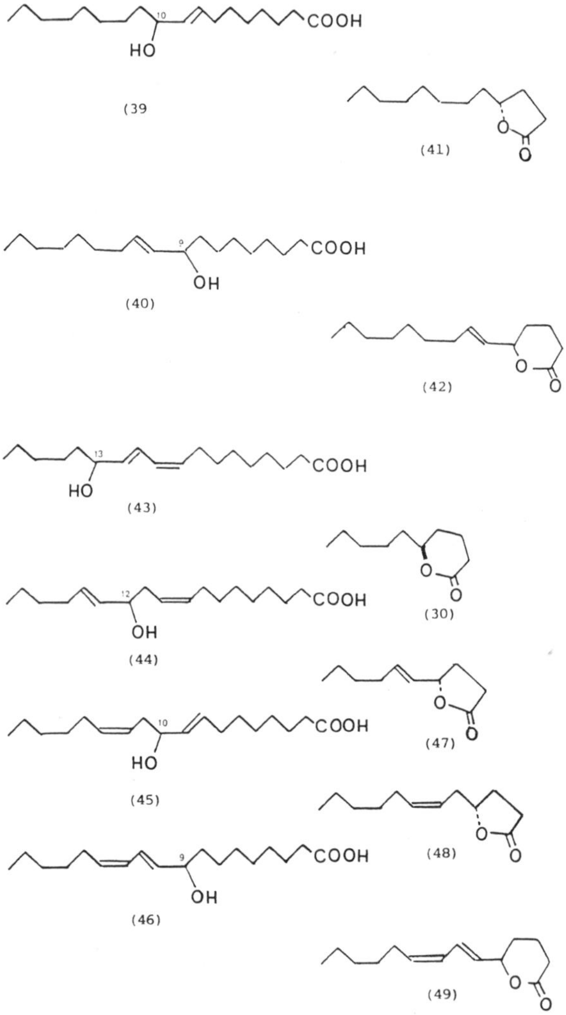

COOH
HO
(39
(41)
O
COOH
OH
(40)
(42)
COOH
HO
(43)
(30)
COOH
OH
(44)
(47)
COOH
HO
(45)
(48)
COOH
OH
(46)
(49)

Scheme 2

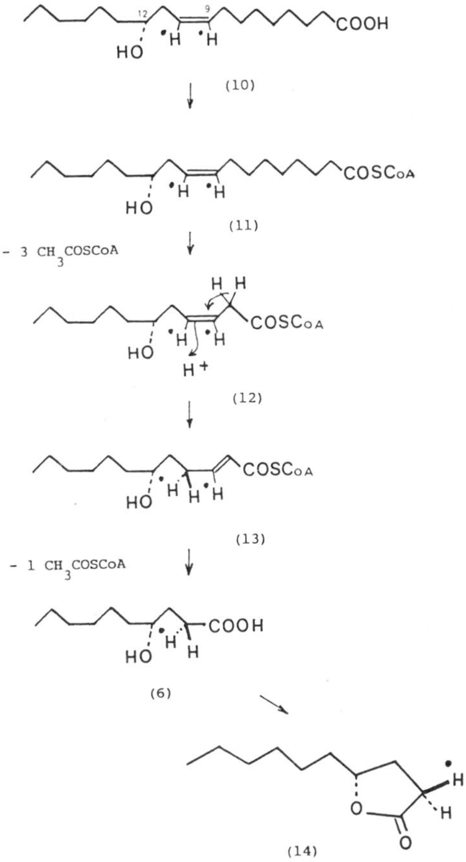

•H = ^{2}H

Scheme 3

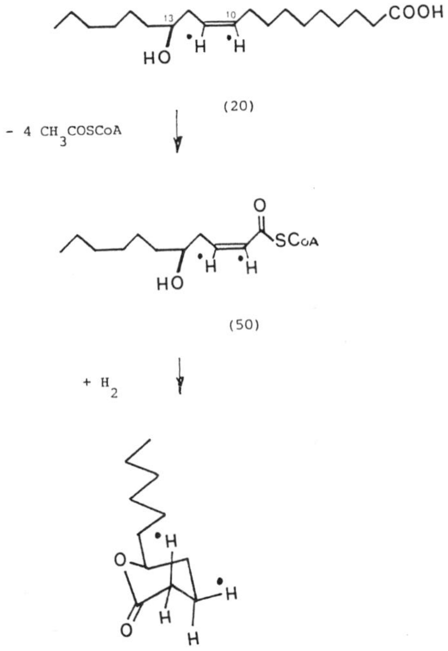

$\bullet H = {}^2H$

Scheme 4

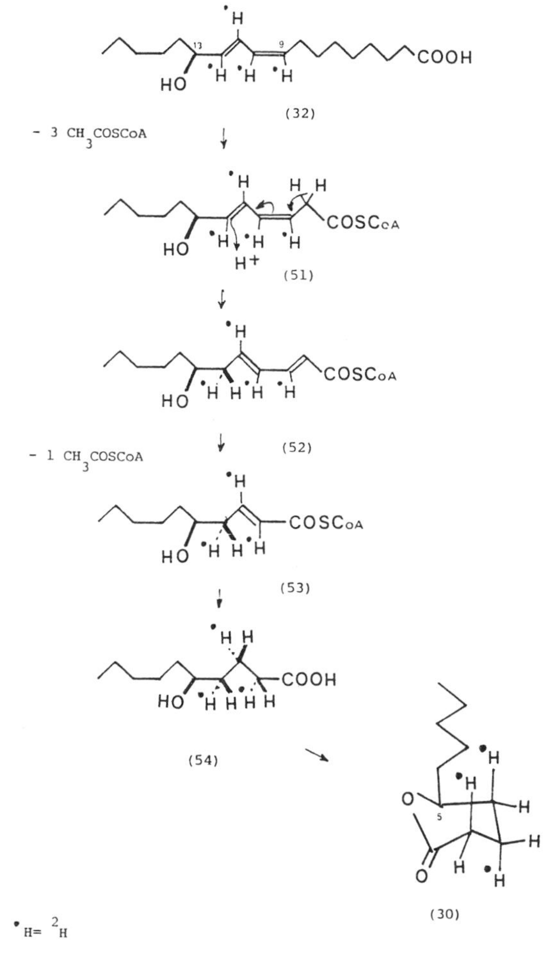

H
HO
COOH
(32)
- 3 CH3COSCoA
H H
COSCoA
H+
(51)
(52)
- 1 CH3COSCoA
(53)
(54)
O
(30)
•H= 2H

Scheme 5

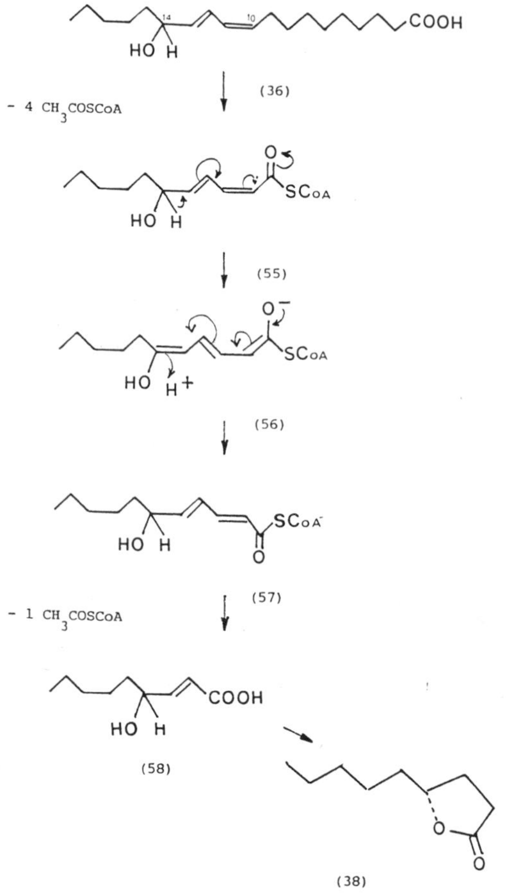
14
10
COOH
HO H
(36)
- 4 CH3COSCoA
O
SCoA
HO H
(55)
O
SCoA
HO H+
(56)
SCoA
HO H
O
(57)
- 1 CH3COSCoA
COOH
HO H
(58)
O
O
(38)

Scheme 6

STEREOCHEMISTRY OF THE BIOGENERATION OF LACTONES

The degradation of fatty acids takes place by a repeating series of steps known as the fatty acid cycle, or the beta oxidation pathway. To enter the fatty acid cycle, a fatty acid has to be coupled to coenzyme A. The first step is dehydrogenation. FAD accepts ($H^- + H^+$) from the alpha and the beta carbons of the fatty acyl unit. The second step is hydration. The third step is another dehydrogenation - a loss of (H^-+H^+). This oxidizes the secondary alcohol to a keto group. The fourth step breaks the bond between the alpha and the beta carbons, releasing one unit of acetyl coenzyme A and the original acyl moiety shortened by two carbons, which is ready to go through the cycle of steps again. Application of the above sequence to the unsaturated hydroxy acids **(10)**, **(20)**, **(32)** and **(36)**, i.e. precursors in C.suaveolens of the lactones **(14)**, **(22)**, **(30)** and **(38)**, respectively, should lead to the acyl intermediates **(12)**, **(50)**, **(51)** and **(55)**. These are indicated in Schemes, 3, 4, 5 and 6 together with the presumed subsequent intermediates on the pathways to the lactones.

We have been studying the steric course of the operations occurring at the double bonds of the above mentioned intermediates, during their metabolism, by means of feeding experiments of deuterated hydroxy acids and determining the stereochemistry of the deuterium retained in the final lactones, via an accurate NMR study. Thus, $|9,10-{}^2H_2|$ **(10)**, $|10,11-{}^2H_2|$ **(20)**, $|9,10,11,12-{}^2H_4|$ **(32)** and $|14-{}^2H_1|$ **(36)** were prepared and fed to C.suaveolens. (S) **(20)** and (S) **(30)**, obtained in these experiments, contained deuterium label located at positions and with the stereochemistry depicted in Schemes 3-5.

The labelling pattern of **(10)** indicates that protonation of the double bond of the (R) intermediate **(12)** occurs on the si face during the conversion into **(13)**, the penultimate product on the pathway from **(10)** to **(14)**. The same stereochemical course is followed during the protonation of the (S) intermediate **(51)**, formed from **(32)** (Schemes). The process results in conjugation of the double bonds and affects the alpha-trans double bond stereochemistry necessary for the subsequent beta oxidation of **(52)** to **(53)**. Finally, the saturation of the cis and trans double bonds in intermediates **(50)** and **(53)**, precursors of **(22)** and **(30)**, respectively, seems to take place by a formal syn hydrogen addition, localing the alpha hydrogen atoms in **(22)** and **(30)** in the same absolute stereochemistry.

Studies on the mode of generation of **(38)** are still in progress. However, experiments with **(36)**, possessing a deuterium atom at position 14, indice that nearly 40% of the deuterium is lost during the formation of (R) nonanolide **(38)** of ca 0.4 ee. A tentative explanation might be the following. The conversion of **(36)** into **(39)** could require the intermediacy of products **(55)** - **(58)**. The (S) enantiomeric form of **(55)** is relayed through the sequence via the intermediate **(56)**, with consequent loss of the hydrogen atom originally present at position 14 in the precursor. The deuterium atom removed from the (R) form of **(55)** is presumably given back during the protonation of **(56)**, once the double bond stereochemistry has been established. However, further experiments are required, to confirm there details.

CONCLUSIONS

The experiments illustrated in Schemes 3-5 indicate that in the biogeneration of enantiomeric gamma and delta lactones from structurally different hydroxy alkene fatty acids some analogous operations at the double bonds takes place with the same absolute stereochemistry. This circumstance might present a biosynthetic link which unifies the array of enantiomeric natural C-6 -- C-12 lactones accessible in C.suaveolens from advanced precursors.

ACKNOWLEDGMENTS

The financial support of Piano Finalizzato CNR Chimica Fine e Secondaria II is acknowledged.

REFERENCES

1. F. Lendl and E. Schleicher, Angew. Chem. Int. Ed. Engl., 29:565 (1990)
2. H. D. Belitz and W. Grosch, "Food Chemistry," Springer Verlag, Berlin, (1987), p 270; J. A. Maja, CRC Crit. Rev. in Food Sci. Nutr., 10:1 (1976)
3. J. Bolding, P. H. Begeman, A. P. De Jonge, and R. S. Taylor, Rev. Fr. Corps Grâs, 13:235, 327 (1977)
4. J. Bolding and R. J. Taylor, Nature, 194:909 (1962)
5. G. Muys Tuynenburg, B. Van der Ven, and A. P. De Jonge, Appl. Microbiol., 11:389 (1963); A. Franke, Biochem. J., 95:633 (1965)
6. J. G. Keppler, J. Amer. Oil Chem. Soc., 54:474 (1977)
7. S. Servi, Synthesis 1 (1990)
8. U. S. Code of Federal Regulations, 21:101.22.a.3
9. "Biogeneration of Aromas," T. H. Parliment and R. Croteau, eds., ACS Symposium Series 317, American Chemical Society, Washington, DC, (1986)
10. R. Tressl, in "Biogeneration of Aromas," T. H. Parliment and R. Croteau, eds., ACS Symposium Series 317, American Chemical Society, Washington, DC, (1986), p 114
11. a) A. Mosandl, C. Günther, M. Gessner, W. Deger, G. Singer, and G. Heusinger, in "Bioflavour'87," P. Scheier, ed., de Gruyter, Berlin, (1988), p 55; b) V. Schuring, ibidem, p 35; c) G. Nitz, H. Kollmansberger, and F. Drawert, Chem. Mikrobiol. Technol. Lebensm., 12:105 (1989)
12. T. Galliard, in "Recent Advances in the Chemistry and Biochemistry of Plant Lipids," T. Galliard and E. I. Mercer, eds., Academic Press, London, (1975), p 318
13. U. S. Patent 4,560,656; Chem. Abstr., 99:4080t (1983)
14. Eur. Pat. 0 258 993; Ital. Appl. 67742
15. S. Okui, M. Uchiyama, M.Mizugaki, and A. Sugawara, Biochim. Biophys. Acta, 70:346 (1963); S. Okui, M. Uchiyama, and M. Mizigaki, J. Biochem., 54:536 (1963)
16. S. Takahara, K. Fujiwara, Ishizaka, J. Mizutani, and Y. Obata, Agric. Biol. Chem., 36:2585 (1972); F. Drawert, R. G. Berger, and K. Neuhauser, Chem. Mikrobiol. Technol. Lebensm., 8:91 (1983), G. -F. Kapfer, R. G. Berger, and F. Drawert, Biotechnol. Lett., 11:561 (1989); J. Sarris and A. Latrasse, Agric. Biol. Chem., 49:3227 (1985)
17. D. Howling, L. J. Morris, M. I. Gurr, and A. J. James, Biochim. Biophys. Acta, 260:10 (1972)

18. F. D. Gunstone, in "The Lipid Handbook," F. D. Gunstone, L. J. Harwood, and F. B. Padley, eds., Chapman and Hall, London, (1986), p 453
19. R. Cardillo, G. Fronza, C. Fuganti, P. Grasselli, V. Nepoti, M. Barbeni, P. F. Guarda, J. Org. Chem., 54:4979 (1989)
20. In shaken cultures of C.suaveolens the maximum concentration of 4-hydroxydecanoic acid derived from the biodegradation of ricinoleic acid is reached in 48 h and corresponds to 40% molar yield
21. D. Dutton and P. K. Stumpf, Archs. Biochem. Biophys., 142:48 (1971)
22. Feeding experiments with $[4-^2H_1]$ 4-hydroxydecanoic acid indicate no loss of deuterium in the recovered lactone, thus suggesting that if oxidation to 4-oxodecanoic acid occurs, this is irreversible (or reversible with retention of the label)
23. W. Stoffel and H. Ceasar, Hoppe Seyler's Z. Physiol. Chem., 342:76 (1965)
24. H. Hatavara, T. Kajiwara, and J. Sekiya, in "Biogeneration of Aromas," T. H. Parliment and R. Croteau, eds., ACS Symposium Series 317, Washington, DC, (1986), p 167
25. W. H. Tallent, J. Hains, I. A. Wolff, and R. E. Lundin, Tetrahedron Letters, 4324 (1966); T. Kato, Y. Yamaguchi, T. Hirano, T. Yokoyama, T. Ugehara, T. Namai, S. Yamada, and N. Harada, Chem. Lett., 409 (1984)
26. C. P. A. Van Os, M. Veute, and J. F. G. Vliegenhart, Biochim. Biophys. Acta, 574:103 (1979); J. C. Andre and M. O. Funk, Anal. Biochem., 158:316 (1986)
27. A. V. Rama Rao, E. R. Reddy, G. V. M. Sharma, P.Yadagiri, and J. S. Yadav, J. Org. Chem., 51:4158 (1986)
28. M. Utaka, H. Watabu, and A. Takeda, J. Org. Chem., 52:4363 (1987)
29. F. D. Gunstone, in "The Lipid Handbook," F. D. Gunstone, L. J. Harwood, and F. B. Padley, eds., Chapman and Hall, London, (1986), p 453
30. Eur. Pat. Appl. 90402217.5
31. R. Cardillo, G. Fronza, C. Fuganti, P. Grasselli, A. Mele, A. Miele G. Allegrone, and M. Barbeni, in press

CATALYTIC ANTIBODIES

Stephen J. Benkovic

The Pennsylvania State University, Department of Chemistry
College Station, Texas 77843-3255, U.S.A.

Summary

The field of catalytic antibodies has made remarkable progress since its inception in 1986/87. The earliest success in the generation of catalytic antibodies were the hydrolyses of esters and carbonates[1,2]. The first nonhydrolytic antibody catalyzed reaction was the formation of a lactone from a hydroxy ester in which one of the hallmarks of enzymic catalysis was also demonstrated. This antibody catalyzed an enantioselective transformation of a racemic substrate to a single enantiomeric product[3].

The fruition of these experiments can be attributed to three developments : i) the increase in our understanding of organic reaction mechanism through advances in physical organic chemistry that led to accurate postulates of the structure of transition-states and high energy reaction intermediates ; ii) the development of monoclonal antibody technology to produce large amounts of a single antibody from a clonal cell ; and iii) the acceptance of transition-state stabilization as a key tenant of catalysis in that catalysts such as enzymes promote reactions by possessing more favorable interactions with the transition-state than with the respective ground states leading to a lowering of the overall free energy of activation[4].

With regard to antibodies one then can imagine : i) inducing an immunological response to a hapten that resembles the transition state or a high energy reaction intermediate for a given reaction ; ii) screening the monoclonal antibodies then for binding of the transition state analog and (iii) finally using those antibodies which satisfy that binding screen as potential catalysts of the reaction being investigated.

Catalytic antibodies now have been reported that catalyze bimolecular amide bond formation[5], stereospecific Claisen rearrangement[6], photochemical thymine dimer cleavage[7], hydrolysis of a p-nitroanilide[8], peptide cleavage[9] and a Diels Alder reaction[10].

It is clear that transition state stabilization alone is insufficient to achieve reaction rates that rival those of enzymes. Many of the existing antibody catalysts increase the rates of reaction approximately 1,000 to 10,000 fold over that of the corresponding spontaneous reaction, although there are a few examples of rate enhancements of ca. one million fold.

Enzymes have at their active sites functional groups that add or substract protons or act as nucleophiles. In the case of some enzymes, there are additional biological cofactors to promote a given reaction. Consequently, a number of genetic as well as chemical modifications have been applied to the binding sites of antibodies in order to improve their reactivity[11].

The genetic approaches have featured protein engineering in order to introduce amino acids capable, for example, of binding metal ions[12] ; the chemical methods have introduced potential acid-base catalysts through the use of cleavable affinity labelling agents[13]. In addition, the potential for finding catalytic antibodies has been enormously expanded by the development of methods to express the

Bioorganic Chemistry in Healthcare and Technology, Edited by U.K. Pandit and F.C. Alderweireldt, Plenum Press, New York, 1991

immunological repertoire in E. colli so that large numbers of antibody clones (up to one million) can be screened for catalytic activity[14].

The antibody that catalyzes the hydrolysis of a p-nitroanilide is unusual in that its rate of reaction over background is enhanced ca. one million fold. Recent investigations of the mechanism used by this antibody, which catalyzes the hydrolysis of an aromatic amide as well as the corresponding p-nitrophenyl ester has revealed that this catalyst operates by a multistep kinetic sequence[15]. The structures of the substrates and transition state analog are numbered **1-4** in Fig. 1.

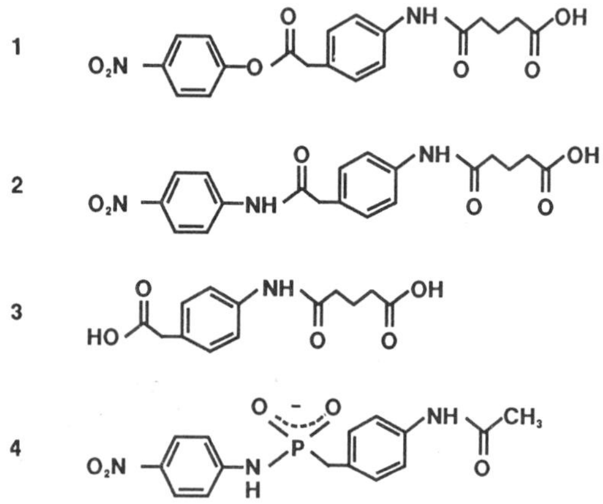

Figure 1

The antibody-catalyzed hydrolysis of the p-nitrophenyl ester **1** and p-nitroanilide **2** have been examined by both pre- and steady-state kinetic techniques. The data have implicated a reaction mechanism that features the formation of a putative acyl intermediate which deacylates through hydroxide ion attack.

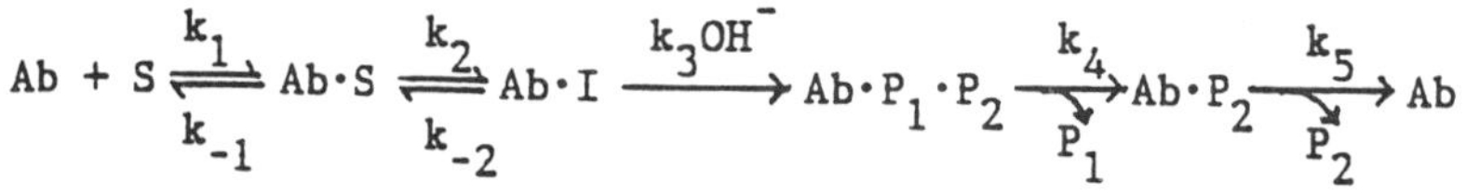

Scheme 1

There are several striking features of this scheme. One is the kinetic consequence of the product complex Ab•P_2 which in the case of the p-nitrophenyl ester controls the turnover of the antibody. A second is the fact that in the formation of the putative acyl antibody intermediate neither the p-nitroaniline nor p-nitrophenol are readily lost from that species. Consequently, unlike the case for serine esterases the acyl antibody intermediate does not accumulate since the close proximity of the bound leaving group reverses the reaction. A third feature is that the breakdown of the putative acyl antibody intermediate is through hydroxide ion attack rather than internal general-acid-base catalysis as is found with the esterase enzymes. Nevertheless, this antibody is within a factor of two hundred (k_{cat}) as effective as chymotrypsin against these substrates (the comparison for the antibody is at pH 9 that of chymotrypsin at pH 7).

Enzymes also hydrolyze peptide substrates by using metal ions, such as carboxypeptidase. By examining the structural motif of a number of enzymes that bind metal ions such as Zn^{2+}, an antibody has been constructed using genetic methods to possess suitable ligands for binding Zn^{2+} and Cu^{2+} metal ions[16]. This antibody indeed shows the same specificity towards those metals as does the carbonic anhydrase model on which it was based. The original hapten was a fluorescein which is still bound. This development represents then the first example of an metallo antibody and presages the creation of antibodies that can act as catalysts for a variety of reactions where the close proximity of metal ion to substrate would be most useful.

References

1. Tramontano, A., Janda, K.D., Lerner, R.A., Science (1986) 234, 1566.
2. Pollack, S.J., Jacob, J.W., Schultz, P.G., Science (1986) 234, 1570.
3. Napper, A.D., Benkovic, S.J., Tramontano, A., Lerner, R.A., Science (1987) 237, 1041.
4. Pauling, L., Am. Sci. (1988) 36, 51.
5. Benkovic, S.J., Napper, A.D., Lerner, R.A., Proceedings Natl. Acad. of Science USA (1988) 85, 5355.
6. Jackson, D.Y., Jacobs, J.W., Sugasawara, R., Reich, S.H., Bartlett, P.A., Schultz, P.G., J. Am. Chem. Soc. (1988) 110, 4841.
7. Cochran, A., Sugasawara, R., Schultz, P.G., J. Am. Chem. Soc. (1988) 110, 7888.
8. Janda, K.D., Schloeder, D., Benkovic, S.J., Lerner, R.A., Science (1988) 241, 1188.
9. Iverson, B.L., Lerner, R.A., Science (1989) 234, 1184.
10. Hilvert, D., Hill, K.W., Nared, K.D., Auditor, M.-T.M., J. Am. Chem. Soc. (1989) 111, 9261.
11. Schultz, P.G., Lerner, R.A., Benkovic, S.J., Chemical & Engineering News Vol. May 28, 1990, pg. 26.
12. Iverson, B.L., Iverson, S.A., Roberts, V.A., Getzoff, E.D., Tainer, J.A., Benkovic, S.J., Lerner R.A., Science (1990) 249, 659.
13. Pollack, S., Nakayama, G. Schultz, P.G., Chem. Eng. News (1990) May 28, 26.
14. Huse, W.D., Sastry, L. Iverson, S.A., Kang, A.S., Alting-Mees, M., Burton, D.R., Benkovic, S.J., Lerner, R.A., Science (1989) 246, 1275.
15. Benkovic, S.J., Adams, J.A., Borders, Jr., C.L., Janda, K.D., Lerner, R.A., Science (in press).
16. Roberts, V.A., Iverson, B.L., Iverson, S.A., Benkovic, S.J., Lerner, R.A., Getzoff, E.D., Proceedings Natl. Acad. Science USA (1990) 87, 6654.

INHIBITION OF STEROL BIOSYNTHESIS BY ANALOGUES OF CARBENIUM ION INTERMEDIATES

Maryse Taton, Pierre Benveniste and Alain Rahier

Institut de Biologie Moléculaire des Plantes du C.N.R.S.
Département d'Enzymologie Cellulaire et Moléculaire
28, rue Goethe, 67083 - Strasbourg Cédex, France

Sterols play at least three basic roles in living cells. First of all they are membrane components and as such they accumulate in the plasma membrane where they can regulate membrane fluidity. This role may be played by the major fraction of sterols ("bulk" sterols) |1,2|. In addition to this structural role, a small fraction of sterols may have a regulatory ("sparking") function. It has been shown that in GL7 mutants of *Saccharomyces cerevisiae* which are auxotrophic to sterols, trace amounts of ergosterol in the medium could reactivate cell division and correlatively stimulate the phosphoinositide cycle |3|. Finally sterols are precursors of compounds having a high physiological activity such as the brassinosteroīds which may be considered as a new class of plant hormones |4|. Therefore the inhibition of sterol biosynthesis may have important physiological consequences and in some circumstances lead to the death of the treated cells or organisms. In this context it is worth noting that several groups of fungicides used in agriculture and in medicine (Table 1) have been recognized to be potent ergosterol biosynthesis inhibitors |5|. It has been even suggested that the inhibition of sterol biosynthesis could be responsible for the fungicidal properties of these molecules |6|. Therefore the biosynthesis of sterols appeared to be a good target for molecules having biocidal (fungicidal and eventually herbicidal) properties. To be used to agricultural or medicinal purposes, such molecules should present two important properties : they should be potent and selective. For example, selective fungicides should control the pathogen without affecting the host. If the putative fungicide is a sterol biosynthesis inhibitor (SBI) this could mean that the molecule would interact much more tightly with its target enzyme in the pathogen than with the homologous enzyme in the host (plant or animal). In addition the

Table 1. Main SBIs used in plant protection and medecine |5|

Azoles, pyrimidines pyridines, etc...	Fungicides Plant growth regulators herbicides ?
Morpholines, piperidines	Fungicides used in Agriculture
Allylamines	Fungicides used in Medicine
Mevilonine, compactine	Hypocholesterolemic drugs

Bioorganic Chemistry in Healthcare and Technology, Edited by U.K. Pandit and F.C. Alderweireldt, Plenum Press, New York, 1991

molecule should not have secondary targets in the host. However it has been reported that some side effects could have beneficial effects as for instance the inhibition by triazole SBIs of absicic catabolism in the plant host |7| and of steroïd hormone biosynthesis in mammalians |8|. For all these reasons, any rational design of inhibitors should take into account :
i) particular features of sterol biosynthesis in plants, animals and pathogenic fungi ; ii) catalytic mechanisms of the target enzymes in the pathogen and in the host. In the first part of this paper we will present a comparative pathway leading to sterol biosynthesis in eukaryotic organisms, then we will discuss some reaction mechanisms involved in several sterol biosynthesis enzymes and finally we will present strategies permitting access to potent and selective inhibitors.

STEROL BIOSYNTHESIS IN MAMMALS, PATHOGENIC FUNGI AND HIGHER PLANTS

Since several reviews have been published on sterol biosynthesis |9-11|, we will only summarize the main features of this process. Figure 1 represents a comparative pathway of sterol synthesis in mammals, fungi and higher plants. Upstream to 2(3)-oxidosqualene (OS) (I) it is generally admitted that the three pathways are very similar. Indeed identical intermediates have been found in all of these organisms. This does not implicate, however, that the enzymes catalyzing these steps are identical. Even if the chemical structures of the substrate(s) or the product(s) involved in a given enzymatic reaction are identical in the three classes of organisms, the enzymatic reaction may be performed with different kinetics or regulation. In this case, it is expected that some structural differences in the enzyme structure may be also found. Indeed it has been recently demonstrated that the primary sequence of HMGCoA reductase in *Arabidopsis thaliana* present important divergences with the primary sequence from the homologeous enzyme in yeast and in mammals |12,13|.

Downstream to OS, profound differences exist between plants on one side, and yeast or mammalian sterol biosynthesis on the other. These differences can be summarized as follows : i) cycloartenol (II) is the first cyclic product in photosynthetic eukaryotes, while lanosterol (III) is formed in non-photosynthetic organisms |14-16|. Exceptions to this rule do not seem to exist. Indeed cycloartenol has been found in *Astasia longa* which is a non-photosynthetic organism |17|, but *Astasia longa* is an apoplastid algae and therefore belongs to a photosynthetic phylum. The occurrence of cycloartenol in *Acantamoeba polyphaga* is more striking |18| because these non photosynthetic organisms cannot be considered at first sight to belong to a photosynthetic phylum. It is worth noting that some recent phylogenetic studies comparing the nucleotidic sequence of ribosomal RNA have proposed an evolutionary tree where amoeba are very close to algae which are photosynthetic organisms and far from non photosynthetic organisms such as protozoans |19| ; ii) a cyclopropane cleaving enzyme, namely cycloeucalenol (IV)-obtusifoliol (V) isomerase, has been found in higher plants only |20|. The presence of such an enzyme in photosynthetic organisms was expected since end-pathway sterols contained in these organisms never possess a cyclopropane ring at C9β,C-19β ; iii) the sequence of C-4 and C-14 methyl group removals is profoundly different : in mammals and in yeast, C-14 methyl is first removed, then the two C-4 methyls are removed stepwise whereas, in higher plants C-4 demethylation of 24-methylene cycloartanol (VIII) occurs first, then C-14 methyl is removed giving 4α-methyl fecosterol (IX) and finally C-4 demethylation of 24-ethylidene lophenol would be performed ; iv) finally, the side chain methylations are restricted to plants and fungi, but exhibit generally opposite stereochemistry at C-24 in these two latter organisms. The situation in higher fungi is more complex than in *Saccharomyces cerevisiae* since in the former the C-24 methylation has been shown to occur at an earlier stage of the biosynthetic pathway than in baker yeast, more precisely at the level of lanosterol, which is converted to

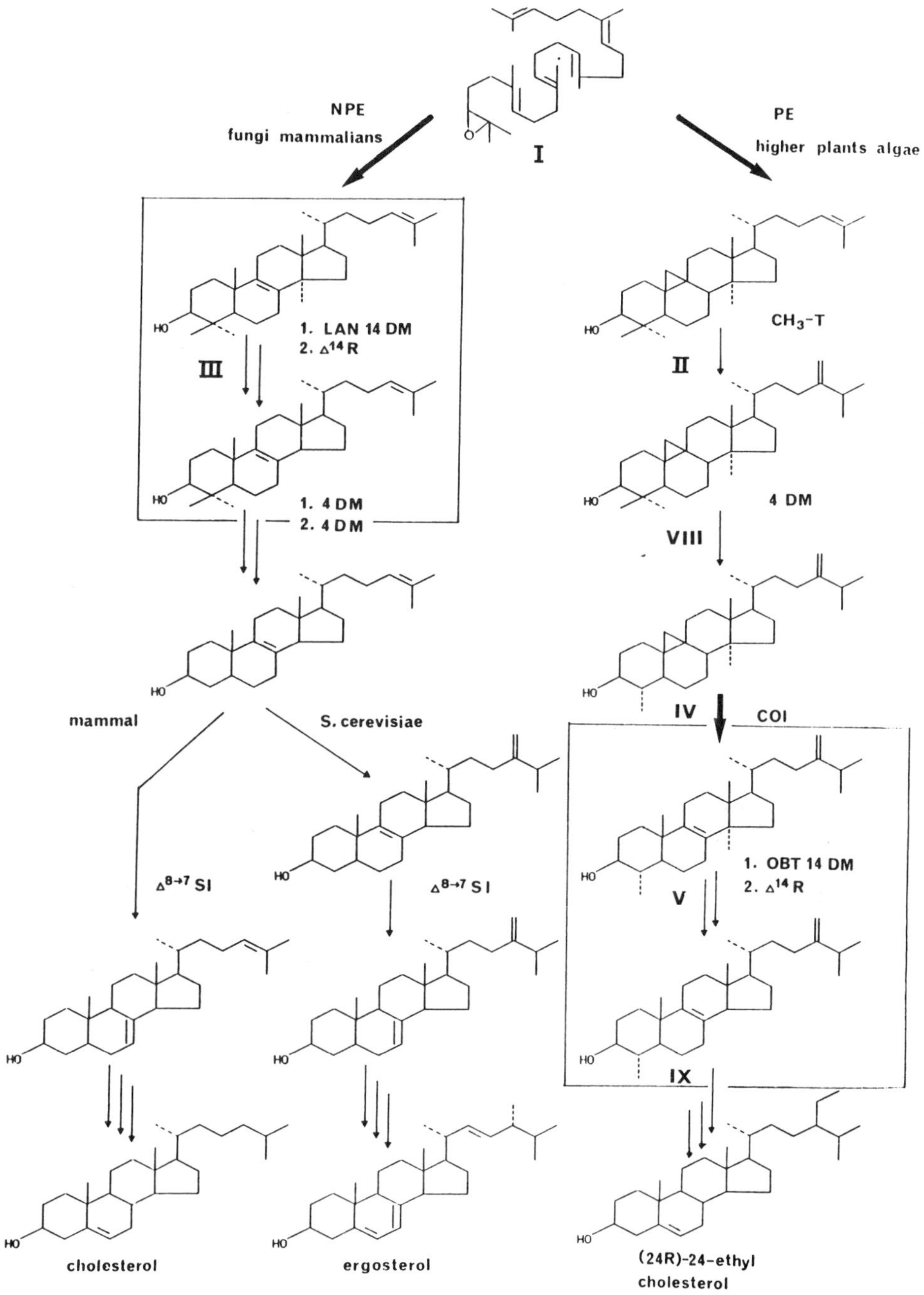

Fig. 1. Comparative biosynthetic pathway leading to cholesterol in mammalian cells, ergosterol in *S. cerevisiae* and sitosterol (24R)-24-ethyl cholesterol in higher plants.

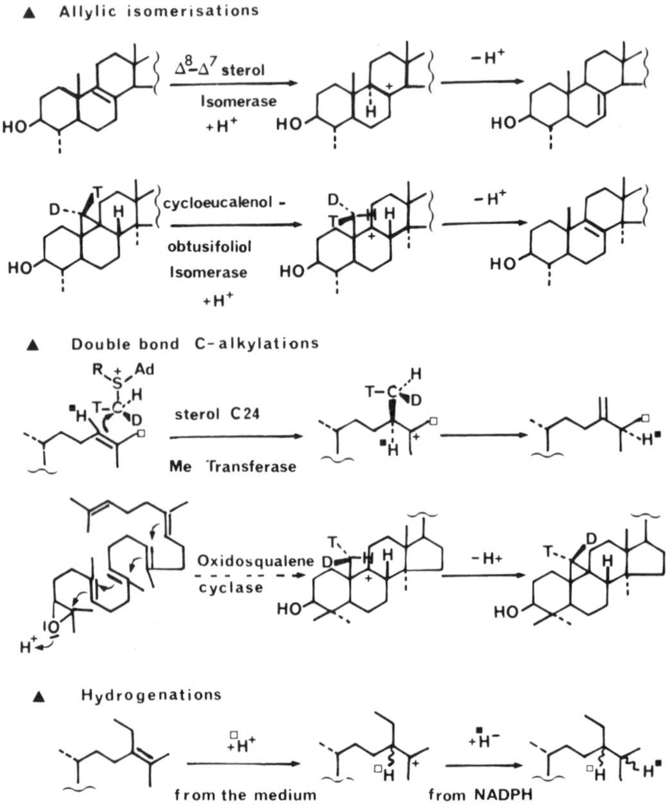

Fig. 2. Enzymic reactions of sterol biosynthesis involving demonstrated (sterol-C-24-methyltransferase) or postulated carbenium ion intermediates. Other reactions involving carbenium ions are the isopentenyldiphosphate : dimethyl-allyl diphosphate isomerase |50,51|, farnesyl-diphosphate-squalene synthetase, $\Delta^{8,14}$-sterol-Δ^{14}-reductase, $\Delta^{5,7}$-sterol-Δ^{7}-reductase.

24-methylene-24-dihydrolanosterol (eburicol). In addition to these four important differences, homologous enzymes in the biosynthetic pathway may feature more subtle differences with regard to changes in substrate specificities or the susceptibility to a given inhibitor. These two distinctions are probably important for the understanding of the selectivity observed with some SBIs. For example, it can be presumed from the biosynthetic scheme (Fig.1) that lanosterol is indeed the best substrate for the 14-demethylase of mammals and yeast, whereas obtusifoliol (V) is expected to be most suitable for the corresponding enzyme in higher plants |21|. Another example of different substrate specificity has been found for the C-4 demethylases. Whereas 4,4-dimethyl zymosterol (VII) was shown to be the best substrate of this enzyme in mammals and baker yeast |22|, some of us have recently demonstrated that 24-methylene cycloartanol (VIII) was the best substrate in maize |23 , S. Pascal, unpublished results|. Other examples have been reported recently |24|. Therefore we suggest that such differences in substrate specificities between homologous animals, fungal and plant sterol biosynthetic enzymes are a common feature.

MECHANISMS OF ACTION OF STEROL BIOSYNTHESIS ENZYMES

The results of mechanistic studies performed on sterol and triterpenoïd biosynthesis enzymes |9,10,25| have shown that several enzymatic reactions involve in their reaction mechanism postulated or demonstrated carbenium ion intermediates. This is especially true in the case of : i) 1,3- allylic rearrangements which have been shown to be of an acido-catalyzed type |26| (Fig. 2) ; ii) double bond C-alkylations and especially those catalyzed by the S-adenosyl-L-methionine (AdoMet)-sterol-C-24-methyltransferases. These last reactions involve a SN_2 type nucleophilic attack of the πorbitals of the Δ^{24} double bond on the sulfonium methyl group of AdoMet with inversion of configuration at the methyl and passage through two carbenium ions |25| ; iii) acido-catalyzed annellation reactions such as those catalyzed by the different 2(3)-oxidosqualene (OS) cyclases which involve electron deficient intermediates with more or less short life time |27,29| ; iv) the acido-catalyzed hydrogenation reactions.

These enzymatic reactions share in common the following properties : all these reactions are triggered by the addition of a proton or another electron-deficient species (the biochemical equivalent of a CH_3^+ carbenium ion in the case of methyltransferases for example). This process leads to an intermediate possessing a carbenium ion at a defined position. The involvment of these carbenium ion intermediates has been demonstrated in the case of the S-adenosyl-L-methionine-cycloartenol-C-24-methyltransferase |28|. In every case the last step of these reactions is the regio- and stereospecific removal of a proton leading to the product. It is expected that the aforementioned enzymes could maintain the carbenium ion intermediates in a hydrophobic and slightly basic environment in such a manner that the structure of the terminal product of the reaction is accurately determined by the position of the active site basic residue which eliminates this proton (kinetic control). Since no racemization occurs during the passage through these carbenium ion intermediates, it probably reflects the existence of strong van der Waals and electrostatic interactions between the charged intermediate species and the enzymatic surface. However the precise nature of these interactions in such an apolar environment is still a matter of debate. In some cases (double bond hydrogenations) the carbenium ion may be neutralized by an hydride ion originating from NADPH.

DESIGN OF STABLE INHIBITORS MIMICKING HIGH ENERGY INTERMEDIATES

According to several authors |30,31| the transition state (TS) analogue concept is a very useful one for the design of potent and selective enzyme inhibitors since TS analogue inhibitors may have a much higher affinity for the active site of the enzyme than traditional ground state analogue inhibitors |30,31| . By definition, the TS for a chemical reaction has a fleeting existence (i.e. the time for a single vibration : 10^{-14}s) and it often leads to molecular metastable High Energy Intermediates (HEI) interacting tightly with the enzyme. According to the Hammond postulate, the structures of these HEI are good approximations of the structure of the true TS. Because of their unstability the carbenium ions involved in the catalytic pathways described above could be considered as HEIs. As discussed elsewhere |35-37|, it is expected that these carbenium ion intermediates would interact with and be stabilized by the active site of the enzyme.

In order to mimic these HEIs, we have synthesized stable analogues possessing a charged heteroatom (generally N, but also S and As) presenting a positive charge at a position identical to this of the sp^2 carbon in the carbenium ion. We were able to show that these analogues mimicked efficiently these HEIs and displayed an affinity for their target enzyme three order of magnitude higher than this of the best substrate |26,33|. A typical molecule is the (24-R,S)-24-methyl-25-aza cycloartanol (Fig.3) (VI) |34|. It appears therefore that a sp^3 tetrahedral ammonium ion (the tertiary amine function of these molecules is protonated at physiological pH) is able to interact tightly with the active site of the target enzyme. According to a thorough study of the molecular features involved in this binding, one can conclude that the existence of a positive charge in the molecule of inhibitor located at a position close or identical to this occupied by the sp^2 carbenium ion in the HEI, is of paramount importance. In other words, the interaction between such species and the enzyme active site is mostly electrostatic.

INHIBITION OF CYCLOEUCALENOL OBTUSIFOLIOL ISOMERASE (COI) Δ^{14}-REDUCTASE (ΔR) AND $\Delta^8 \rightarrow \Delta^7$-STEROL ISOMERASE (SI)

To illustrate the application of the preceding concepts we have focused our attention on the inhibition of COI, ΔR and SI. A comparative mechanism of action of these three enzymes is depicted in the Fig. 4.

In the case of the COI, results of mechanistic studies performed on an enzymatic preparation from maize (*Zea mays*) are consistent with a general acid-catalyzed cyclopropane ring opening with incorporation of one proton from the medium at C-19 |35,36|. The cleavage of the C9-C19β bond of the cyclopropane occurs with retention of configuration on the C19 carbon atom. This result suggests that the enzyme subsite which is involved in the acid-catalyzed cyclopropane ring opening is located above the C ring of the substrate |36,37|. Moreover, since the cleavage of the C9-C19β bond and the elimination of the C8-Hβ are cis, the reaction cannot be concerted and has to go through the intermediate carrying a carbenium ion at C-8.

Although little is known about the mechanism of the SI, it has been shown that in rat liver the reaction involves an antarafacial mechanism, i.e. 9α-protonation and 7β-elimination, whereas in yeast the 7α hydrogen atom is eliminated |38|. Antarafacial transfers are mostly encountered for allylic isomerizations of non-activated double bonds (for which allylic deprotonation is not facilitated), and must involve a two-base mechanism |39| which is most probably non concerted, i.e. there must be protonation of the double bond, leading to a carbenium ion intermediate from which a proton is eliminated. Such an antarafacial allylic isomerization involving a carbenium ion has been demonstrated in the case of isopentenyl-diphosphate isomerase|39,50|.These considerations and the findings of different groups which studied the mechanism of this reaction |40-42| led us to consider that the most probable mechanism for the plant $\Delta^8 \rightarrow \Delta^7$-sterol isomerase would involve a carbenium ion at C-8.

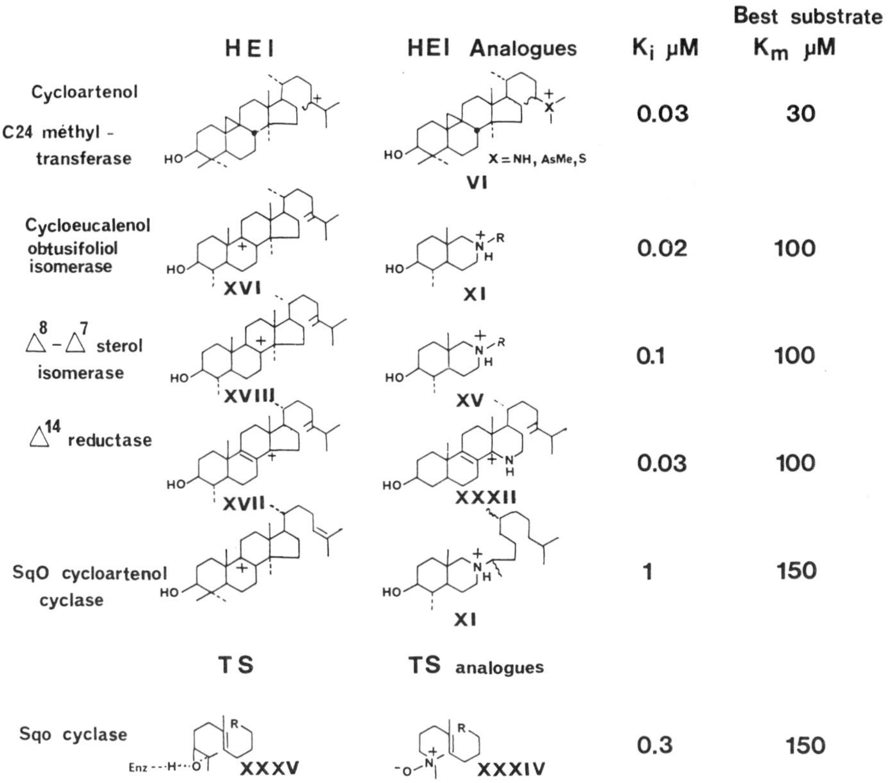

Fig. 3. Inhibition of enzymes of sterol biosynthesis by stable analogues of high energy intermediates (HEI) or transition states (TS) involved during their catalytic mechanism. These analogues possess a positive charge at a position identical to this of the sp^2 carbon in the carbenium ion. The positive charge is confered by a charged heteroatom (generally N, but also S and As). The tertiary amine N-oxide derivative (XXXIV) would present electronic and structural similarities with a possible dipolar transition state (XXXV) resulting from the polarisation of the oxiranne ring of 2,3-oxido-squalene by an enzyme proton.

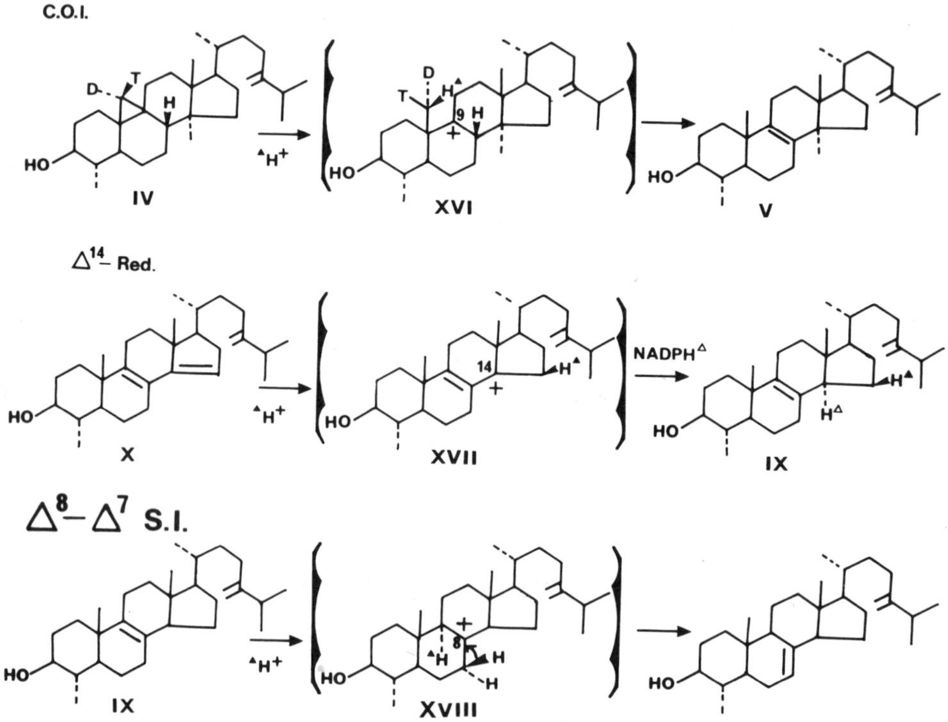

Fig. 4. Comparative mechanism of action of COI, $\Delta^{8,14}$-sterol- Δ^{14}-reductase (ΔR) and $\Delta^{8} \rightarrow \Delta^{7}$-sterol isomerase (SI) involving high energy carbenium ion intermediates. IV : cycloeucalénol ; V : obtusifoliol ; X : 4α-methyl-5α-ergosta-8,14,24(28)-trien-3β-ol ; IX : 4α-methyl-fecosterol ; XVI, XVII and XVIII : carbenium ion carrying high energy intermediates involved during catalytic mechanisms of COI, ΔR and SI respectively.

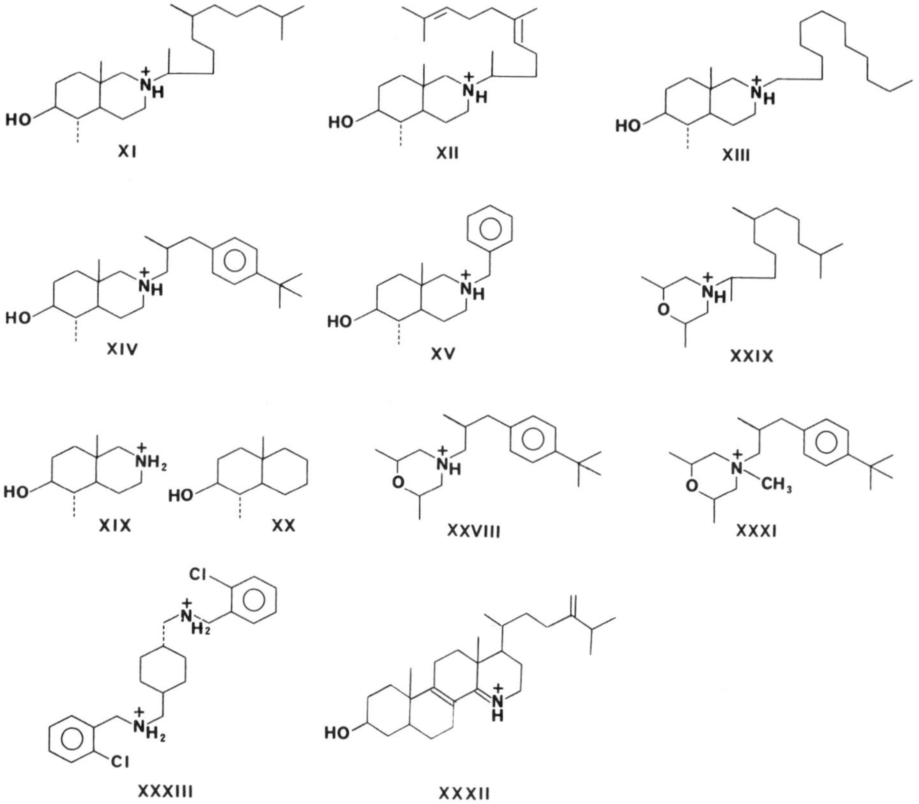

Fig. 5. Chemical structures of some inhibitors considered in the present work. XI : N-|(1,5,9)-trimethyldecyl|-4α,10-dimethyl-8-aza-trans-decal-3β-ol ; XII : N-|trans-1,5,9-trimethyl-5,9-decadien| 4α,10-dimethyl-8-aza-trans-decal-3β-ol ; XIII : N-lauryl-4α,10-dimethyl-8-aza-trans-decal-3β-ol ; XIV : N-|3-(4-tert-butyl phenyl)-2-methyl|-propyl-8-aza-4α,10-dimethyl-trans-decal-3β-ol ; XV : N-benzyl-4α,10-dimethyl-8-aza-trans-decal-3β-ol ; XIX : 4α,10-dimethyl-8-aza-trans-decal-3β-ol ; XX : 4α,10-dimethyl-trans-decal-3β-ol ; XXIX : tridemorph ; XXVIII : fenpropimorph ; XXI : N-methyl-fenpropimorph ; XXXII : 15-aza-24-methylene-D-homocholestadiene-3β-ol (A 25822B) ; XXXIII : AY9944.

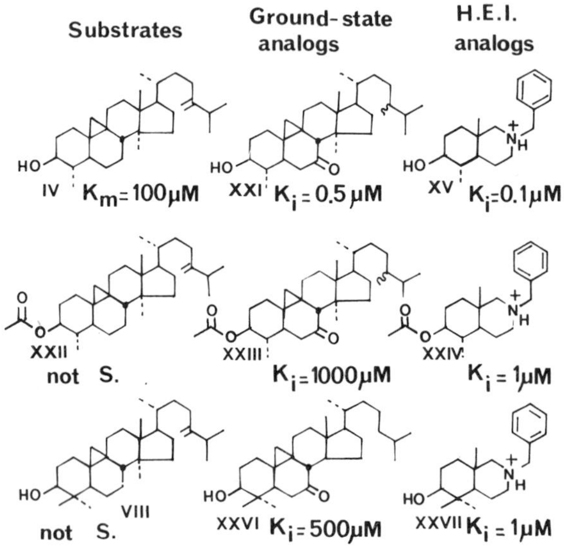

Fig. 6. Structure-Activity relationship for substrates, ground state analogues and High Energy Intermediate (HEI) analogues. The structural features which are compulsory for binding of substrates (IV, XXII and VIII) or ground state analogues (XXI, XXIII and XXVI) affect the binding of the related analogues in the azadecalin series (XV, XXIV and XXVII) to a much lesser extent. IV : cycloeucalenol ; XXII : cycloeucalenyl acetate ; VIII : 24-methylene cycloartanol ; XXI : 7-ceto-24-dihydro cycloeucalenol ; XXIII : 7-ceto-24-dihydro cycloeucalenyl acetate ; XXVI : 7-ceto-cycloartanol ; XV : N-benzyl-8-aza-4α,10-dimethyl trans-decal-3β-ol ; XXIV : N-benzyl-8-aza-4α,10-dimethyl-trans-decal-3β-acetoxy ; XXVII : N-benzyl-8-aza-4,4,10-trimethyl-trans-decal-3β-ol.

Table 2. Inhibition of the cycloeucalenol-obtusifoliol isomerase (COI), the $\Delta^{8,14}$-sterol-Δ^{14}-reductase (ΔR) and the $\Delta^{8}\rightarrow\Delta^{7}$-sterol isomerase (SI) by the 8-azadecalins (XI-XV) and XIX, XX.

	XI	XII	XIII	XIV	XV	XIX	XX
				ID50 (μM)			
COI	0.025	0.035	0.10	0.10	0.10	NI	17.5
ΔR	0.30	ND	0.09	0.07	0.08	NI	4.5
SI	0.20	ND	ND	ND	0.13	NI	10

NI : non inhibitory
ND : not determined

$\Delta^{8,14}$-sterol Δ^{14}-reductase has been described in animals |43| and fungi |44|. The results of mechanistic studies performed in the rat liver system clearly show that the first step of the reaction is the stereospecific protonation of the double bond generating the most stable carbenium ion, which is then neutralized stereospecifically in a trans manner by a hydride ion from NADPH |43|.

N-Substituted Azadecalins

The N-substituted 8-azadecalins (XI-XV) (Fig.5) have electronic and structural similarities with the carbenium ion carrying HEIs (XVI-XVIII) involved in the three enzymatic steps depicted in Fig.4 and could, therefore, be considered as potential inhibitors of COI, ΔR and SI |33|. As shown in the Table 2, we were able to show that a positive charge at C-8 is the major cause of the affinity, since an electrostatic neutral isosteric analogue (XIX) of (XX) does not inhibit all three reactions. The affinities measured for the different analogues in this series also show the importance of the presence of a N-substituent. For example the presence of a benzyl substituent (XV) lead to an increase of affinity of about two order of magnitude for all three enzymes |33|. The behaviour of the azadecalin carrying a trimethyl decyl substituent (XI) is unique in the sense that XI inhibits the COI with an affinity ten times higher than the two other enzymes. In the case of the COI a ID50/Km ratio of 2×10^{-4} was measured |45-47|. The great potency of XI on the COI could be explained by the presence of the trimethyldecyl substituent able to mimic the hydrophobic part of the C,D rings and lateral chain of the steroïd nucleus. Moreover the selectivity of this molecule could result from the presence of a methyl group at C'-10 of the N-substituent since it could be argued that this methyl is present in HEI (XVI) but absent in both HEIs (XVII and XVIII) could hamper the binding of (XI) with the active site of the enzymes in conformations competent to bind the HEIs (XVII and XVIII).

It is interesting to consider how the same ammonium derivative is able to mimic three different HEI possessing a carbenium ion AT C-9, C-14 and C-8. As already discussed |33,46,47,49|, this might be due to the delocalized charge of such ammonium ions. The relative freedom within the active site of model analogues of HEI with a flexible N-substituent could also account for the inhibition of these three enzymes by the same N8-ammonium derivatives, stressing again the importance of the electrostatic interactions during the binding of these stable HEI analogues or the stabilization of the unstable carbenium ion intermediates.

Table 3. Inhibition of the cycloeucalenol-obtusifoliol isomerase (COI), the $\Delta^{8,14}$-sterol-Δ^{14}-reductase (ΔR) and the $\Delta^8 \rightarrow \Delta^7$-sterol isomerase (SI) by fenpropimorph (XXVIII, tridemorph (XXIX) and fenpropidine (XXX).

	XXVIII	XXIX	XXX
		ID50 (µM)	
COI	0.4[a]	0.4[a]	0.2[a]
ΔR	0.8	25	3
SI	0.4[a]	0.4[a]	0.08

[a] Data from |24|

Table 4. Inhibition of the cycloeucalenol-obtusifoliol isomerase (COI), the $\Delta^{8,14}$-sterol-Δ^{14}-reductase (ΔR) and the $\Delta^8 \rightarrow \Delta^7$-sterol isomerase (SI) by AY9944, (XXXIII), 15-azasterol (XXXII) and the trimethyldecyl-8-aza-decalin (XI).

	XI	XXXII	XXXIII
		ID50 (µM)	
COI	0.025	NI	NI
ΔR	0.30	0.030	40
SI	0.20	0.8	0.5

NI : non inhibitory

In the case of the COI, comparison of the structure-affinity relationship for substrates and for compounds of the azadecalin series (Fig.6) |49|, shows that structural features which are compulsory for binding of substrates (IV, XXII and XXV) or ground state analogues (XXI, XXIII and XXVI), affect the binding of the related analogues in the azadecalin series (XV, XXIV and XXVII) to a much lesser extent. Thus, substrates and ground state analogues inhibitors on one hand and azadecalins on the other probably interact with two different conformations of the active site of COI ; this would be in accordance with the nature of HEI analogues postulated for the azadecalins |33|.

N-Substituted Morpholines

One important group of fungicides that act as inhibitors of sterol biosynthesis are the morpholine derivatives such as fenpropimorph (XXVIII), tridemorph (XXIX) and a piperidine, fenpropidine (XXX). Using an enzyme assay in a microsomal preparation isolated from maize embryos, we have shown that XXVIII, XXIX and XXX strongly inhibit both COI and $\Delta^8 \rightarrow \Delta^7$-sterol isomerase (Table 3) |24,48|. The $\Delta^{8,14}$-sterol Δ^{14}-reductase is also inhibited by compounds XXVIII-XXX with less potency |49|. Based on our previous results obtained with the azadecalin series we investigated the importance of the charge on the nitrogen atom of the morpholine ring in these molecules. We first showed that the quaternary ammonium derivative XXXI of fenpropimorph displays a strong activity relative to the parent amine probably binding in its protonated form. We then demonstrated that the potency of inhibition of COI and SI is strongly pH dependent |24|. Assuming that the observed strong affinity is due mainly to the interaction of the morpholine cation with a complementary negatively charged subsite of the enzyme, the variation in its affinity with pH is not only dependent on the fraction of inhibitor in the charged form but also on the protonation state of the residue(s) of the enzyme interacting with it. A study of the variation of affinity with pH of fenpropimorph and its ammonium derivative, for both COI and SI, has shown that changing pH alters the affinity of the enzyme for fenpropimorph in a manner consistent with exclusive binding of the morpholinium cation to both COI and SI. The results suggest the existence of an electrostatic interaction between this cation and a negatively charged residue of the enzyme with a more basic pKa (<8) in the case of COI than in the case of SI (pKa <6). An attractive possibility would be that these residues are those stabilizing the HEIs involved in both reactions. N-substituted alkyl morpholines could then be considered as in the case of azadecalins, as transition-state analogues inhibitors |24|.

Miscellaneous

The 15-aza-24-methylene-D-homocholestadiene-3β-ol (A25822B) (XXXII) has been shown to inhibit strongly $\Delta^{8,14}$-sterol, Δ^{14}-reductase in yeast |44|. We have recently shown that XXXII is also a potent inhibitor of the ΔR from maize embryos (Table 4) and we have rationalized this remarkable inhibition by the assumption that after protonation of XXXIII, the iminium derivative would mimic the C-14 carbocationic HEIs (XVIII and XXXIX) implicated in the catalytic pathway. In contrast to the morpholines, XXXII was shown to be strongly specific for ΔR since it did not inhibit COI at all |49|. In a more systemic search for active inhibitors of the three enzymes, we discovered that AY9944 (XXXIII) inhibited the SI but had little or no effect on the R and the COI. It is remarkable that AY9944, A25822B and the trimethyldecyl-8-aza-decalin (XI) are specific inhibitors of the SI, ΔR and COI, respectively.

DISCUSSION

Results from our and other laboratories |50-55| show that compounds possessing a positively charged heteroatom at a position identical to that of the sp^2 carbon atom in the putative HEI involved in the catalytic pathway are in most cases strong enzyme inhibitors. However, this result alone does not prove that the designed inhibitors are really TS or HEI analogue inhibitors. The following list of likely criterions have been proposed for tentative characterization of TS or HEI analogue inhibitors : i) transition state (high energy intermediate) resemblance |30| ; ii) tight binding |56| ; iii) correlation between Ki values for inhibitors and Km/kcat values for the corresponding substrates |31| ; iv) structural features essential for binding of the substrate affect the binding of TS or HEI analogue inhibitors to a much lesser extent (present work) ; v) slow binding enzyme inhibition |32|. Only criterions i, ii and iv were met in our own work. Tests for criterions iii and v would require thorough kinetic studies which are difficult to accomplish with heterogenous enzymatic systems composed of membrane-bound enzymes and water insoluble substrates and inhibitors. These tests might become feasible with solubilized and extensively purified enzymes.

Finally, the results described above shed light on the paramount importance of electrostatic interactions for the binding of HEI analogues possessing a positively charged heteroatom to active sites. The nature of the negatively charged group(s) present in the enzyme and interacting with these inhibitors is unknown, but it could be a delocalized carboxylate anion as shown in glycosidases |57|, or a thiolate as suggested for isopentenyl diphosphate isomerase |50|. It is worth mentioning that morpholine fungicides have been shown to inhibit bacterial cholesterol oxidase, an enzyme characterized by an aspartate residue essential for catalytical activity |58|.

As stated in the introduction, one goal of these studies was to obtain SBIs fungicides more potent and selective than those presently on the market. As shown in Tables 2-4, it appears that the N-substituted-8-aza decalins are more potent inhibitors of the COI, ΔR and SI from maize embryos than the classical N-alkyl morpholines. Several of the aza decalines were tested also in enzymes from *Saccharomyces cerevisiae*, the best of them was the trimethyl decyl-8-azadecaline (XI). In comparison to fenpropimorph (XXVIII) it was a slightly better inhibitor of ΔR but 20 times a weaker inhibitor of SI (P. Masner, personal communication). This may explain the fact that in controlling phytopathogenic fungi, N-substituted-8-azadecalines are no better than the commercially available fenpropimorph (XXVIII). In recent months, inhibitors of the yeast $\Delta^{8,14}$-sterol Δ^{14}-sterol reductase have been tentatively designed |55| using the principles developed in our laboratory. Compounds possessing a charge analogy and structural similarities with the HEI (XVII) were synthesized and optimized by using a ΔR enzymatic assay. Several compounds with IC50 values significantly lower than that of fenpropimorph were found. Most of which displayed good activity against pathogenic fungi in green-house trials. Under field conditions, however, none of these compounds have surpassed the efficacy of fenpropimorph so far |55|.

ACKNOWLEDGMENTS

We are indebted to Drs. C. Anding and P. Place (Rhône-Poulenc Agrochimie) for their participation in the synthesis of the 8-azadecalins |33|. We acknowledge BASF Agrochemical Station (Limburgerhof, F.R.G.) for having provided the N-alkyl-morpholines used in the present work. Many thanks to B. Bastian for kindly typing the manuscript.

REFERENCES

1. R.S. Burden, D.T. Cooke and G.A. Carter, Phytochemistry 28:1791 (1988).
2. R.J. Rodriguez, C. Low, C.D.K. Bottema and L.W. Parks, Biochim. Biophys. Acta, 837:336 (1985).
3. C. Dahl, H.P. Biemann and J. Dahl, Proc. Natl. Acad. Sci. USA, 89: 4012 (1987).
4. N.B. Mandava, Ann. Rev. Plant Physiol. Plant Mol. Biol. 39:23 (1988).
5. F.J. Schwinn, Pestic. Sci., 15:40 (1984).
6. H. Van den Bossche, Willemsens, W. Cools, F. Cornelissen, W.F. Lauwers and Van Cutsem, Antimicrob. Agents Chemother. 17:922 (1980).
7. D.F. Gillard and D.C. Walton, Plant Physiol., 58:790 (1976).
8. H. Van den Bossche, Mode of action of pyridine, pyrimidine and azole antifungals, in Sterol Biosynthesis Inhibitors, Berg. D., and Plempel M., Eds, Ellis, Horwood, Chichester, 79 (1988).
9. P. Benveniste, Annu. Rev. Plant Physiol., 37:275 (1986).
10. T.W. Goodwin, Annu. Rev. Plant Physiol., 30:369 (1979).
11. W.R. Nes, in Isopentenoïds in Plants : Biochemistry and Function (W.D. Nes, G. Fuller and L. Tasi eds), p. 325. Dekker, New York (1984).
12. C. Caelles, A. Ferrer, L. Balcells, F.G. Hegardt and A. Boronat, Plant Molec. Biol., in press.
13. R.M. Learned and G.R. Fink, Proc. Natl. Acad. Sci. USA, 86:2779 (1989).
14. P. Benveniste, L. Hirth and G. Ourisson, Phytochemistry, 5:31 (1966).
15. M. Ardenne, G. Osske, K. Schreiber, K. Steinfelder and R. Tümmler, Die Kulturpflanze, 13:115 (1965).
16. P.D.G. Dean, Steroidologia, 2:143 (1971).
17. M. Rohmer and R.D. Brandt, Eur. J. Biochem., 36:446 (1973).
18. D. Raederstorff and M. Rohmer, Eur. J. Biochem., 164:427 (1987).
19. J.H. Gunderson, H. Elwood, A. Ingold, K. Kindle and M.L. Sogin, Proc. Natl. Acad. Sci. USA, 84:5823 (1987).

20. R. Heintz and P. Benveniste, J. Biol. Chem., 249:4267 (1974).
21. A. Rahier and M. Taton, Biochem. Biophys. Res. Commun., 140:1064 (1986).
22. E.I. Mercer, Pestic. Sci., 15:133 (1984).
23. S. Pascal, M. Taton and A. Rahier, Biochem. Biophys. Res. Commun., 172:98 (1990).
24. M. Taton, P. Benveniste and A. Rahier, Pestic. Sci., 21:269 (1987).
25. D. Arigoni, Ciba Found. Symp., 60:243 (1978).
26. A. Rahier, M. Taton and P. Benveniste, Inhibition of sterol biosynthesis in Higher plants by analogues of high energy carbocationic intermediates, in Biochemistry of Cell walls and Membranes in Fungi, P.J. Kuhn et al., Eds. Springer-Verlag, Berlin, 206 (1990).
27. E.E. Tamelen, Pure and Applied Chem., 53:1259 (1981).
28. M.M. Mihailovic, Ph. Dissertation Eidgenosische Technische Hochshule, Zürich (1984).
29. S.W. Johnson, S.D. Lindell and J. Steele, J. Am. Chem. Soc., 109:2517 (1987).
30. R. Wolfenden, Ann. Rev. Biophys. Bioeng., 5:271 (1976).
31. P.A. Bartlett and C.K. Marlowe, Biochemistry, 22:4618 (1983).
32. J.V. Schloss, Acc. Chem. Res., 21:348 (1988).
33. A. Rahier, M. Taton, P. Schmitt P. Benveniste, P. Place and C. Anding, Phytochemistry, 24:1223 (1985).
34. A. Rahier, J.C. Génot, F. Schuber, P. Benveniste and A.S. Narula, J. Biol. Chem., 259:15215 (1984).
35. A. Rahier, Biosynthèse des stérols chez les plantes. PhD Thèse Université de Strasbourg, (1980).
36. A. Rahier, L. Cattel and P. Benveniste, Phytochemistry, 16:1187 (1977).
37. W.A. Blättler, Sterischer Verlauf der Bildung und Offnung des Cyclopropanringes in der Biosynthese von Phytosterinen. PhD Thesis, Eidgenossische Technische Hochschule, Zürich (1978).
38. M. Akhtar, A.D. Rahimtula and D.C. Wilton, Biochem. J., 117:539 (1970).
39. J.W. Cornforth, Tetrahedron, 30:1515 (1974).
40. M. Akhtar, A.D. Rahimtula, J. Chem. Soc., Chem. Commun., 259 (1968).
41. W.H. Lee, R. Kammereck, B.N. Lusky, J.A. Mc Closkey and G.J. Schroepfer J. Biol. Chem., 244:2033 (1969).
42. L. Canonica, A. Fiecchi, M. Galli Kienle, A. Scala, G. Galli, E. Grossi Paoletti and R. Paoletti, Steroïds, 11:287 (1968).
43. D.C. Wilton, I.A. Watkinson, M. Akhtar, Biochem. J., 119:673 (1970).
44. C.K. Bottema and L.W. Parks, Biochim. Biophys. Acta, 531:301 (1978).
45. A. Rahier, M. Taton and P. Benveniste, Biochem. Soc. Trans., 18:48 (1990).
46. A. Rahier, M. Taton and P. Benveniste, Eur. J. Biochem., 181:615 (1989).
47. M. Taton, P. Benveniste and A. Rahier, Pure and Appl. Chem., 59:287 (1987).
48. A. Rahier, P. Schmitt, B. Huss, P. Benveniste and E.H. Pommer, Pestic. Biochem. Physiol., 25:112 (1986).
49. M. Taton, P. Benveniste and A. Rahier, Eur. J. Biochem., 185:605 (1989).
50. J.E. Reardon and R.H. Abeles, Biochemistry, 25:5609 (1986).
51. M. Muehlbacher and C. Dale Poulter, Biochemistry, 27:7315 (1988).
52. C.D. Poulter, T.L. Capson, M.D. Thompson and R.S. Bard, J. Am. Chem. Soc., 111:3734 (1989).
53. A.C. Oehlschlager, R.H. Angus, A.M. Pierce, H.D. Pierce and R. Srinavasan, Biochemistry, 23:3582 (1984).
54. M.A. Ator, S.J. Schmidt, J.L. Adams and R.E. Dolle, Biochemistry, 28:9633 (1989).
55. A. Akers, E. Ammermann, E. Buschmann, N. Götz, W. Himmele, G. Lorenz, E.H. Pommer, C. Rentzea, F. Röhl, H. Siegel, B. Zipperer, H. Sauter and M. Zipplies, Pestic. Sci., in press.
56. K.T. Douglas, Chemistry and Industry, 311 (1983).
57. P. Lalégerie, G. Legler and J.M. Yon, Biochemistry, 64:377 (1982).
58. P.G. Hesselink, A. Kerkenaar and B. Witholt, Pestic. Biochem. Physiol., 33:69 (1989).

MOLECULAR MODELLING and STRUCTURAL DATABASES IN PHARMACEUTICAL RESEARCH

Klaus Müller

Central Research Units
Hoffmann-LaRoche Ltd
Basel, Switzerland

There is a vast and rapidly growing amount of structure and biosequence information available to pharmaceutical research. For practical purposes such information must be quickly accessible. This requires specially designed databases with facilities for interactive and focussed search of the relevant data as well as a large variety of sophisticated tools for interactive data analysis and processing. The basic concepts of our in-house database systems and data handling tools have been outlined elsewhere[1,2]. Here, we wish to discuss some applications to illustrate typical aspects of the interplay between molecular modelling and structural databases in structure-based chemical research.

Constraining the conformational flexibility of biologically active peptides by suitable structural modifications is an often encountered problem. This task can be accomplished in basically two different ways. The first involves structural modifications within the peptide fragment responsible for biological activity, e.g., by incorporating peptidomimetics or rigid nonpeptidic building blocks. The second leaves this essential peptide domain untouched, but attempts to induce or fix the putative bioactive conformation by incorporating the peptide unit into a macrocyclic system.

A number of interesting mono-, di-, and tricyclic building blocks have been proposed, and used with varying success, as (partial) substitutes for specific peptide folds[3,4]. This approach can be expected to be useful in cases where a receptor recognizes a specific pattern of amino acid side chains, but does not interact with the peptide backbone itself. However, from many crystal structures of protease inhibitor complexes and tightly bound oligomeric proteins, there is growing evidence that molecular recognition generally involves specific interactions with both amino acid side chains and backbone amide units. A striking example is provided by the refined crystal structure of glycosomal triosephosphate isomerase (TIM)[5], in which

Bioorganic Chemistry in Healthcare and Technology, Edited by U.K. Pandit and F.C. Alderweireldt, Plenum Press, New York, 1991

the two identical subunits are interlocked by exposed β-turns fitting tightly into recognition pockets. As can be seen from the hydrogen bonding network analysis (Figure 1a) for one of the two β-turns containing a Thr-Gly unit in a turn of type II, all heteroatoms of this turn, in particular those of the exposed backbone amide unit, are engaged in specific hydrogen bonding interactions with functional groups of the recognition pocket. Any attempt to arrive at a potential TIM dimerization antagonist by designing a (rigid) peptide analog to this Thr-Gly β-turn unit would have to take into account its specific interaction pattern. Hence, there is virtually no room for the design of peptidomimetics or rigid building blocks within the β-turn domain. Rather, we have to resort to the design of a suitable macrocyclic system. Obvious candidates may be Thr-Gly containing cyclic hexapeptides for their known tendency to adopt β-turn containing conformations, or derivatives thereof by incorporating less flexible building blocks outside the Thr-Gly domain. Smaller ring systems, in particular ten- to twelve-membered lactam derivatives[1,6,7], may also be considered as interesting substitutes for the Thr-Gly β-turn domain.

It is at this stage where searches through the structural database become most revealing. For example, we may want to collect all well resolved crystal structures of standard cyclic hexapeptides in order to gain information about their conformational preferences. A typical protocol of such a search in our ROCSD (ROche Cambridge Structural Database) system[1,2] is reproduced in Figure 2. It illustrates that relevant structural or substructural information can be retrieved almost immediately. A number of tools are then available for interactive examination of geometric and energetic aspects of the retrieved structures. The majority of structures adopt conformations with two β-turns joined by short antiparallel strands (Figure 3). They are essentially devoid of torsional strain. However, while most structures exhibit trans-annular hydrogen bonds, characteristic for standard β-turns, one structure[8] shows no such intramolecular hydrogen bonding. This unusual peptide conformation can be related to other 18-membered ring systems. Fourier-analysis of ring shapes[9] for all 18-membered macrocyclic structures from the database, followed by a cluster analysis, reveals that this unusual peptide conformation is very similar to those observed for a tetralactone[10] as well as three 18-crown-6 ethers[11-13]. Rigid-body superposition of the cyclic hexapeptide to the four nonpeptidic macrocyclic structures confirms the close conformational similarity (Figure 4). Interestingly, theoretical conformational analysis on cyclic hexaglycyl using our MOLOC-model building system[1,2] indicates that this conformation is a torsionally unstrained, but high-energy local minimum (ca. 7.5 kcal/mol above the global minimum) due to the lack of intramolecular hydrogen bonding. Instant generation of the crystal packing and graphic-assisted analysis of the intermolecular interactions (Figure 5) then reveals that this unusual peptide conformation is stabilized by tight packing of hydrophobic side chains, on the one hand, and by two orthogonal strictly intermolecular hydrogen bonding networks, involving also crystal water molecules. In this way, all heteroatoms of the peptide backbone are engaged in hydrogen bonds. This structure is of particular interest since it

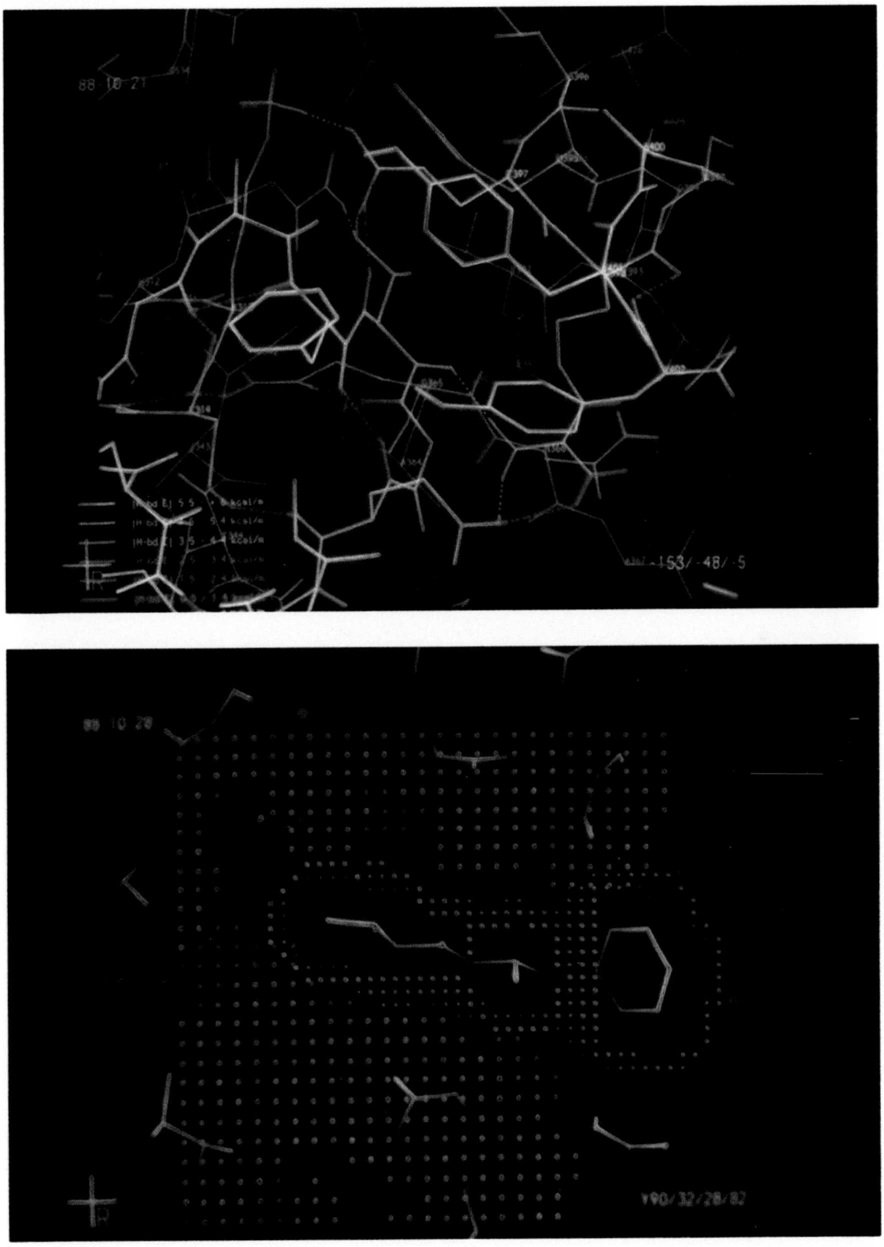

Figure 1: A (top): Color-coded display of putative hydrogen bonds between two subunits of glycosomal TIM involving the Thr-Gly β-turn of the Phe-Thr-Gly-Glu interface loop of one subunit tightly bound in a recognition pocket of the other subunit. B (bottom): Cubic-raster analysis of the packing of the same interface loop (raster representation of the van der Waals surface with a grid size of 0.4A; yellow dots) in the recognition pocket of the other TIM subunit (raster representation of the van der Waals surface with a grid size of 0.6A; blue dots). A narrow cross section through the C_{α}-atoms of Thr and the preceding Phe as well as parts of the threonine side chain is displayed.

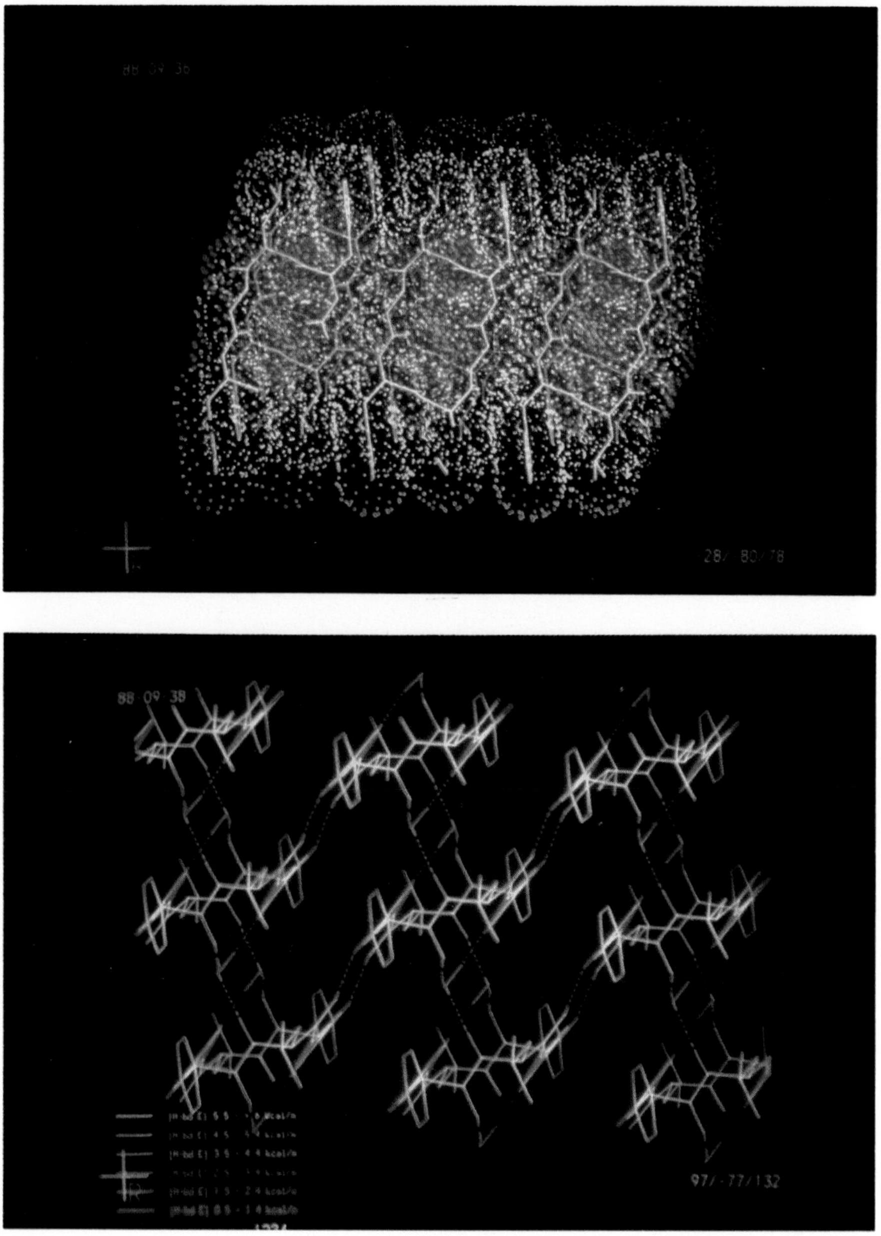

Figure 5: Computer-graphic analysis of the crystal packing of the cyclic hexapeptide cyclo-(-L-Phe-D-Leu-Gly-D-Phe-L-Leu-Gly-)[8]. Top: van der Waals dotted surface for an array of nine packed molecules, yellow and blue colors indicating hydrophobic and hydrophilic surface domains, respectively. Note that the leucine and phenyl-alanine side chains are tightly packed into hydrophobic layers. Bottom: Color-coded display of the hydrogen bond network for the same array of molecules. Note that all hydrogen bonds are intermolecular, the β-strand segments forming short antiparallel β-sheet domains, the remaining polar groups being engaged by water molecules.

R O C S D   S E A R C H   for   C Y C L I C   H E X A P E P T I D E S

command	meaning	set	no.hits	CPU-sec
SF 248	18-memb. ring	1	811	0.05
SF 127	cyclic -CO-X-	2	7776	0.05
MER 1 A 2	set 1 AND 2	3	196	0.03
SF 156	dipeptide	4	643	0.05
MER 3 A 4	set 3 AND 4	5	68	0.03
EC N 6-*	at least 6 N	6	4954	0.45
EC O 6-*	at least 6 O	7	13313	0.43
MER 5-7/A	sets 5,6,7/AND	8	65	0.04
FRAG O 'cyclohexa peptide'	retrieve fragment from library	-	-	0.17
CONN 8	connectivity search on set 8	9	41	0.87
SDP 16	str.det.param. R-factor < 10% no disorder no part.disorder coord. available	10	41151	0.06
MER 9 A 10	well resolved 'standard' cyclic hexapeptides	11	15	0.03

Figure 2. Protocol of a search for well resolved X-ray structures of cyclic hexapeptides in ROCSD (CPU-times are for a VAX-8700).

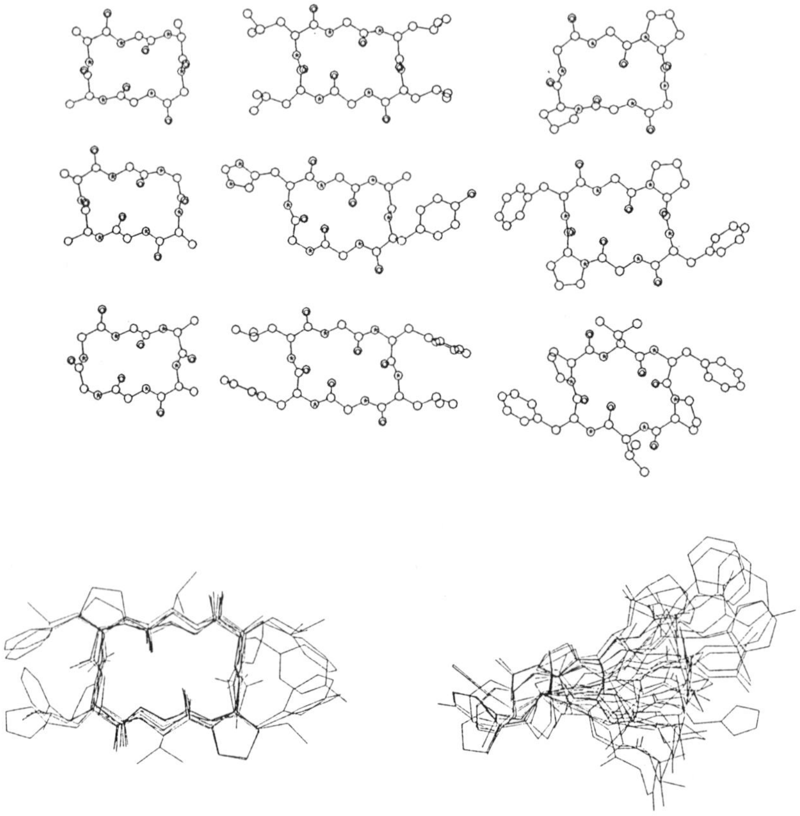

Figure 3. A (top): Juxtaposition of the crystal structures of nine selected cyclohexapeptides as retrieved from ROCSD. The data base reference codes are (from left to right and top to bottom): AAGAGG10[19], CGDLLL10[20], CGLPGL[21], AAGGAG10[19], BAMLIK[22], CYBGPP[23], GGAAGG[24], CAHWEN[8], PAPRVA[25]. B (bottom left): Rigid-body superposition of the nine structures with respect to their C_α atoms. C (bottom right): Rigid-body superposition of the nine structures with respect to either of the two β-turn amide units (note that mirror images were used for some structures so that all macrocyclic rings extend to the right).

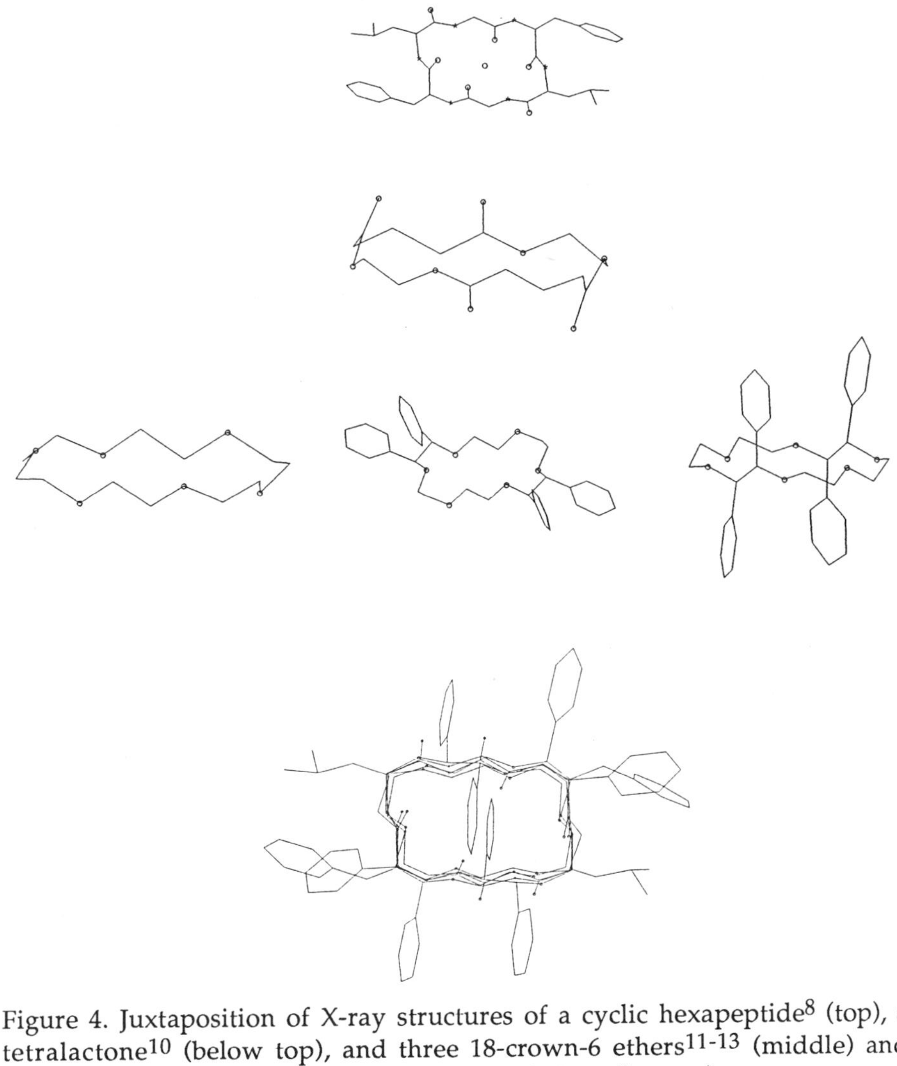

Figure 4. Juxtaposition of X-ray structures of a cyclic hexapeptide[8] (top), a tetralactone[10] (below top), and three 18-crown-6 ethers[11-13] (middle) and rigid-body superposition of the 18-membered rings (bottom).

provides a telling illustration of how a high-energy conformation can be stabilized by crystal packing forces, i.e., by intermolecular interactions with a structured molecular environment as is also available at protein binding sites.

Although the overall ring shapes of the selected cyclic hexapeptides are quite similar, they show considerable conformational diversity. This can be revealed by detailed secondary structure analyses for the cyclic peptide backbones, but is most easily seen by rigid-body superpositions of the ring systems with respect to either their six C_α atoms (Figure 3b) or the central peptide unit of a selected β-turn (Figure 3c). The latter mode of superposition is particularly relevant in view of a design of cyclic analogs to the β-turn of the TIM interface loop. In fact, packing analyses for this loop shows that it is tightly sandwiched between two domains of the other subunit (Figure 1b). Thus, appropriate superposition of the experimental cyclohexapeptide structures directly on top of the interface loop allows us quickly to sort out unsuitable conformations and to use the remaining ones as templates for the modelling of cyclohexapeptides with amino acid side chain patterns that are spatially and functionally complementary to the topology of the recognition pocket. This procedure allows us to concentrate on a relatively small number of particularly promising loop analog candidates out of an unmanageably large number of possibilities (160'000 different Thr-Gly containing cyclohexapeptides could be constructed with only the twenty regular amino acids of proteins).

The exposed and interlocked Thr-Gly β-turn appears to be a highly conserved structural motif of TIM. Thus, it is of interest to examine whether the exposed Thr-Gly β-turn is a unique structural motif of the dimeric TIM enzyme or whether it is also found in other protein structures. A protocol of related searches for primary and secondary structure information in ROPDB (ROche Protein structure Data Base)[1,2] is reproduced in Figure 6. Again, relevant information is readily gathered. A number of occurrencies of Thr-Gly or Ser-Gly β-turns are uncovered in a variety of proteins and enzymes. The observation that this structural motif is also found well exposed on the surface of some antibodies[14] and viral coat proteins[15] indicates interesting opportunities for further potential applications of cyclic Thr-Gly containing loop analogs.

Cyclic hexapeptides are still quite flexible molecules. Theoretical conformation analyses for the prototype system cyclo-hexaglycyl in the gas-phase using our MOLOC model builder system produces over 100 conformations within an energy band of 0-10 kcal/mol. If the topological symmetry is reduced by incorporation of one or more non-Gly amino acids, the number of possible low-energy conformations may increase up to about six-fold, not counting the conformational diversity arising from side chain flexibilities. On first site, this large number may come as a surprise. However, it implies that there are on average some 1.7 different orientations per flexible bond (the backbone of a cyclic hexapeptide contains twelve flexible bonds), thus conforming to the earlier findings of

R O P D B   S E A R C H   for   Phe-Thr-Gly-Glu

	command	set	occs/hits	CPU-sec
PE	peptide sequence			
	FTGE	1	2 occur	0.3
	F[TS]G[DE]	2	6 occur	0.5
	[FLIMY][ST]G[DE]	3	18 occur	0.9
	[FLIMY][ST]G[DENQ]	4	24 occur	1.0
	[ST]G	5	885 occur	0.3
KA	Kabsch & Sander sec.str.assignments			
	T* (b-turn, any type)	6	12181 hits	0.05
	T1 (b-turn, type I)	7	5011 hits	0.05
	T2 (b-turn, type II)	8	2120 hits	0.05
	T3 (b-turn, type III)	9	2441 hits	0.05
MER	Boolean set operations			
1 A 8	FTGA in b-turn II	10	4 hits	0.05
5 A 8	[ST]G in b-turn II	11	143 hits	0.05

Figure 6. Protocol of a search in ROPDB for short peptide fragments related to the tetrapeptide Phe-Thr-Gly-Glu of the TIM interface loop (the CPU-times are for a VAX-8700).

conformation analyses on other macrocyclic systems[9]. Although solvation effects and specific interactions involving amino acid side chains may be expected to reduce the number of potentially observable co-existing conformations, we may wish to constrain conformational flexibility by incorporating rigid building blocks.

The fact that the majority of cyclic hexapeptides, in their crystal states, adopt conformations with two β-turns on opposite sites provides the interesting possibility to incorporate a rigid β-turn analog at one site, while keeping the other β-turn domain in the macrocyclic system untouched. If properly designed, such a building block may not only reduce conformational flexibility of the macrocyclic system, but also act as a template inducing the formation of a β-turn in the attached peptide unit. The concept may be illustrated by the incorporation of the bicyclic lactam system shown in Figure 7, which has been proposed by Sato & Nagai[16] as an analog to the D-Phe-L-Pro β-turn of type II' found in the cyclic decapeptide antibiotic gramicidin S. In Figure 7, we use the enantiomer of the proposed bicyclic lactam to substitute for a dipeptide β-turn type II in a cyclic hexapeptide. This bicyclic structure fits well on top of the peptide β-turn unit. Furthermore, energy minimization of a tetrapeptide attached to this template in a preformed turn conformation results in an unstrained low-energy conformation with a β-turn opposite to the bicyclic template. While this represents the standard procedure by which novel oligocyclic templates can be assessed as potential analogs of a peptide β-turn, a full conformational analysis of the template-fixed peptide is required if we are to assess the potential of this template to induce the desired β-turn at the opposite site in the macrocyclic system.

Such conformational analyses can be performed within reasonable CPU-times using the generic ring shape algorithm for loops[9]. Again, a considerable number of conformations are collected in the energy range of 0-10 kcal/mol. Gratifyingly, the desired conformation is among the low-energy structures. However, there are many other conformations of similar or even lower energies, most of which do not exhibit the desired conformational characteristics (Figure 8a). One of the reasons for the occurrence of undesired low-energy conformations is the very presence of the lactam unit in the bicyclic template, the carbonyl group of which can engage in hydrogen bond interactions with backbone amide groups, resulting in differently folded macrocyclic structures (Figure 8b). Based on this finding, analogs to the bicyclic lactam system, lacking the carbonyl group, can be proposed immediately and their template properties assessed as described above. It is our experience that the conformational consequences of the incorporation of rigid templates into macrocyclic ring systems can be assessed properly only by means of exhaustive searches for all low-energy conformations, and by suitable screening facilities that allow us interactively to analyze relevant conformational aspects.

Another approach to the design of a β-turn analog may start with a search for all macrocyclic lactam or lactone derivatives with nine- to eighteen-membered rings in the structural database. Several hundred well resolved crystal structures can be retrieved from ROCSD. Such a large body of structural information requires special tools for efficient interactive screening of the structures. A particularly useful way consists in a superposition of all structures on top of selected sets of peptide β-turns retrieved from the protein structure database. This then allows us to page through all structures interactively on the screen while keeping the sets of peptide β-turns as visible references in the background. In this way, potentially interesting macrocyclic systems become immediately evident. Two examples are provided in Figure 9.

The first consists of a phthaloyl diamide of bis-aminoethylether[17]. This structure is remarkable in view of its potential analogy to a Pro-X β-

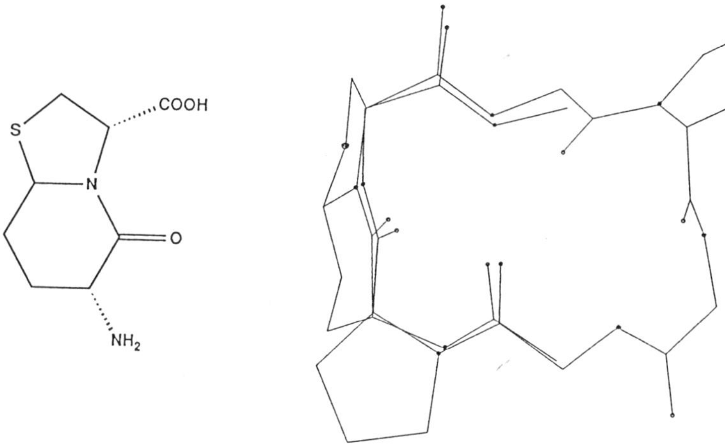

Figure 7. Bicyclic template as a potential substitute for a β-turn[16] (left) and superposition of this template to an X-ray structure of a cyclic hexapeptide taken from ROCSD (reference code CGLPGL[21]).

turn, its synthetic availability, and its possibilities for derivatization. Furthermore, inspection of the structural superposition immediately suggests a number of potentially interesting analogs, most of them not in the database, in which the benzene ring, the nonfunctional amide group, or the saturated ether linkage are replaced by related structural units. This example illustrates that there are generally a variety of criteria, apart from purely geometric ones, that are relevant in assessing the analog potential of a given structure. Often such criteria arise on an ad hoc basis from a visual inspection of structures and structural superpositions. The human interface is indispensable here. This requires sophisticated tools for efficient structure examination and evaluation.

The second compound is caprinolactam[18] with an eleven-membered ring system. We note that the s-trans amide unit is essentially planar and that the macrocyclic ring system nicely fills the space spanned by the set of β-turns. In particular, the ring is similarly folded at the corresponding C_α positions where amino acid side chains would have to be incorporated. Unfortunately, saturated eleven-membered lactam derivatives will be very flexible. For the unsubstituted prototype system, we calculate over 200 conformations within a 10 kcal/mol energy band. Again, some reduction of this number may be expected upon introduction of side chains. However, further structural modifications to constrain conformational flexibility are clearly desired. Of course, there could be countless attempts to achieve this goal. Three typical structural modifications, annelation of ring units, embedding of rigid unsaturated moieties, incorporation of bridging units, are exemplified in Figure 10.

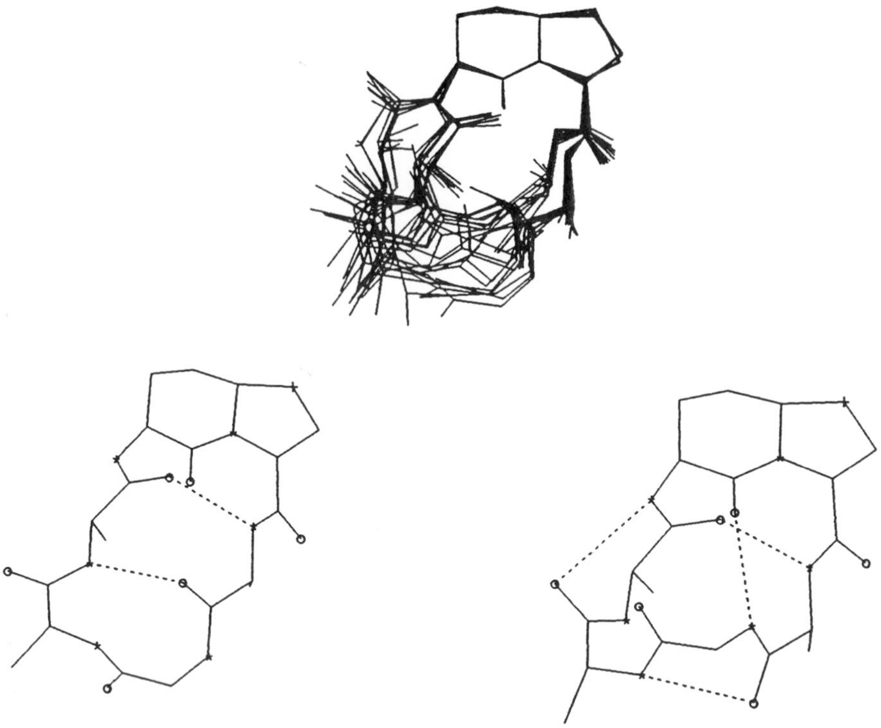

Figure 8. Conformation analysis for a tetrapeptide Ala-Gly-Ala-Ala attached to the bicyclic template shown in Fig. 7. A (top): rigid-body superposition of the first 19 conformations lying within 5 kcal/mol relative to the lowest-energy conformation. B (bottom): hydrogen bond analysis (dotted lines) for the lowest-lying conformation with the desired β-turn conformation for the central Gly-Ala dipeptide unit (left; 17-th conformation) and one representative lower-energy conformation (right, 5-th conformation) being stabilized by additional hydrogen bonding to the lactam carbonyl group of the template.

Based on the experimental structure of caprinolactam, annelation of two cyclohexane rings at two ring sites with synclinal conformations might appear to be a viable route to reduce ring flexibility. However, a conformational analysis suggests that such derivatives may be almost as flexible as the parent system itself. A complete set of low-energy conformations is of considerable value, even if not too much meaning can be attributed to the exact energy level ordering of the individual conformations due to inadequacies of simple force field methods and the fact that calculations refer to isolated molecules. Screening the complete set of possible conformations allows us to identify those domains in the macrocycle which are particularly suitable for most consequential structural modifications. For example, compiling the distribution of antiperiplanar

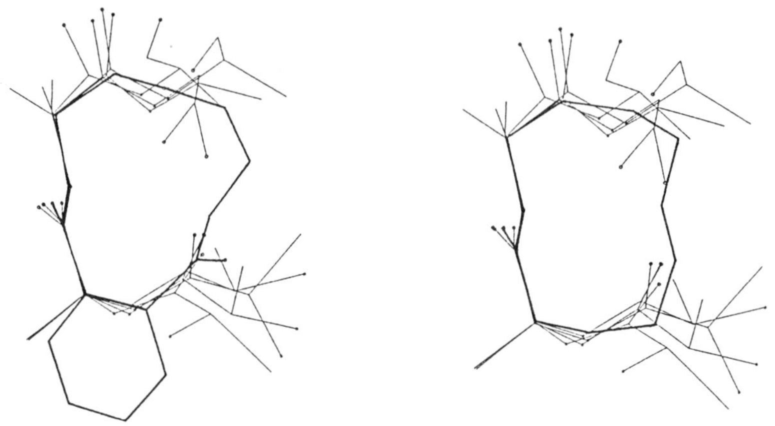

Figure 9. Rigid-body superposition of the X-ray structures of a macrocyclic phthaloyl diamide[17] (left) and caprinolactam[18] (right) to a set of representative tetrapeptide β-turn type I conformations retrieved from ROPDB.

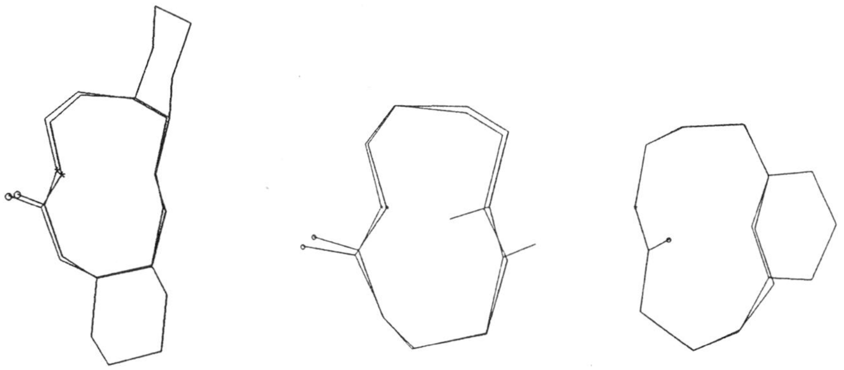

Figure 10. Superposition of the X-ray structure of caprinolactam[18] with a low-energy conformation of a modelled derivative containing two annelated cyclohexane rings (left) and a low-energy conformation of an unsaturated derivative containing a dimethyl-substituted trans double bond at C_5-C_6 (middle). The superposition at right shows one low-energy conformation calculated for caprinolactam, having two consecutive antiperiplanar arrangements for the bonds C_5-C_6 and C_6-C_7, and a low-energy conformation of a modelled bicyclic lactam derivative of the meta-cyclophane type.

single-bond arrangements in the aliphatic domain for all calculated caprinolactam conformations reveals remarkable patterns that suggest the incorporation of unsaturated or bridging units at specific locations. Thus, the comparatively high propensity of antiperiplanar arrangements found for the C_5-C_6 bond in the low-energy band of all s-trans lactam conformations calls for the incorporation of a trans-configured double bond at this location. The occasional occurrence of two successive antiperiplanar arrangements, e.g., for the bonds C_5-C_6 and C_6-C_7, suggests the incorporation of a bridging unit such as a meta-substituted benzene ring to form a meta-cyclophane. In fact, molecular modelling indicates that both structural modifications are fully compatible with an unstrained s-trans lactam system of the given ring size. Moreover, complete conformational analyses for these two derivatives reveal that the structural modifications exert dramatic filtering effects, in each case eliminating more than 90% of all conformations of the parent lactam system, while keeping the desired conformations within the remaining set of low-energy conformations.

In principle, full conformational analyses have to be performed for each new derivative, if the conformational consequences of structural modifications are to be assessed more quantitatively. However, for most design applications this level of sophistication is not required. It is then sufficient, and more convenient, to start from the complete set of conformations of a given parent system lying within a reasonably wide energy band (e.g., 0-15 kcal/mol), to define a structural modification prototypically for the first member of this set, and to have the same modifications applied automatically to all members of the set, followed by unconstrained energy minimizations. Even if this procedure does not in general generate a complete set of conformations for the modified ring system, it provides a satisfactory basis for a qualitative assessment of conformational consequences of structural modifications. It is a vital tool for the design of conformationally constrained macrocyclic systems.

It is beyond the scope of this short account to detail specific applications. Rather, its intention is to illustrate some of the potentials of combining molecular modelling techniques with compounded three-dimensional structure information. Furthermore, we wish to emphasize that computer-assisted molecular modelling systems should not be considered as potential substitutes for, but rather as tools to guide experimental work. Molecular modelling should always be performed in close relation to experimental data. In fact, one of the virtues of our modelling systems is their capabilities to gather relevant and usually widely distributed structural information quickly and to indicate new research opportunities by establishing novel relationships between such data through appropriate structure processing.

References

1. K. Mueller, H. J. Ammann, D. M. Doran, P. R. Gerber, K. Gubernator and G. Schrepfer, 'Complex Heterocyclic Structures - A Challenge for Computer-Assisted Molecular Modeling', Bull. Soc. Chim. Belg. 97:655 (1988).

2. K. Mueller, H. J. Ammann, D. M. Doran, P. R. Gerber, K. Gubernator and G. Schrepfer, 'Use of Computer Modeling and Structural Databases in Pharmaceutical Research', in: "Trends in Medicinal Chemistry '88", H. van der Goot, G. Domany, L. Pallos and H. Timmerman, eds., Elsevier, Amsterdam (1989).
3. V. J. Hruby, F. Al-Obeidi and W. Kazmierski, 'Emerging approaches in the design of receptor-selective peptide ligands: conformational, topo-graphical and dynamic considerations', Biochem. J. 268:249 (1990).
4. J. B. Ball and P. F. Alewood, 'Conformational Constraints: Nonpeptide β-Turn Mimics', J. Mol. Recogn 3:55 (1990).
5. R. K. Wierenga, K. H. Kalk and W. G. J. Hol, 'Structure Determination of the Glycosomal Triosephosphate Isomerase from Trypanosoma brucei brucei at 2.4A Resolution', J. Mol. Biol. 198:109 (1987).
6. M. Kahn, S. Wilke, B. Chen, K. Fujika, Y.-H. Lee and M. E. Johnson, 'The design and synthesis of mimetics of peptide β-turns', J. Mol. Recogn. 1:75 (1988).
7. D. S. Kemp and W. E. Stites, 'A convenient preparation of derivatives of 3(S)-amino-10(R)-carboxy-1,6-diazacyclodeca-2,7-dione, the dilactam of L-α-γ-diaminobutyric acid and D-glutamic acid: a β-turn template', Tetrahedron Lett. 29:5057 (1988).
8. C. L. Barnes and D. van der Helm, 'Conformation and Structures of Two Cycloisomeric Hexapeptides: cyclo(L-Phe-D-Leu-Gly-L-Phe-L-Leu-Gly) Tetrahydrate and cyclo(L-Phe-D-Leu-Gly-D-Phe-L-Leu-Gly) Dihydrate', Acta Cryst. B38:2589 (1982).
9. P. R. Gerber, K. Gubernator and K. Müller, 'Generic Shapes for the Conformation Analysis of Macrocyclic Structures', Helv. Chim. Acta 71:1429 (1988).
10. A. Shanzer, N. Mayer-Shochet, F. Frolow and D. Rabinovich, 'Synthesis with Tin Templates: Preparation of Macrocyclic Tetralactones', J. Org. Chem. 46:4662 (1981).
11. E. Maverick, P. Seiler, W. B. Schweizer and J. D. Dunitz, '1,4,7,10,13,16-Hexaoxaoctadecane: Cryctal Structure at 100 K', Acta Cryst. B36:615 (1980).
12. E. Blasius, R. A. Rausch, G. D. Andreetti and J. Rebizant, 'Darstellung und Komplexbildung phenyl- und (4-phenylbutyl)-substituerter Kronenether', Chem. Ber. 117:1113 (1984).
13. G. Weber, G. M. Sheldrick, T. Burgemeister, F. Dietl, A. Mannschreck and A. Merz, 'Crystal Structure and Solution Conformation of the meso-Forms of 2,3,11,12-Tetraphenyl-[18]crown-6', Tetrahedron 40:855 (1984).
14. M. Marquart, J. Deisenhofer, R. Huber and W. Palm, 'Crystallographic Refinement and Atomic Models of the Intact Immunoglobulin Molecule Kol and its Antigen-binding Fragment at 3.0 A and 1.9 A Resolution', J. Mol. Biol. 141:369 (1980).
15. E. Arnold and M. G. Rossmann, 'The Use of Molecular-Replacement Phases for the Refinement of the Human Rhinovirus 14 Structure', Acta Cryst. A44:270 (1988).

16. K. Sato and U. Nagai, 'Synthesis and Antibiotic Activity of a Gramicidin S Analogue containing Bicyclic β-Turn Dipeptides', J. Chem. Soc. Perkin Trans. I, 1231 (1986).
17. Yu. G. Ganin, E. V. Ganin, Yu. A. Simonov, V.F. Anikin and G.L. Kamalov, 'Structure of macrocyclic phthalic acid diamides', Zh. Strukt. Khim 23:103 (1982).
18. F. K. Winkler and J. D. Dunitz, 'Medium-Ring Compounds. XXVII. Caprinolactam Hemihydrochloride', Acta Cryst. B31:286 (1975).
19. M. B. Hossain and D. van der Helm, 'Conformation and Crystal Structures of Two Cycloisomeric Hexapeptides: cyclo-(L-Alanyl-L-alanylglycylglycyl-L-alanylglycyl) Monohydrate (I) and cyclo-(L-Alanyl-L-alanylglycyl-L-alanylglycylglycyl) Dihydrate (II)', J. Am. Chem. Soc. 100:5191 (1978).
20. K. I. Varughese, G. Kartha and K. D. Kopple, 'Crystal Structure and Conformation of cyclo-(Glycyl-D-leucyl-L-leucyl)$_2$', J. Am. Chem. Soc. 103:3310 (1981).
21. E. C. Kostansek, W. E. Thiessen, D. Schomburg and W. N. Lipscomb, 'Crystal Structure and Molecular Conformation of the Cyclic Hexapeptide cyclo-(Gly-L-Pro-Gly)$_2$', J. Am. Chem. Soc. 101:5811 (1979).
22. C.-H. Yang, J. N. Brown and K. D. Kopple, 'Crystal Structure and Solution Studies of the Molecular Conformation of the Cyclic Hexapeptide cyclo-(Gly-L-His-Gly-L-Ala-L-Tyr-Gly)', J. Am. Chem. Soc. 103:1715 (1981).
23. J. N. Brown and C. H. Yang, 'Crystal and Molecular Structure of the Cyclic Hexapeptide cyclo-(Gly-Pro-d-Phe)$_2$', J. Am. Chem. Soc. 101:445 (1979).
24. I. L. Karle, J. W. Gibson and J. Karle, 'The Conformation and Crystal Structure of the Cyclic Polypeptide cyclo-(Gly-Gly-D-Ala-D-Ala-Gly-Gly) . $3H_2O$', J. Am. Chem. Soc. 92:3755 (1970).
25. J. L. Flippen-Anderson, 'Conformation of the cyclic hexapeptide (D-Phe-Pro-Val)', Pept., Struct. Biol. Funct., Proc. Am. Pept. Symp., 6th, 145 (1979).

Supramolecular Reactivity and Catalysis of Phosphoryl Transfer

Jean-Marie Lehn

Université Louis Pasteur, Strasbourg and
Collège de France, Paris

Reactivity and catalysis represent major features of the functional properties of supramolecular systems. Molecular receptors bearing appropriate functional groups may bind selectively to a substrate, react with it, and release the products. Supramolecular reactivity and catalysis thus involve two main steps: recognition of the substrate followed by transformation of the bound species into products. The design of efficient and selective molecular catalysts may give mechanistic insight into the elementary steps of catalysis, provide new types of chemical reagents, and produce models of reactions effected by enzymes that reveal factors contributing to enzymatic catalysis.

Bond cleavage reactions have been extensively studied in this respect. A further step lies in the design of systems capable of inducing bond formation, which would thus effect synthetic reactions as compared to degradative ones. In order to realize "bond-making" rather than "bond-breaking" processes, the presence of several binding and reactive groups is essential. Such is the case for coreceptor molecules in which subunits may cooperate for binding and transformation of the substrates; as a consequence, they should be able to perform co-catalysis by bringing together substrate(s) and cofactor(s) and mediating reactions between them within the supramolecular complex.

These reactional properties of supramolecular systems will be illustrated by several processes: - activation of ester cleavage by binding; - hydrogen transfer to a bound substrate; - cleavage of activated esters bound to functionalized macrocycles; - hydrolysis of adenosine triphosphate (ATP) by macrocyclic polyamines. In particular, bond formation has been realized in a system performing catalytic substrate phosphorylation and pyrophosphate synthesis via cocatalysis by a macrocyclic polyamine. These reactions are represented on the following figures.

Bioorganic Chemistry in Healthcare and Technology, Edited by U.K. Pandit and F.C. Alderweireldt, Plenum Press, New York, 1991

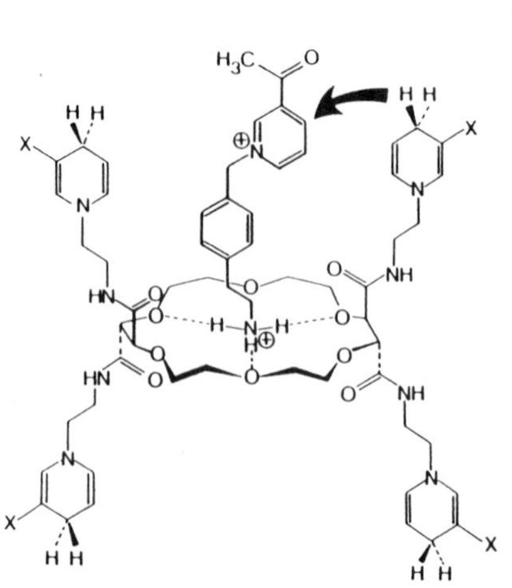

Hydrogen Transfer

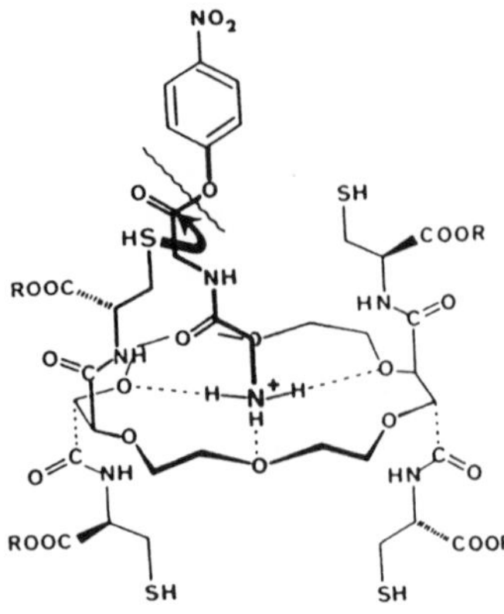

Activated Ester Cleavage

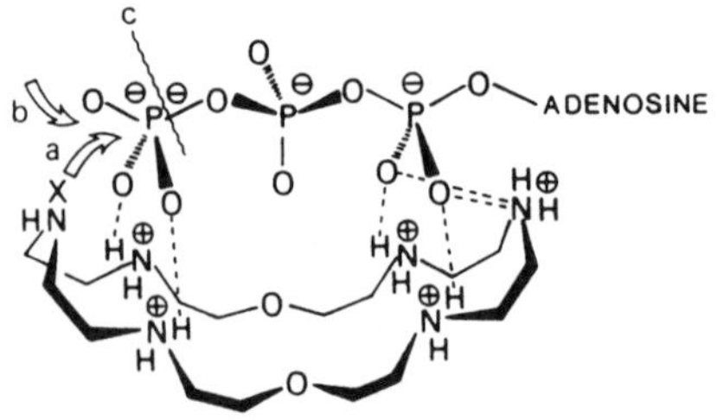

ATP Hydrolysis

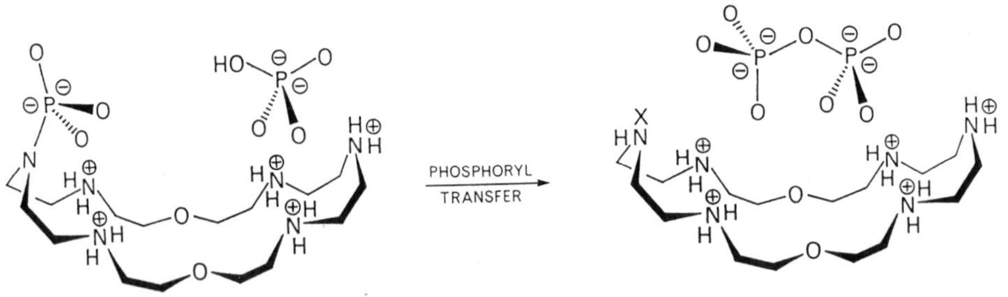

Pyrophosphate Synthesis

POLYMER ASSISTED MOLECULAR RECOGNITION: THE CURRENT UNDERSTANDING OF THE MOLECULAR IMPRINTING PROCEDURE

Günter Wulff

Institute of Organic and Macromolecular Chemistry
Heinrich-Heine University of Düsseldorf
4000 Düsseldorf F.R.G.

INTRODUCTION

Molecular recognition is of great importance in many areas in the biological world. The most well-known are enzyme-substrate recognition, receptor-hormone recognition, and antibody-antigen recognition. In all these cases the receptor consists of a void or a cleft in a biopolymer in which suitably arranged binding site groups effect binding. With this method the shape of the void as well as the position of the binding sites have considerable influence on the selectivity of recognition. In enzymes the binding of the right substrate in a stereochemically controlled arrangement is a substantial prerequisite for catalysis. In addition, enzyme active sites usually contain nucleophilic, electrophilic, basic or acidic groups originating from amino acid side chains. They are precisely positioned to react with a bound substrate.

Researchers have been trying for a considerable time to synthesize artificial enzyme analogues which possess a similarly high substrate-, reaction-, and stereoselectivity and which are at the same time better accessible, more stable, and catalyze a larger variety of reactions[3,4]. Furthermore, such enzyme analogues offer the opportunity to study the characteristics of enzyme catalysis in greater detail and thereby to gain a better understanding of the whole process.

The crucial problem of the construction of enzyme models is the synthesis of the active center. The following conditions have to be created for the construction of enzyme models in order to approach enzyme similarity[5]:

a) A cavity or a cleft must be prepared with a defined shape corresponding to the shape of the substrate of the reaction. For catalysis it is preferable if the shape of the cavity corresponds to the shape of the transition state of the reaction. If the transition state of the reaction is more strongly bound than the substrate or the product the activation energy of the reaction is lowered and turnover numbers of the reaction are increased. In many enzyme models turnover numbers are low, and sometimes inhibition is observed since it is not the transition state which is more firmly bound, but the product or a reaction intermediate.

Bioorganic Chemistry in Healthcare and Technology, Edited by U.K. Pandit and F.C. Alderweireldt, Plenum Press, New York, 1991

b) A procedure must be developed in which the functional groups acting as coenzyme analogues, binding sites or catalytic sites are introduced into the cavity in the right stereochemistry.

c) Since binding of the substrate or the transition state of the reaction in the active center of the enzyme is a rather complex overlap of different interactions which are not known in detail, simplified binding mechanism have to be developed. For each substance binding site group should be found exhibiting fast and reversible interactions.

d) In many enzymes coenzymes or prosthetic groups play an important role. Analogues of such groups should be developed.

e) An important part of the preparation of an enzyme model is the choice of a suitable catalytically active environment. In enzymes the cooperativity of different catalytically active groups is of importance. In models one normally has to choose simplified systems without omitting the essential characteristics. In enzyme models, in contrast to enzymes themselves, a large variety of different catalytically active groups can be choosen since one is not restricted to the functional groups of amino acid side chains.

In the past many approaches have been used to achieve the construction of an enzyme model. The results in many cases were rather poor and only low catalytic activities were observed. Successfull experiments in the past used low-molecular-weight ring systems as the binding site, like crown ethers[6], cryptates[7], cyclophanes[8], cyclodextrines[8,9], concave molecules[10] etc. and attached catalytically active functional groups. An extremely interesting approach of recent years used monoclonal antibodies[11,12]. These antibodies were generated against the transition state analogue of a reaction and they showed remarkably high activity in catalysis. In other experiments antibodies with defined binding sites and a catalytically active group have been generated which show high catalytic activity as well.

In principle a procedure similar in approach to the generation of an antibody against a reactant was introduced quite some time ago by our group[13-16]. In our case polymerizable binding site groups are bound by covalent or non-covalent interaction to a suitable template molecule [see Scheme I]. This template monomer is copolymerized in the presence of a high amount of crosslinking agent. After splitting off the template molecule from the polymer, microcavities are obtained as imprints which have a shape and an arrangement of functional groups corresponding to that of the template. In this review the principle will be briefly discussed, followed by an analysis of the new results which facilitate the understanding of the whole procedure. Finally, there will be a report of the first results in stereoselective reactions using these polymers.

THE PREPARATION OF CHIRAL CAVITIES IN CROSSLINKED POLYMERS

An example for the imprinting procedure is the polymerization of the template monomer 1[15-17]. In this case phenyl-α-D-mannopyranoside acts as the template. Two molecules of 4-vinylbenzeneboronic acid are bound by diester linkages to this template. In this case the binding sites (the boronic acids) are bound by a covalent interaction. The monomer was copolymerized by free radical initiation in the presence of an inert solvent with a large amount of bifunctional crosslinking agent. Polymers thus obtained are macroporous possessing a permanent pore structure and a

Scheme I. Schematic representation of the preparation of microcavities with functional groups by an imprinting procedure using template molecules.

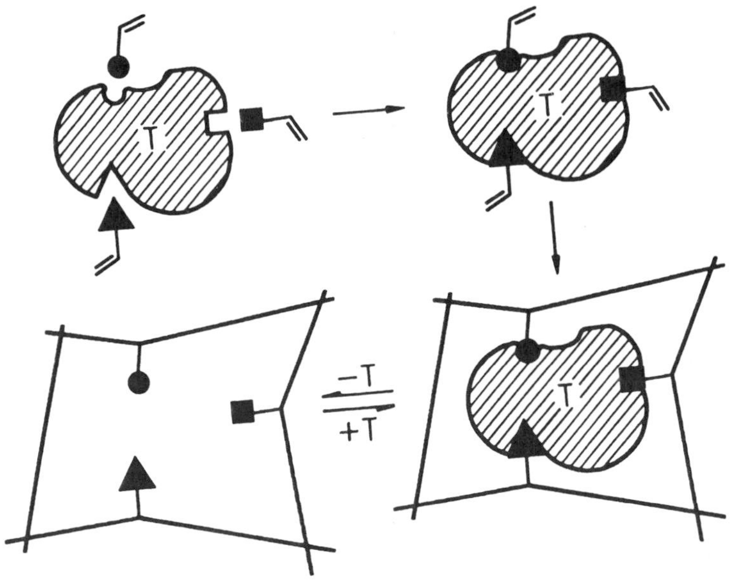

high inner surface area. Polymers of this type possess good accessibility on the surface of the pores and a rather rigid structure with low mobility of the polymer chains.

The template can be split off by water or methanol to an extent of up to 95% (see Scheme II). The accuracy of the steric arrangement of the binding sites in the cavity can be tested by the ability of the polymer to resolve the racemate of the template, in this case of phenyl-α-D,L-mannopyranoside. Therefore the polymer is equilibrated in a batch procedure with a solution of the racemate under conditions under which rebinding in equilibrium is possible. The enrichment of the antipodes on

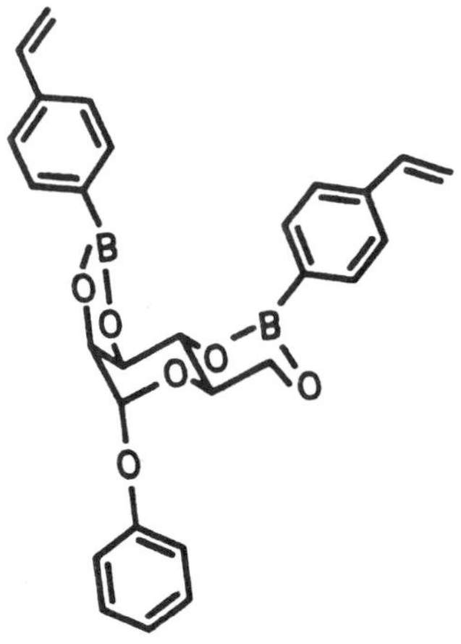

Scheme II. Schematic representation of the removal and uptake of the template molecule phenyl-α-D-mannopyranoside from a highly crosslinked copolymer of 1.

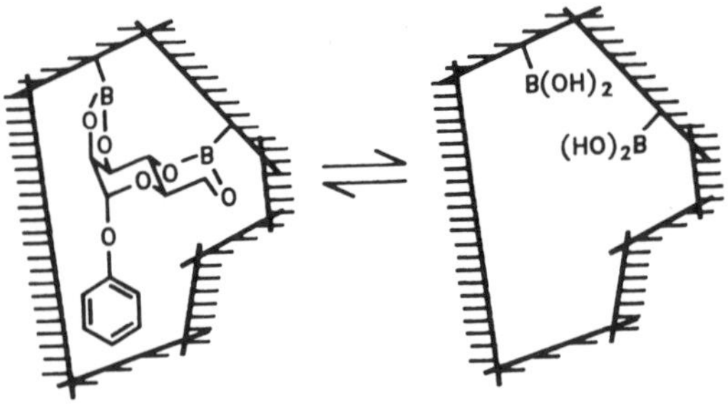

the polymer and in solution is determined and the separation factor α, i.e. the ratio of the distribution coefficients of D and L compound between polymer and solution is calculated. After extensive optimization of the procedure, α-values between 3.5 - 6.0 were obtained[17,18]. This is an extremely high selectivity for racemic resolution that cannot be reached by most other methods.

Polymers obtained by this procedure can be used for the chromatographic separation of the racemates of the template molecules[1,19-21]. The selectivity of the separation process is fairly high (separation factors up to α = 4.56) and at higher temperatures with gradient elution, resolution values of R_S = 4.2 with base line separation have been obtained (see Fig. 1). These sorbents can be prepared conveniently and possess excellent thermomechanical stability. Even when used at 80°C under high pressure for a long time, no leakage of the stationary phase nor decrease of selectivity during chromatography was observed.

Apart from using sugar derivatives as templates with this method amino acids, hydroxy carboxylic acids, diols and other racemates have been separated[16]. Furthermore it has been possible to localize two amino groups in a defined distance at the surface of silica with this method[22]. These materials bind strongly by a cooperative binding dialdehydes or diacids in those cases where the functional groups have the correct distance from each other.

Other research groups in the meanwhile have also been working on molecular imprinting in crosslinked polymers[16,23-26]. By using a variety of different templates and binding site groups the scope of the method has been enlarged. The group led by Mosbach[27,28] concentrated on non-covalent interactions during polymerization and chromatography. In this way they were able to completely resolve derivatives of amino acid racemates.

FUNDAMENTALS OF THE IMPRINTING PROCEDURE

Origin of the Chirality of the Imprinted Polymers

The asymmetry of the empty cavities can be analyzed by the excellent racemate resolution ability but it can also be directly detected by

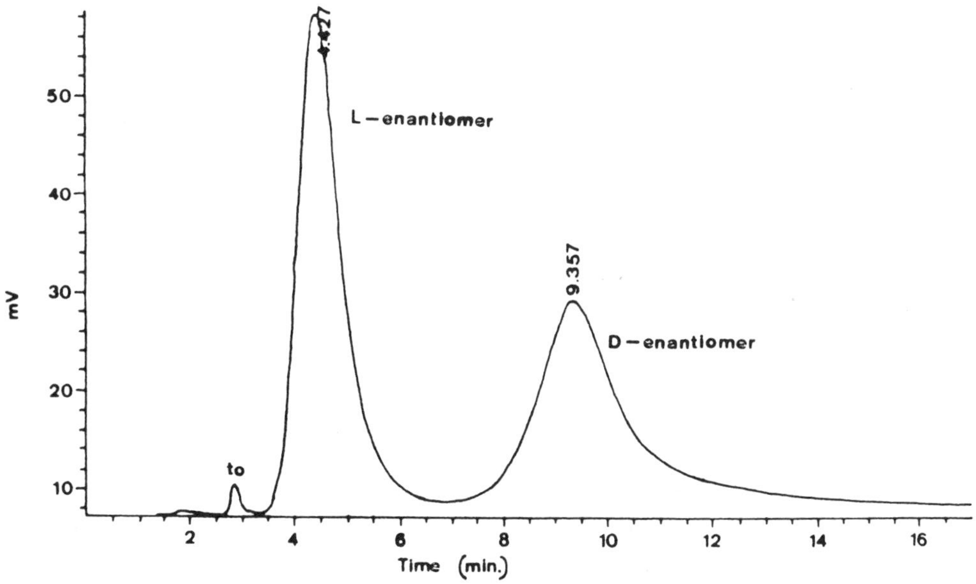

Fig. 1. Chromatographic separation with gradient elution of phenyl-α-D,L-mannopyranoside on a polymer prepared from **1**. Temp. 90°C[21].

measuring the optical activity[29]. This is measured by suspending the polymer in a solvent which has the same refractive index as the polymer, a technique which was developed for other types of insoluble polymers. The values of molar optical rotation measured are shown in Table I.

If we compare the value -61.7° for polymer P1 containing the template, with the value -448.9° for the template monomer 1 it becomes apparent that the molar optical rotation value has decreased considerably as a result of the polymerization. This could have several causes, one being the influence of the polymer matrix. Its effect can be determined by splitting off the optically active template. If the boronic acids are converted with an achiral diol to the ethyleneglycol ester, the polymer gives a positive molar rotation $[M]^{20}_{546}$ +110.0°. This shows that in P1' the imprints generated in the polymer make a positive contribution to the optical rotation value. Measuring the optical rotation in the solid phase thus allows the properties of chiral cavities in the polymer to be directly determined, like the binding situation of a bound substrate, different swelling situations etc.[29].

Table I. Molar optical rotation values for polymers with chiral cavities

	Template monomer 1	Polymer P1 with template	Polymer P1' template split off[a)]
$[M]^{20}_{546}$	-448.9°	-61.7°	+110.0°

a) as the ethyleneglycolester

In this case the optical rotation is not caused by individual chiral centers, as is usual, but by the empty imprints as a whole. Their chiral construction is stabilized by means of the crosslinking of the polymer chains. This type of chirality can arise from the asymmetric configuration of the crosslinking points, as well as from asymmetric conformations of the polymer chains which are stabilized by crosslinking. The extent to which these two factors contribute is not known.

It is unlikely that the chiral configurations of the linear portions of the chains contribute to the asymmetry of the cavity since no asymmetric cyclocopolymerization is possible for the template monomer 1. With other types of template monomers, though, such type of contribution of backbone-chiral portions of the polymer might be expected[30,31]. Another possible reason for the observed racemic resolution could be the existance of the few optically active templates that could not be split off. Careful investigations showed that racemic resolution is not brought about by the interaction of the racemate with the remaining templates[16].

Temperature Dependence of Selectivity

There is a considerable influence of temperature on the selectivity of racemic resolution with polymers prepared by imprinting[18-21]. It is both influenced by the thermodynamics and by the kinetics of the binding process. If the equilibration of the polymer with the racemate is performed at different temperatures[18] the α-values increase with increasing temperature. Whereas an α-value of 4.7 is observed at 20°C, the same polymer under otherwise identical conditions shows an α-value of 5.5 at 60°C. The reason for this strong increase might be twofold. The monomeric mixture with the template monomer by initiation with azo-bis(isobutyronitrile) was polymerized at 65°C for 48 hrs.. Therefore the imprinting occurred at this temperature, and after splitting off the template the cavity might be better preserved at 60°C than at 20°C. To elucidate this question a second polymer was prepared with the initiator azo-bis(2,4-dimethyl-valeronitrile) that could be used for polymerization at 20°C for 12 days. If this polymer is equilibrated at 20°C and 60°C the difference is much less pronounced (20°C α = 3.1; 60°C α = 3.2)showing that the assumption is, at least partially, correct[18]. A similar result was obtained by Mosbach et al.[32]

A second reason for this temperature dependence could be that at a higher temperature the embedding process is faster and more cavities, especially the most selective ones, are loaded. At higher temperature the swelling of the polymer is stronger and the size of the cavity will be enlarged. Therefore the embedding is facilitated. Even highly crosslinked polymers freed from the template can possess a swellability of more than 100% in the solvent used for the equilibration in the batch procedure. On adding the template, the original volume is restored by deswelling. Assuming that the swelling occurs with solvation of the functional groups in the microcavities, the microcavities apparently restore their original form on binding, as the high selectivity of these polymers indicate. This could be regarded as an analogy to the "induced fit" in enzymes. Furthermore, under these conditions the substrate will be preferably bound by the more selective two-point binding. In our example this means that phenyl-α-D-mannopyranoside is bound by two boronic ester bonds. Calorimetric measurements showed that at room temperature the correct enantiomer is bound to less than 50% by a two-point binding, the other part being bound by a one-point binding. At a higher temperature the two-point binding increases strongly[29]. The "wrong" enantiomer does not fit into the cavity with a two-point binding and is preferably bound by a one-point binding.

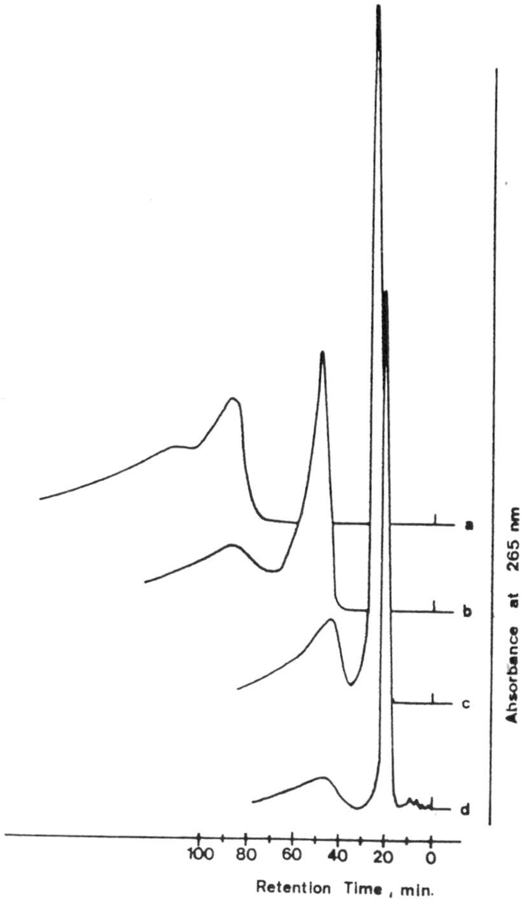

Fig. 2. Chromatographic separation at different temperatures of racemic phenyl-α-mannopyranoside with a support prepared from 1. a. 25°C; b. 40°C; c. 65°C; d. 80°C[19].

Using these polymers as chromatographic supports for h.p.l.c. separations the influence of both the temperature on the thermodynamics (i.e. on the separation factor) and on the kinetics (i.e. on the number of theoretical plates) can be directly inspected. Fig. 2 shows a chromatographic separation at different temperatures[19]. At a higher temperature the mass transfer is strongly increased and the separations become much better (see also Fig. 1)[21]. The D-form (the "right" enantiomer), in particular, shows stronger spreading. The difference may be ascribed to the hindrance imposed on the D-enantiomer on its entrance and subsequent two-point binding into highly selective cavities. The association of the D-form within a selective cavity in a cooperative two-point binding fashion and its subsequent dissociation process might be expected to be more hindered kinetically than in a one-point binding.

Influence on Selectivity of the Arrangement of Functional Groups versus Shape Selectivity

Another important question which still remains unsolved in the molecular imprinting procedure is the precise mechanism of molecular recognition. The high selectivity observed is believed on the one hand to be mainly due to the shape of the cavity. In this case the binding sites within the cavity are primarily responsible for the driving force to bring the substrates inside the cavity. On the other hand, molecular recognition could also be due to the spatial arrangement of the functional groups (binding sites) within the cavity. A recent observation by Shea and Sasaki[33] showed that shape selectivity was the most important factor for molecular recognition in the examples investigated by them.

We adressed this problem by preparing a number of polymers with similar but different template monomers[2,34,35]. These polymers were then tested for their ability for racemic resolution for a variety of different racemates. The racemates were on the one hand of different shape and on the other hand of different arrangement of the groups interacting with the binding sites.

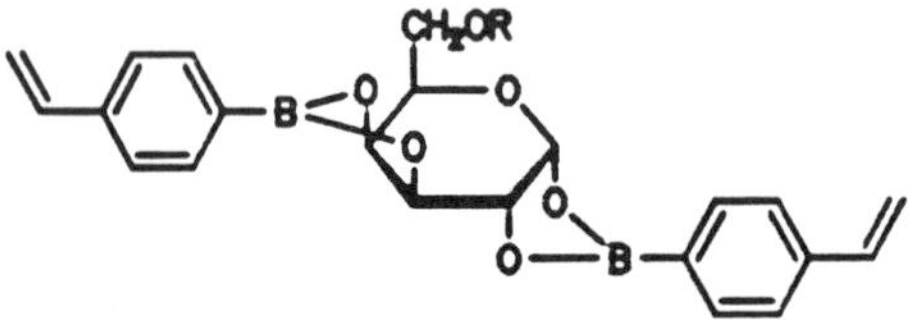

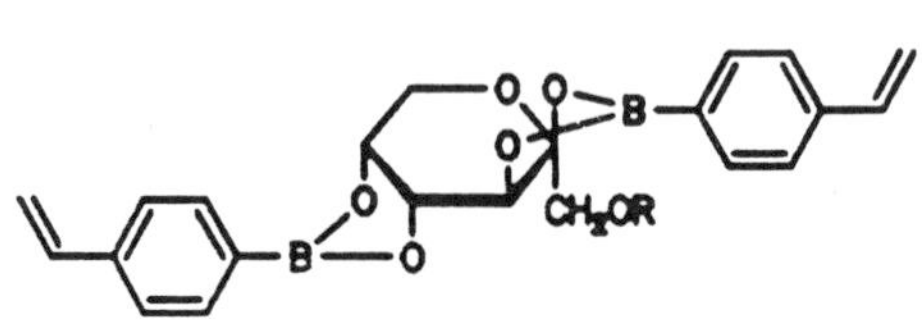

2a R=H
2b R=Methyl
2c R=Benzyl

3a R=H
3b R=Methyl
3c R=Benzyl

Polymers were compared, prepared from the template monomer 1, 2a-c, and 3a-c in which phenyl-α-D-mannopyranoside, ß-D-fructopyranose, 1-O-methyl- and 1-O-benzyl-ß-D-fructopyranose, α-D-galactopyranose, 6-O-methyl- and 6-O-benzyl-α-D-galactopyranose are the templates. In all cases two molecules 4-vinyphenyl boronic acid were bound by ester linkages. These polymers were prepared in the same manner as described for 1 and P1 and P1' and the templates could be split off to a high percentage. Equilibration of the corresponding racemates of the templates with their polymers showed selectivity for racemic resolution but to a lower extent, than in the case of phenyl-α-D-mannopyranoside (see Table II). The reason for the lower selectivity might be that these racemates exist in a mutarotational equilibrium in which the isomer with the correct arrangement of the diol groupings is only present to a certain degree.

Table II Selectivity for racemic resolution of polymers P3, P4, and P5

Polymer	Template monomer	Racemate	Separation factor α
P1	1	phenyl-α-D,L-mannoside	5.0
P2a	2a	D,L-fructose	1.63
P2b	2b	1-O-methyl-D,L-fructose	1.76
P2c	2c	1-O-benzyl-D,L-fructose	1.64
P3a	3a	D,L-galactose	1.38
P1	1	D,L-mannose	1.60
P1	1	D,L-fructose	1.34
P3b	3b	D,L-galactose	1.39
P3c	3c	D,L-galactose	1.58
P2a	2a	D,L-galactose	0.85 (1.17)
P3a	3a	D,L-fructose	0.80 (1.25)
P3b	3b	D,L-fructose	0.71 (1.41)

Nevertheless, it is possible to separate the racemates of the free sugars. This is more convenient than preparing suitable derivatives of the racemates, as was necessary in the past.

D,L-Mannose can be separated on a polymer prepared from 1. This shows that a considerably smaller racemate than the original one can also be separated by P1. Polymer P3c prepared from 6-O-benzylgalactose showed a much better capacity of resolution for D,L-mannose (with a separation factor $\alpha = 1.58$ which is nearly 50% better) than on P3a. In this case the cavity is enlarged by the bulky benzyl group and now apparently the embedding of the correct enantiomer is facilitated.

In these cases the racemates were of the same type as the template but had a different shape. If other sugar racemates are used, D,L-fructose can be separated on a polymer prepared with phenyl-α-D-mannopyranoside as the template. All other sugar racemates tried (arabinose, glucose, galactose, xylose, ribose) could not be separated on this polymer. Polymers imprinted with D-galactose and D-fructose showed surprising behaviour when equilibrated with the other sugar racemate. Thus the polymer imprinted with D-fructose binds preferably L-galactose from D,L-galactose and that imprinted with D-galactose binds preferably L-fructose. D,L-Mannose on the contrary is not separated on either polymers. The inverse selectivity observed on polymers P2a and P3a can be explained by inspecting molecular models of the preferred conformation of the free sugars (see Fig. 3). Both ß-D-fructopyranose and α-L-galactopyranose, which are both preferably bound by P2a, show that they don't differ in the orientation of the interacting hydroxyl groups; only the hydroxy methyl groups are at different places in the molecules. This causes these molecules to possess considerably different shapes. The selectivity, however, did not change due to this fact, but was only reduced to some extent. This clearly suggests that for selectivities in these cases it is the orientation of the functional groups inside the cavity which is primarily responsible for molecular recognition. Shape selectivity is only of secondary importance. The same is true if phenyl-α-D-mannopyranoside as the template is compared with D-mannose and D-fructose both of which can be incorporated in P1. In both cases the orientation of the diol-groupings is such that they can be incorporated in the cavities of polymer P1.

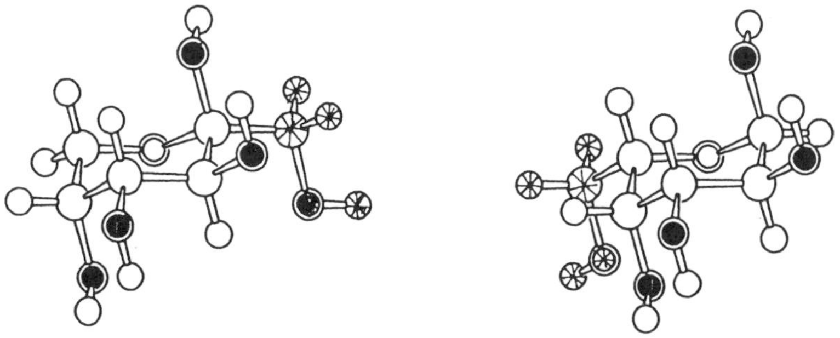

Fig. 3. Ball and stick models of the preferred conformation of ß-D-fructopyranose (left) and α-L-galactopyranose (right). The hydroxyl groups interacting with the boronic acids are marked black.

These findings have a number of consequences for the concept of molecular imprinting with templates[34]:

(a) If the orientation of the interacting groups inside the cavity is the principal factor governing the selectivity, other molecules than the templates but with a similar orientation of their functional groups can be separated. In this case it is possible to prepare a well-defined monomer as in the case of 1, and separate D,L-mannose or D,L-fructose afterwards.

(b) In contrast to imprinting with non-covalent interactions, in our system the functional groups are present in a defined number and orientation per active center. With non-covalent interactions an excess (usually four fold) of binding sites need to be incorporated to obtain reasonable selectivity. This implies that (at best) only a quarter of the functional groups is arranged in an oriented manner, while the rest are irregularly arranged all over the polymer. This may be an important shortcoming for the construction of active centers behaving in a catalytic mode. With covalent interactions on the other hand, certain functionalities can be introduced in a defined number and orientation into the active site and these groups can be transformed to other functionalities. This is especially true for the boronic acid moiety that can be easily replaced by a variety of other functional groups.

(c) It is possible to prepare active centers with a defined orientation of functional groups and with a particular shape which selectively bind certain substances or catalyze specific reactions. Moreover, it is also possible to introduce additional pockets or niches at the active centers. This can be achieved by attaching bulky substituents at the template molecule. Thus, in the case of 1 the phenyl ring provides additional space in the active centers. This principle can find application in the construction of catalytically active polymers.

REACTIONS INSIDE IMPRINTED CAVITIES

Up to now imprinted polymers with different types of templates were primarily employed for the resolution of the racemate of the template as a principal diagnostic criterion for evaluating their molecular recognition abilities. On the other hand, as outlined in the introduction, the final aim of these investigations is to use this principle for the construction of catalysts which work in a similar fashion as enzymes do. Initial attemps to use these polymers for performing enantioselective or diastereoselective organic synthesis were made in the laboratories of Shea[23] and Neckers[24], who carried out the synthesis of chiral cyclopropane derivatives and a diastereoselective dimerization of cinnamic acid, respectively. In later investigations, Sarhan et al.[26] observed an enantioselective protonation-deprotonation and Leonhardt and Mosbach[36] noticed a substrate selective hydrolysis using suitable imprinted polymers. The results of these attempts, however, were not very satisfying. The same is true for an attempt by Mosbach's group to use a transition-state analogue imprinted polymer to catalyze esterolysis in analogy to the catalytic antibodies. Only a 60% increase in rate was observed[37].

Our attempts first concentrated on an enantioselective C-C bond formation inside the chiral cavities with the objective of synthezizing optically active amino acids[38,39]. Our synthetic approach is outlined in Scheme III. The template monomer 4 is polymerized to yield a macroporous polymer **A**, following standard experimental procedures. After splitting off

Scheme III. Schematic representation of an asymmetric synthesis of α-amino acids using imprinted polymers.

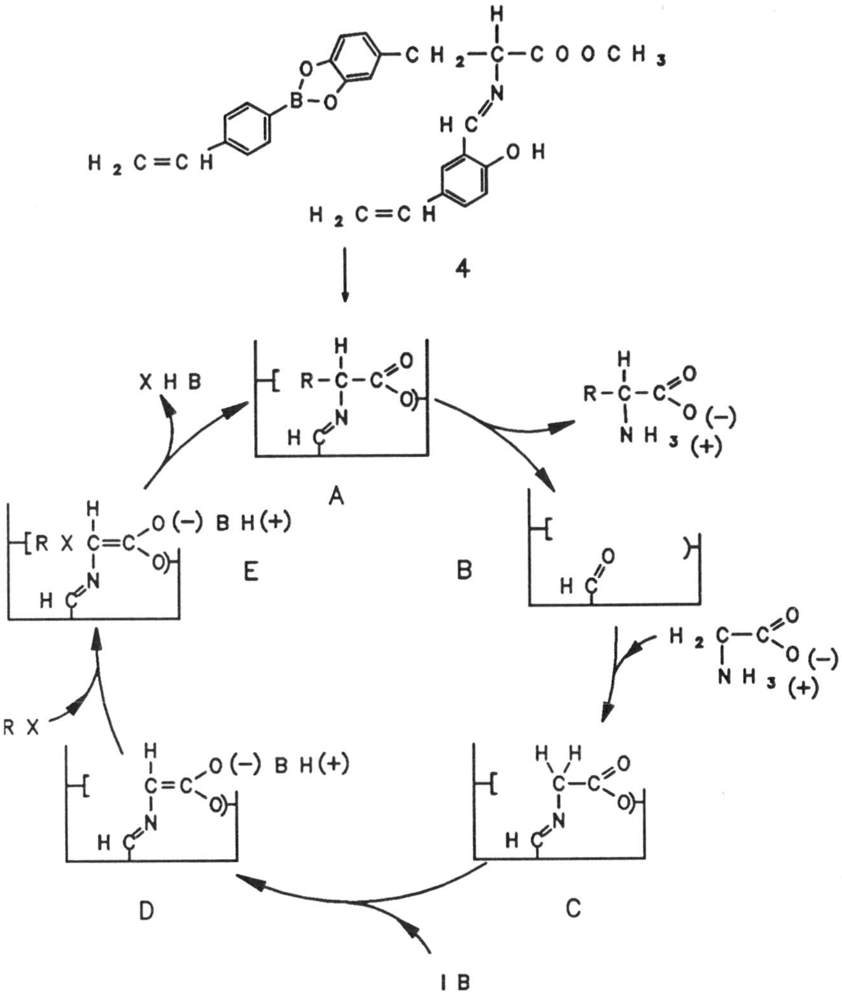

the template, a glycine molecule is bound, forming C inside the resulting free cavities of B. By deprotonation with a base Bl, the resulting ester enolate D possesses enantiotopic sites, which become diastereotopic inside the chiral cavity. The electrophilic reagent RX is subsequently bound in the cavity to yield E. Reaction of RX at one of the diastereotopic sites of the ester enolate gives the bound chiral amino acid A which can be isolated in the next step.

By following this route polymer P4 was prepared. It turned out, however, that the planned reaction between 3,4-dihydroxy benzyl halides (bound as boronic ester in the cavity, see E) and the polymer bound ester enolate of glycine was not feasible, since 3,4-hydroxy benzyl halides gave rise to a large number of by-products. Therefore the reaction between the polymer-bound-glycine enolate and acetaldehyde was studied which yielded threonine and allo-threonine. While the reaction of the bound glycine ester enolate with acetaldehyde showed poor enantioselectivity, the corresponding transition metal complexes were proven to be superior. Asymmetric inductions in this aldol type reaction can be obtained with

Scheme IV. Synthesis of L-threonine inside a chiral cavity of a polymer prepared by copolymerization of 4 and removal of the template.

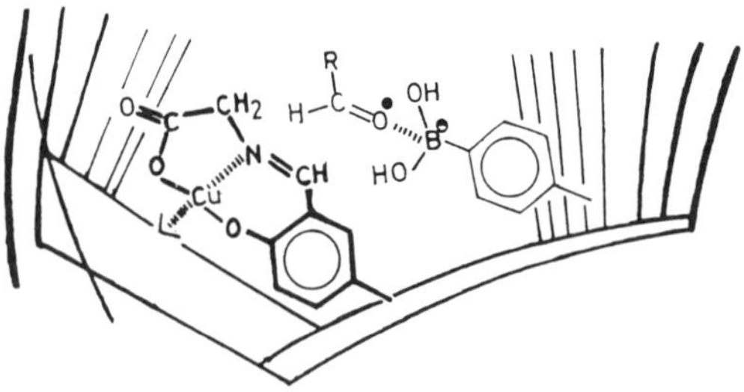

36 per cent of ee for L-threonine by performing the reaction inside the cavity. This enantioselectivity can be explained by the fixation via a transition metal complex and by the presence of a coordination of the aldehyde to the boronic acid groups. By this interaction the incoming aldehyde is oriented at the Si-face of the enolate, as shown in Scheme IV, and can preferably attack from this face. It appears from these observations that the transition state of the reaction is forced into an asymmetric arrangement by the chiral cavity.

The asymmetric induction observed in this case is by far the highest among the reported results on the asymmetric synthesis performed inside chiral cavities. In comparison with other methods of asymmetric synthesis the extent of asymmetric induction in the present case is somewhat modest. Nevertheless, this approach provides the possibility of combining the cavity effect with the conventional asymmetric synthesis, whereby the selectivity could be significantly increased, as has been observed by Fuji et al.[25] during racemic resolution with such types of systems.

CONCLUSION

Molecular imprinting using template molecules is a convenient procedure to prepare microcavities in crosslinked polymers. These microcavities possess a shape and an arrangement of functional groups inside the cavity corresponding to that of the template. These polymers show high selectivity in chromatographic racemic resolution. It is possible to resolve a variety of different racemates with this method. Usually for each racemate a separate polymer has to be prepared. If another racemate than the template is to be separated, it has to have a similar orientation of the functional groups as the template molecule. These investigations constitute important prerequisites for the construction of catalysts which are expected to act in a similar manner as natural enzymes.

ACKNOWLEDGMENT

These investigations were supported by financial grants from Minister für Wissenschaft und Forschung des Landes Nordrhein-Westfalen, Fonds der Chemischen Industrie, and Deutsche Forschungsgemeinschaft.

LITERATURE

1. Enzyme-analogue built polymers, Part 31; Part 30: cf[2]
2. G. Wulff, M. Minárik, and S. Schauhoff, Chromatograhische Racemattrennung an Adsorbentien mit chiralen Hohlräumen, GIT Fachz. Lab. in press.
3. M. L. Bender, R. J. Bergeron, and M. Komiyama, "The Bioorganic Chemistry of Enzymatic Catalysis", Wiley, New York (1984).
4. H. Dugas, "Bioorganic Chemistry - A Chemical Approach to Enzyme Action", Springer Verlag, New York (1989).
5. G. Wulff and A. Sarhan, Models of the receptor sites of enzymes in: "Chemical Approaches to Understanding Enzyme Catalysis", B. S. Green, Y. Ashani, and D. Chipman, eds., Elsevier, Amsterdam (1982), p. 106.
6. D. J. Cram, Molecular hosts and guests, and their complexes, Angew. Chem. 100, 1041 (1988); Angew. Chem. Int. Ed. Engl. 27, 1009 (1988).
7. J. M. Lehn, Supramolecular chemistry - molecules, supermolecules, and molecular functional units, Angew. Chem. 100, 91 (1988); Angew. Chem. Int. Ed. Engl. 27, 89 (1988).
8. I. Tabushi and Y. Kuroda, Cyclodextrins and cyclophanes as enzyme models, Adv. Catal. 32, 417 (1983).
9. R. Breslow, J. Chmielewski, D. Foley, and B. Johnson, Optically active amino acid synthesis by artificial transaminase enzymes, Tetrahedron 44, 17, 5515 (1988).
10. J. Rebek, Jr., Molecular recognition using concave model compounds, Angew. Chem., 102, 261 (1990); Angew. Chem. Int. Ed. Engl. 29, 245 (1990).
11. P. G. Schultz, Catalytic antibodies, Angew. Chem. 101, 1336 (1989); Angew. Chem. Int. Ed. Engl. 28, 1283 (1988).
12. P. G. Schultz, R. A. Lerner, and S. J. Benkovic, Catalytic antibodies, Chem. Eng. News 68, (22) 26 (1990).
13. G. Wulff and A. Sarhan, Use of polymers with enzyme-analogous structures for the resolution of racemates, Angew. Chem. 84, 364 (1972); Angew. Chem. Int. Ed. Engl. 11, 341 (1972).
14. G. Wulff, A. Sarhan, and K. Zabrocki, Enzyme-analogue built polymers and their use for the resolution of racemates, Tetrahedron Lett. (1973), 4329.
15. G. Wulff, W. Vesper, R. Grobe-Einsler, and A. Sarhan, Enzyme-analogue built polymers, IV. On the synthesis of polymers containing chiral cavities and their use for the resolution of racemates, Makromol. Chem. 178, 2799 (1977).
16. For a comprehensive review see: G. Wulff, Molecular recognition in polymers prepared by imprinting with templates, in: W. T. Ford (ed.), "Polymeric Reagents and Catalysts", ACS Symp. Ser. 308, Washington (1986), p. 186.
17. G. Wulff, J. Vietmeier, and H.-G. Poll, Influence of the nature of the crosslinking agent on the performance of imprinted polymers in racemic resolution, Makromol. Chem. 188, 731 (1987).
18. G. Wulff and K. Jakobi, unpublished results.
19. G. Wulff and M. Minárik, Pronounced effect of temperature on racemic resolution using template-imprinted polymeric sorbents, J. High Resolut. Chromatogr. Chromatogr. Commun. 9, 607 (1986).
20. G. Wulff and M. Minárik, Tailor-made sorbents. A modular approach to chiral separation; in: M. Zief and L. J. Crane (eds.), "Chromatographic Chiral Separations", Marcel Dekker, New York (1988), p. 15.
21. G. Wulff and M. Minárik, Template-imprinted polymers for HPLC separation of racemates, J. Liqu. Chromatogr. in press.

22. G. Wulff, B. Heide, and G. Helfmeier, Molecular recognition through the exact placement of functional groups on rigid matrices via a template approach, J. Am. Chem. Soc. 108, 1089 (1986).
23. K. J. Shea, E. A. Thompson, S. D. Pandey, and P. S. Beauchamp, Template synthesis of macromolecules. Synthesis and chemistry of functionalized macroporous polydivinylbenzene. J. Am. Chem. Soc. 102, 3149 (1980).
24. J. Damen and D. C. Neckers, Stereoselective syntheses via a photochemical template effect, J. Am. Chem. Soc. 102, 3265 (1980).
25. Y. Fujii, K. Matsutani, and K. Kikuchi, Formation of a specific coordination cavity for a chiral amino acid by template synthesis of a polymer Schiff base cobalt(III) complex, J. Chem. Soc. Chem. Commun. 1985, 415.
26. A. Sarhan and M. A. El-Zahab, Racemic resolution of mandelic acid on polymers with chiral cavities. Enzyme-analogue stereospecific conversion of configuration, Makromol. Chem. Rapid Commun. 8, 555 (1987).
27. B. Sellergren, M. Lepistö, and K. Mosbach, Highly enantioselective and substrate-selective polymers obtained by molecular imprinting utilizing noncovalent interactions. NMR and chromatographic studies on the nature of recognition, J. Am. Chem. Soc. 110, 5853 (1988).
28. For a review see: B. Ekberg and K. Mosbach, Molecular imprinting: A technique for producing specific separation materials, TIBTECH 7, 92 (1989).
29. G. Wulff and G. Kirstein, Measuring the optical activity of chiral imprints in insoluble highly crosslinked polymers, Angew. Chem. 102, 706 (1990); Angew. Chem. Int. Ed. Engl. 29, 684 (1990).
30. G. Wulff, R. Kemmerer, and B. Vogt, Optically active polymers with structural chirality in the main chain prepared through an asymmetric cyclocopolymerisation, J. Am. Chem. Soc., 109, 7449 (1987).
31. G. Wulff, Main-chain chirality and optical activity in polymers consisting of C-C-chains, Angew. Chem. 101, 22 (1989); Angew. Chem. Int. Ed. Engl. 28, 21 (1989).
32. D. J. O'Shannessy, B. Ekberg, and K. Mosbach, Molecular imprinting of amino acid derivatives at low temperature (0°C) using photolytic homolysis of azobisnitriles, Anal. Biochem. 177, 144 (1989).
33. K. J. Shea and D. Y. Sasaki, On the control of microenvironment shape of functionalized network polymers prepared by template polymerization, J. Am. Chem. Soc. 111, 3442 (1989).
34. G. Wulff and S. Schauhoff, Racemic resolution of free sugars with macroporous polymers prepared by molecular imprinting. Selectivity dependence on the arrangement of functional groups versus spatial requirements, J. Org. Chem. in press.
35. G. Wulff and J. Haarer, The preparation of defined chiral cavities for the racemic resolution of free sugars, Makromol. Chem. in press.
36. A. Leonhardt and K. Mosbach, Enzyme-mimicking polymers exhibiting specific substrate binding and catalytic functions, React. Polym. 6, 285 (1987).
37. D. K. Robinson and K. Mosbach, Molecular imprinting of a transition state analogue leads to a polymer exhibiting esterolytic activity, J. Chem. Soc., Chem. Commun. 1989, 969.
38. G. Wulff and J. Vietmeier, Synthesis of macroporous copolymers from α-amino acid based vinyl compounds, Makromol. Chem. 190, 1717 (1989).
39. G. Wulff and J. Vietmeier, Enantioselective synthesis of amino acids using polymers possessing chiral cavities obtained by an imprinting procedure with template molecules, Makromol. Chem. 190, 1727 (1989).

THE CHEMISTRY OF LIGNIN BIODEGRADATION

CATABOLISM OF VERATRYL ALCOHOL AND ITS METHYL ETHER

H.E. Schoemaker[1], U. Tuor[2], A. Muheim[2],
H.W.H. Schmidt[3], and M.S.A. Leisola

[1]DSM Research, Bio-Organic Chemistry Section
P.O. Box 18, 6160 MD Geleen, The Netherlands
[2]Swiss Federal Institute of Technology
Department of Biotechnology, ETH-Hönggerberg
CH-8093 Zürich, Switzerland
[3]Givaudan Forschungsgesellschaft AG,
CH-8600 Dübendorf, Switzerland
[4]Cultor Ltd., Research Center, 02460 Kantvik, Finland

INTRODUCTION

Lignin is the most abundant natural aromatic polymer in the biosphere. It comprises 20-30% of woody plant cell walls and by forming a matrix surrounding the cellulose, it provides strength and protection, a.o. against biodegradation. Lignin is a highly branched and heterogeneous three-dimensional structure made up of phenylpropanoid units which are interlinked through a variety of different bonds. In figure 1 some characteristic lignin model substructures are depicted. In the lignin nomenclature normally used, α and β refer to, respectively, the Cα and Cβ carbon atoms of the phenylpropane units, O-4 indicates the oxygen atom of the *p*-hydroxy group. The numbers refer to the carbon atoms in the aromatic ring, with C-1 connected to the propyl side-chain. The polymer is formed via a peroxidase-catalyzed polymerization of (methoxy)-substituted *p*-hydroxycinnamyl alcohols. The benzylic hydroxy groups are introduced via quinone methide intermediates (Higuchi, 1990). The biological significance of lignin combined with the commercial importance of lignocelluloses has generated widespread interest in understanding the biochemistry of lignin biodegradation. The degradation by the white-rot fungus *Phanerochaete chrysosporium* has been widely used as a model system to understand the processes of lignin biodegradation. The isolation of ligninase from ligninolytic cultures of this fungus in 1983 (Tien and Kirk, 1983; Glenn et al, 1983) has brought lignin biodegradation research into the realm of (bio)chemistry and molecular biology. Since then the biodegradation process could be studied on a molecular level. Although at first described as an H_2O_2-requiring oxygenase, in

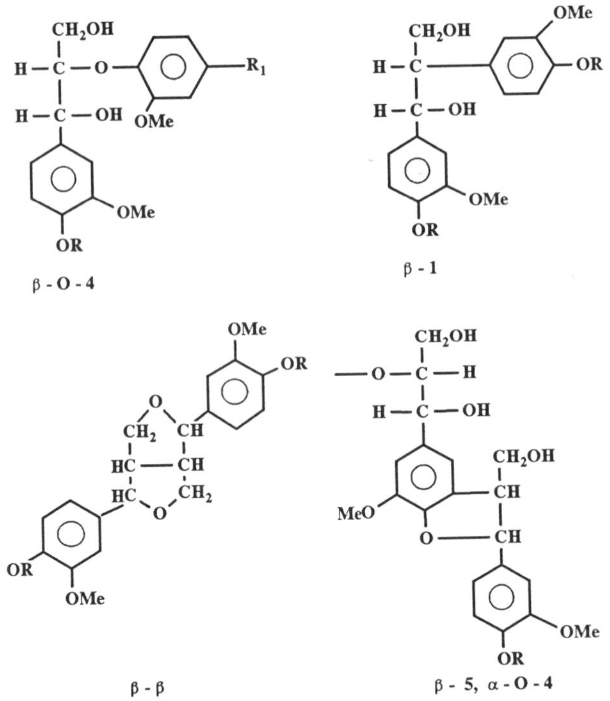

Figure 1

1985 it was established that ligninase should be designated a peroxidase (Harvey et al, 1985; Kuila et al, 1985; Renganathan and Gold, 1986). At present, the enzyme is generally referred to as lignin peroxidase (LiP). In fact, a number of isoenzymes with molecular weights of approximately 41000 have been isolated from *P. chrysosporium* (Leisola et al, 1987). The physical and kinetic properties of these isoenzymes have been described (Farrell et al, 1989; Glumoff et al, 1990). Also, LiP has been crystallized (Glumoff et al, 1989).
LiP will oxidize non-phenolic electron-rich aromatic rings to the corresponding radical cations at the low optimum pH of 3 (Kersten et al, 1985; Schoemaker et al, 1985). Recently, a number of review papers on lignin biodegradation (Kirk and Farrell, 1987; Tien, 1987; Buswell and Odier, 1987; Umezawa, 1988; Schoemaker and Leisola, 1990, Schoemaker, 1990) have appeared. In this paper, therefore, only a general outline of the insights gained in this exciting field will be given. Emphasis is focussed on aspects of the mechanism that have received little attention and on some results obtained recently.

CHEMICAL MECHANISM OF LIGNIN PEROXIDASE CATALYSIS

LiP will oxidize non-phenolic electron-rich aromatic compounds by a single-electron transfer mechanism that entails formation of radical cations. The ensuing reactions of the radical cations include Cα-Cβ-cleavage, demethoxylation and other ether bond cleaving reactions, hydroxylation of benzylic methylene groups, oxidation of benzylic alcohols, decarboxylation, formation of phenols and quinones and aromatic ring cleavage. In these processes reaction of oxygen with intermediate radicals will occur, resulting either in oxygen incorporation or in oxygen activation (i.e. reduction to superoxide anion). These reactions have been comprehensively reviewed (Schoemaker, 1990 and references cited).

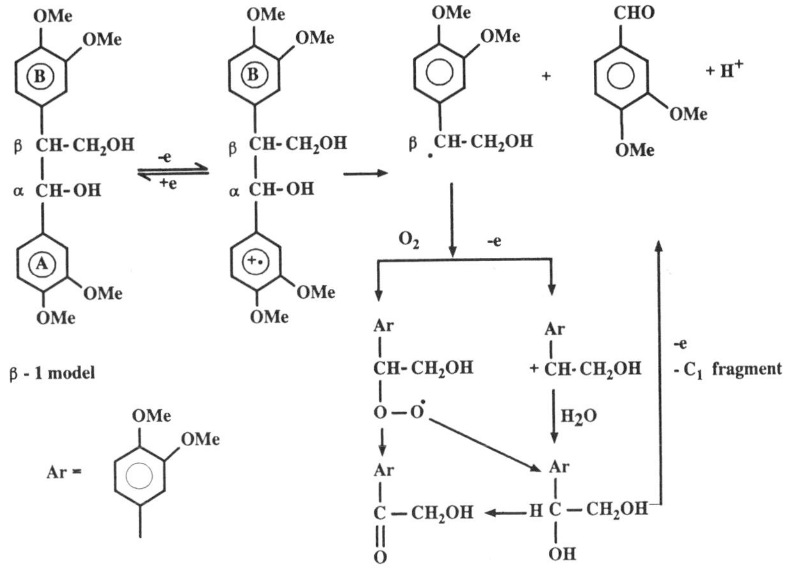

Scheme 1

As examplified for the oxidation of a β-1 lignin model, in scheme 1 the result of one-electron oxidation of a lignin substructure is given. Cα-Cβ cleavage of the radical cation results in the formation of a radical at the Cβ carbon atom, which will react in an almost diffusion controlled manner with molecular oxygen (thus, the enzyme is not an oxygenase). The presence of the Cα-hydroxy substituent facilitates the cleavage reaction by the formation of an aromatic aldehyde (in this case veratraldehyde) as the product. This Cα-Cβ cleavage can be considered as a model for a depolymerizing reaction.
At this point it is interesting to note that the biosynthesis of lignin is initiated by plant peroxidase-induced radical formation of phenolic cinnamyl alcohols (at almost neutral pH). In this process, the benzylic Cα-hydroxy groups are introduced chemically via quinone methide intermediates. In the biodegradation process, the white rot fungi will also use peroxidases, most notably LiP, this time to initiate

a depolymerization reaction via one-electron oxidation. However, this reaction is not the microscopically reversed reaction of the polymerization. The Cα-Cβ cleavage that now takes place is highly facilitated due to the presence of the Cα-hydroxy substituent. Other differences are the pH of the reaction (biodegradation is a slightly acidic process) and the fact that LiP has a somewhat higher redox potential than other peroxidases, which is manifested in its ability to oxidize non-phenolic electron-rich aromatic rings to the corresponding radical cations.

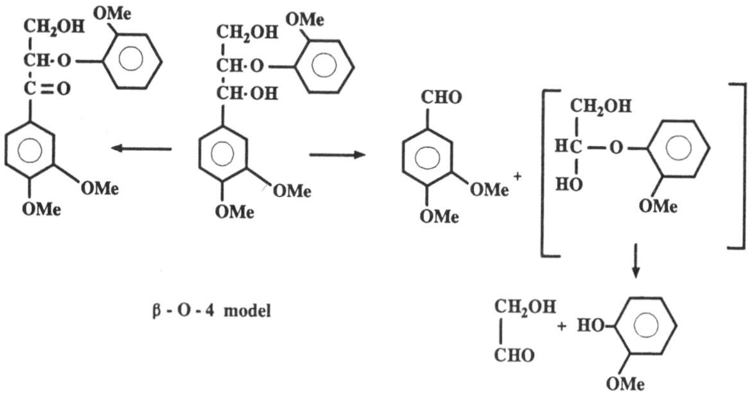

Scheme 2a

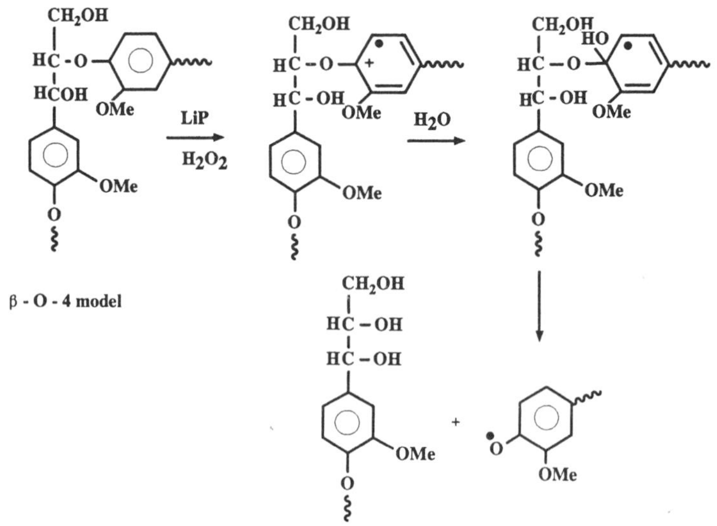

Scheme 2b

In scheme 2a and 2b, other consequences of such a peroxidase induced radical cation formation are depicted. In the oxidation of β-O-4 substructure models, phenols are formed as the products. As stated before, peroxidases will polymerize phenols and phenolic lignin and these reactions indeed are also observed with the isolated LiP (Haemmerli et al, 1986; Odier et al, 1987; Kern et al, 1989). Still, addition of LiP to ligninolytic cultures of *P.chrysosporium* will stimulate the degradation of lignin (Leisola et al, 1988). Obviously, LiP is only part of the ligninolytic system of the white-rot fungi. As one possible mechanism by which the fungus will degrade polymeric lignin, we have suggested that by the extracellular oxidants - peroxidases, phenoloxidases and active oxygen species - a dynamic system is established (a so-called polymerization-depolymerization 'equilibrium') which is shifted towards further degradation by fungal uptake of smaller fragments (a.o. quinones and ring opened products). In this scheme, reductive enzymes also play a role (Schoemaker et al, 1989; Schoemaker and Leisola, 1990; Schoemaker, 1990). In the sequel some results with model systems will be discussed, which form the basis for the hypothesis.

REDOX CYCLE OF LIGNIN PEROXIDASE

In its most basic form the redox cycle of LiP can be represented as depicted in scheme 3:

$$\text{Enz-Fe}^{III}\text{P} + H_2O_2 \longrightarrow \text{Enz-Fe}^{IV}=\text{O P}^{+\cdot} + H_2O$$

compound I

$$\text{Enz-Fe}^{IV}=\text{O P}^{+\cdot} + H^+ + e \longrightarrow \text{Enz-Fe}^{IV}=\text{O P}$$

compound II

$$\text{Enz-Fe}^{IV}=\text{O P} + H^+ + e \longrightarrow \text{Enz-Fe}^{III}\text{P} + H_2O$$

Enz = apo-enzym; P = protoporhyrin IX

Scheme 3

A complication in the redox cycle is the reaction of compound II with excess H_2O_2, yielding compound III (Fe^{III}-$O_2^{-\cdot}$P or Fe^{II}-O_2P), a species with catalytic limited ability (Cai and Tien, 1989, 1990; Wariishi and Gold, 1989, 1990; Wariishi et al, 1990). Despite recent studies using stopped flow techniques, still little is known on the exact mechanism of LiP-catalysis. Especially, the role of veratryl alcohol is unclear. Veratryl alcohol is oxidized to veratraldehyde, presumably in two consecutive one-electron oxidation steps. However, in contrast to the oxidation of dimethoxybenzene (Kersten et al, 1985) no spectral evidence for the intermediacy of a radical cation has been obtained sofar (see, however, Harvey et al, 1989a). Still, the formation of side-products (arising from aromatic ring oxidation) is best rationalized by attack on an intermediate radical cation by water and dioxygen or reduced oxygen species (*vide infra*). If one assumes lignin degradation to be initiated by one-electron oxidation, efficient reduction of compound II of LiP should occur. It has been postulated that the conversion of compound II to the native enzyme is one of the physiological roles of the secondary metabolite veratryl alcohol (Schoemaker, 1990); it has also been suggested that a veratryl alcohol radical cation modified form of compound II plays a role in this process (Harvey et al, 1989b). Alternatively, it has been postulated that compound II is first converted into compound III or compound III* by reaction with H_2O_2, followed by a ligand displacement initiated by veratryl alcohol with the formation of superoxide anion (Wariishi and Gold, 1989, 1990; Wariishi et al, 1990). The process is still under active investigation in a number of laboratories (Tien, Gold, Harvey). Sofar, most kinetic studies have been performed with veratryl alcohol. However, since the major product of the reaction is veratraldehyde, formed after a two-electron oxidation, the results of these studies are not easy to analyze, because it is rather difficult to distinguish between two consecutive one-electron oxidation steps and a direct conversion to the aldehyde. Therefore, substrates, including phenols, that are clearly oxidized via one-electron oxidation should also be investigated.

OXIDATION OF VERATRYL ALCOHOL AND ITS METHYL ETHER

Veratryl alcohol 1 (3,4-dimethoxybenzyl alcohol) plays a pivotal role in studies on lignin biodegradation. It is a secondary metabolite of the white-rot fungus, *P.chrysosporium*. Remarkably, veratryl alcohol is also catabolized by ligninolytic cultures of this fungus. It is an inducer for LiP and it seems to protect the enzyme against H_2O_2. In general, the formation of veratraldehyde from veratryl alcohol is used as an assay for LiP activity.
In this paper, the oxidation and further degradation of veratryl alcohol and its derivatives are described to illustrate the complexity of the ligninolytic system. Also, an attempt will be made to extrapolate the data obtained with simple monomeric (and dimeric) lignin-like compounds and/or substructures to the polymeric system. Evidently, such an extrapolation should be done with the utmost care, since the two systems vary widely and still little is known from studies with the actual polymer. However, due to the complexity of the problem, in our opinion, progress can only be made with the aid of model studies.

Oxidation of veratryl alcohol by LiP

At first, it was believed that veratraldehyde was the only product and that oxygen did not play a role in the process (Tien et al, 1986). However, subsequent research indicated that molecular oxygen is used both as an electron acceptor and as a reagent. Dioxygen is also converted to H_2O_2 via intermediacy of the superoxide anion (Palmer et al, 1987; Haemmerli et al, 1987; Schmidt et al, 1989). Thus, when veratryl alcohol 1 was oxidized by LiP in the presence of oxygen, a number of products were observed (scheme 4). Most of them have been isolated and identified (Schmidt et al, 1989). Veratraldehyde 2 was the major product (over 70%), the quinones accounted for about 10% of the reaction products (2-methoxy-*p*-quinone 3: 3.5%; 2-methoxy-5-(hydroxymethyl)-2,5-cyclohexadiene-1,4-dione 4: 7.0%; 4,5-dimethoxy-3,5-cyclohexadiene-1,4-dione 5: 2.5%; and the five-membered ring lactones (mainly 4-(2-(methoxycarbonyl)-ethenyl)-2(5H)-furanone 6), formed after aromatic ring cleavage, for about 15-20%.

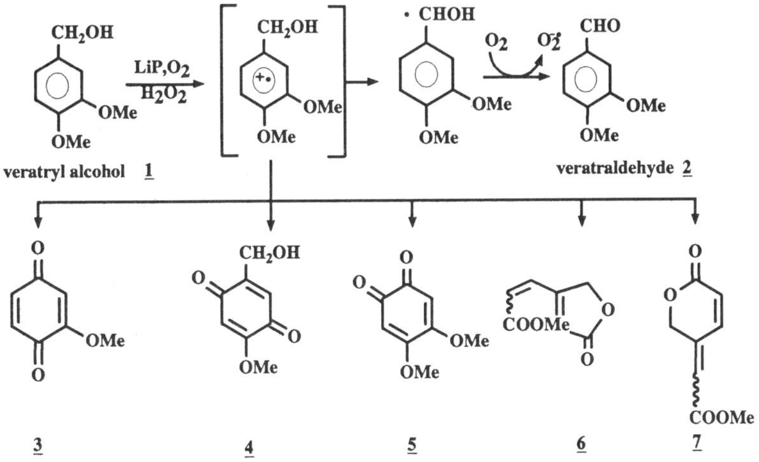

Scheme 4

When the reaction was carried out under an argon atmosphere veratraldehyde was practically the only product. The one-electron oxidant cerium ammonium nitrate gave essentially the same products as LiP under both conditions.

Effect of pH and Mn(II) on veratryl alcohol oxidation

The relative amounts of the various oxidation products in the enzyme reaction were dependent on the reaction pH. The highest amount of quinones was formed at pH 3.0, while formation of the ring cleavage lactones was more favoured between pH 4.0 and 4.5. Above pH 5.0, veratraldehyde was the only product. When Mn(II) was added to the reaction mixtures, the amount of ring cleavage lactones decreased considerably (pH 3.0). The effect was even more pronounced at pH 4.5. A similar phenomenon was observed in the formation of quinone 4. Oxidation of veratryl alcohol to the aldehyde and, remarkably, formation of the quinone 3 were not significantly affected by Mn(II). Only Mn(II) concentrations of 5 mM and more had an inhibitory effect on the enzyme reaction.
In 1985, Leisola and coworkers described for the first time LiP-catalyzed ring opening of veratryl alcohol. Palmer et al proposed that intermediate radical cations of 1 react with perhydroxyl radicals and or superoxide anion, generated in the oxidation process, with formation of a dioxetane (Palmer et al, 1987).

Subsequent decomposition of the latter compound should then yield the observed ring-opened lactones (see scheme 5, route a). A corrollary of this mechanism is the incorporation of two oxygen atoms derived from dioxygen into the final product. However, Shimada recently showed, with labeling studies, that in the ring-opened lactone 6 one oxygen is derived from water and one oxygen atom is derived from molecular oxygen (see scheme 5, route b). The oxygen atom derived from water is regioselectively incorporated at the 3-position of 1, in contrast to the non-enzymatic (tetraphenylporphyrinato)-iron(III)/*tert*-BuOOH system or other heme-protein models. Originally it was proposed that the intermediate radical cation of 1 reacts with water at the 3-position and with molecular oxygen at the 4-position (Shimada et al, 1987; Hattori et al, 1988). A modification of this mechanism has been proposed in which the intermediate radical cation of 1 reacts with water and with perhydroxyl radical, generated during LiP catalysis, via reaction of the intermediate hydroxy-substituted 3,4-dimethoxybenzyl radical with molecular oxygen. Reaction of the radical cation of 1 with perhydroxyl radical was postulated to explain the inhibition of aromatic ring opening by Mn(II). Also, upon addition of Mn(II) more equivalents of veratraldehyde are formed, compared to the equivalents of H_2O_2 added as oxidant. A possible explanation for these phenomena is the rapid reduction of superoxide anion or perhydroxyl radical by Mn(II) with formation of H_2O_2 (see scheme 5). The mechanism by which the primarily formed *Z*-lactone is converted into the *E*-isomer has not yet been elucidated (Haemmerli et al, 1987). The mechanism of quinone formation has not yet been studied in great detail. No data on oxygen incorporation or the possible intermediacy of the corresponding hydroquinones are available.

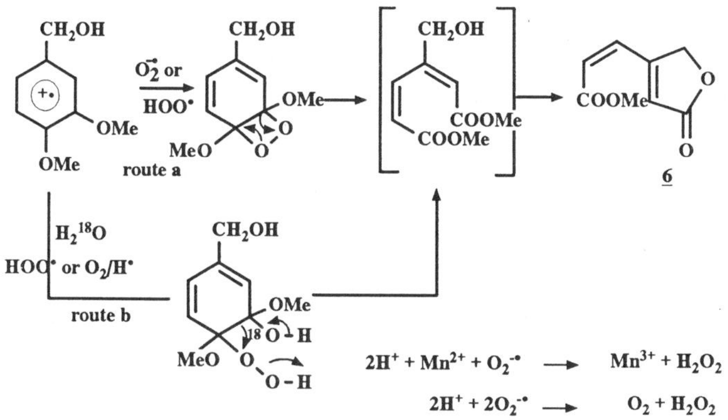

Scheme 5

Oxidation of 3,4-dimethoxybenzyl methyl ether (DMME) by LiP

LiP catalyzed oxidation of 3,4-dimethoxybenzyl methyl ether (DMME) afforded at least six products (see scheme 6). Under anaerobic conditions only trace amounts of products other than veratraldehyde could be detected. Oxidation of DMME with the one-electron oxidant cerium(IV)ammonium nitrate gave comparable results. The products of LiP-catalyzed DMME oxidation were isolated, purified and characterized by analytical techniques and eventually by independent synthesis (Schmidt et al, 1989). At pH 3.0, veratraldehyde 2 was the most prominent product (53%). Two quinones were formed which were also products of the veratryl alcohol oxidation; these being 2-methoxy-*p*-quinone 3 (3.3%) and 4,5-dimethoxy-3,5-cyclohexadiene-1,2-dione 5 (2.5%). The third rather prominent quinone was identified as the 2-methoxy-5-(methoxymethyl)-2,5-cyclohexadiene-1,4-dione 9 (9.2%), a compound analogous to quinone 4 formed in the veratryl alcohol oxidation. Another prominent oxidation product was identified as dimethyl 3-(methoxymethyl)-(*Z*,*Z*)-muconate 11 (30.3%). NMR analysis, including NOE-DIFF experiments indicated the *Z*,*Z*-configurations of the double bonds. As a minor product the methyl ester of veratric acid 10 (1.2%) could be identified. This is the first time that a carboxylic acid derivative has been detected in a LiP-catalyzed reaction.

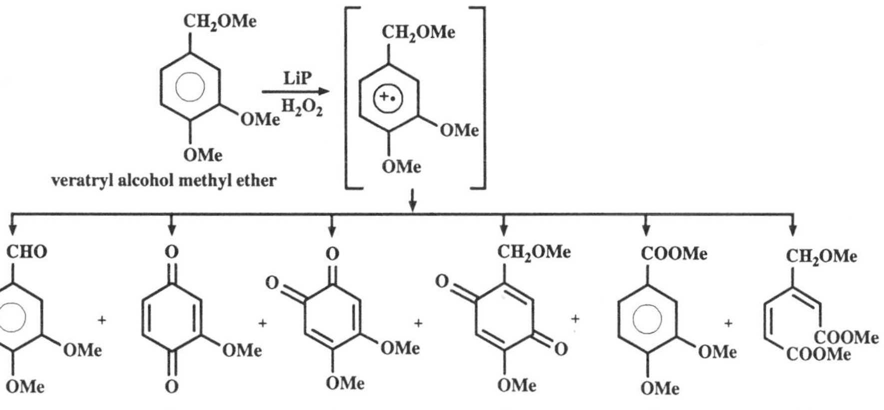

Scheme 6

Effect of pH and Mn(II) on DMME oxidation

The effect of pH and Mn(II) addition in the oxidation of DMME is far more pronounced compared to similar studies with veratryl alcohol. The effect of pH between 3.0 and 5.0 has been investigated. With increasing pH there was a slight decrease in the formation of the quinones 3, 5 and 9. The decrease in the formation of the ring cleavage product was more pronounced whereas the relative production of veratraldehyde remained more or less the same. However, there was a dramatic increase in formation of the ester 10 upon increasing the pH.

In fact, the ester became the most prominent product at pH 5.0 (aldehyde/ester ratio: pH 3.0 = 1:0.37; pH 4.0 = 1:0.80; pH 5.0 = 1:1.18). Possible reaction mechanisms have been discussed (Schmidt et al, 1989), but a detailed description of the process has to await extensive labeling and kinetic studies.

CATABOLISM OF VERATRYL ALCOHOL AND ITS METHYL ETHER

As already stated in the introduction, the oxidative system of *P. chrysosporium* will polymerize phenolic lignin. The mechanism to prevent polymerization is currently unknown. The secondary metabolite veratryl alcohol (non-phenolic!) is both synthesized and degraded by ligninolytic cultures of *P. chrysosporium*. In the lignin peroxidase catalyzed oxidation of veratryl alcohol quinone type structures, formally derived from phenoxy-radical type intermediates are formed. Thus, even in the oxidation of the non-phenolic veratryl alcohol, products formed are prone to polymerize under the oxidative reaction conditions. Therefore, the metabolism of this compound was studied as a simple model system.

Veratryl alcohol was rapidly and almost completely (>80%) degraded to CO_2 in carbon-limited cultures compared to slow and incomplete (ca.40%) degradation in nitrogen-limited cultures. In those studies uniformly labeled ^{14}C-veratryl alcohol, ^{14}C-quinones and ^{14}C-lactones were used.

From these studies the metabolic scheme depicted in scheme 7 was proposed (Leisola et al, 1988). Veratryl alcohol is oxidized by lignin peroxidase to a mixture of veratraldehyde (ca 70%), quinones and ring opened products. The quinones are rapidly metabolized by the fungus. In N-limited cultures, the ring-opened products accumulate as lactones. The latter compounds are slowly metabolized by the fungus. In C-limited cultures no accumulation of ring opened products is observed. Presumably, the intermediate muconic ester derivative is rapidly metabolized under these carbon-limited conditions. Veratraldehyde is rapidly reduced by the fungus to veratryl alcohol and then further degraded. The aryl aldehyde reductase involved in this conversion has been isolated and purified (Muheim et al, in press). A metabolic pathway via veratric acid could be discounted because both veratraldehyde and veratric acid are not substrates for the lignin peroxidase. Instead, veratric acid is also rapidly reduced by ligninolytic cultures of *P. chrysosporium*.

In a search for degradation products further down the catabolic pathway ^{14}C-labeled veratryl alcohol was fed to nitrogen limited cultures of *P. chrysosporium*. In addition to the evolution of $^{14}CO_2$ the formation of polar metabolites was also monitored using HPLC. Two unknown metabolites were found to contain radioactivity. They were only slowly metabolized to CO_2 by the fungus. However as soon as glucose was depleted in the cultures, these intermediates were rapidly degraded further and finally disappeared (Tuor et al, 1990). At first it was thought that these polar intermediates were ring opened structures. Subsequent studies, however, indicated that the products were

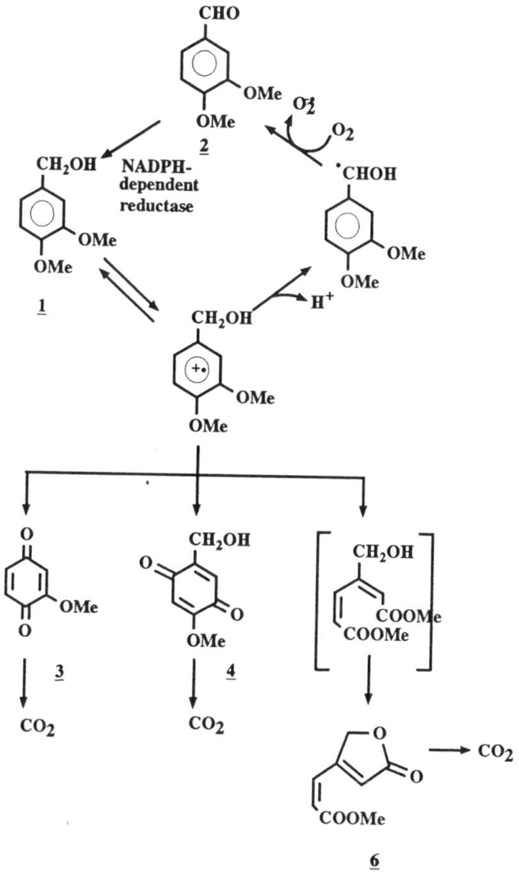

Scheme 7

derived from the quinones. Also, it was demonstrated that all products from the oxidation of DMME were further degraded by intact mycelium from *P.chrysosporium*. The first step in the degradation of the quinones is reduction to the corresponding hydroquinones (Schoemaker et al, 1989). It has been proposed for 2-methoxy-*p*-quinone that the next step is hydroxylation (Ander et al, 1980) or demethoxylation (Ander et al, 1983), followed by ring opening. However, sofar we have not obtained evidence for such a pathway with the hydroxymethyl or methoxymethyl substituted 2-methoxy-hydroquinones. Instead, the isolated products were 4-hydroxy-cyclohex-2-enones,i.e. the formally reduced hydroquinones (*vide infra*). To the best of our knowledge this constitutes a unique conversion in the microbial degradation of aromatic compounds.
Thus, addition of quinones 3, 4 and 9 to three separate, nitrogen-limited cultures of *P.chrysosporium* gave at first an intense yellow color to the respective supernatants. This color gradually disappeared, due to reduction to the corresponding hydroquinones. The reactions were monitored by HPLC analysis of the culture samples. In the course of the process polar

metabolites from both quinone 4 and 9 were formed. The concentration of these metabolites increased in the first 24 hrs, remained constant for 2-3 days and then the compounds gradually disappeared when the glucose levels of the cultures reached depletion. In the case of quinone 3 only reduction to 2-methoxy-*p*-hydroquinone 3a was observed. Still, the latter compound is eventually degraded to CO_2.

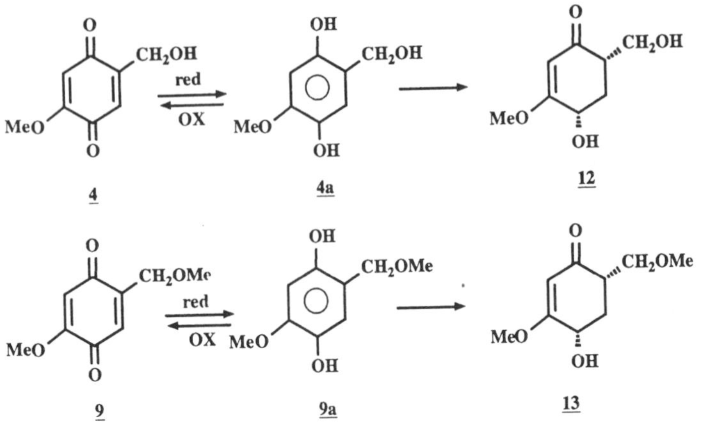

Scheme 8

It turned out to be rather difficult to isolate the polar metabolites from the cultures, because they are very hydrophilic and rather labile. Only isolated yields of about 10% could be obtained. The rest of the material polymerized during work-up. HPLC analysis indicated that quinone 4 and the hydroquinone 4a are rapidly interconverted and subsequently slowly further reduced to *cis*-4-hydroxy-6-hydroxymethyl-3-methoxy-cyclohex-2-en-one 12. Analogously, from the equilibrium of quinone 9 and the corresponding hydroquinone 9a the slow formation of *cis*-4-hydroxy-6-methoxymethyl-3-methoxy-cyclohex-2-en-one 13 was observed (scheme 8). Thus, it could not be unambiguously established that the hydroquinones are the immediate precursors of these novel metabolites. Remarkably, the two hydroxycyclohexenones had different conformations. Presumably, due to internal hydrogen bond formation the 4-hydroxy and the 6-methoxymethyl substituents in compound 13 were in pseudo-axial positions. In contrast, the 4-hydroxy and the 6-hydroxymethyl groups in compound 12 occupied pseudo-equatorial positions, as depicted in figure 2.

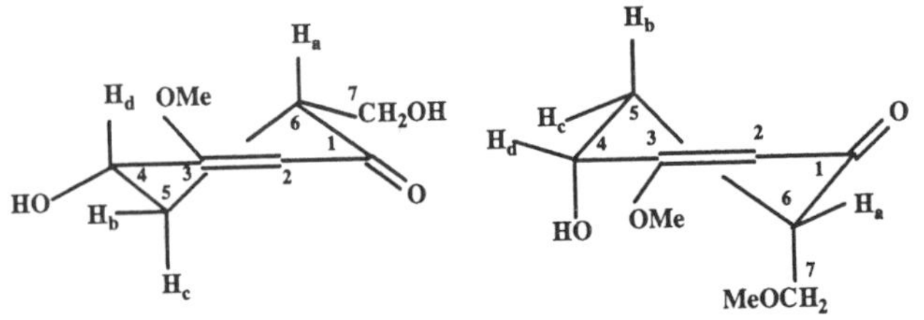

Figure 2

DISCUSSION

To the best of our knowledge the formation of 4-hydroxy-cyclohex-2-enones - formally reduced hydroquinones - from the corresponding quinones or hydroquinones is unprecedented in the literature. However, the physiological role of this novel type of reductive conversion is not clear at the moment. Evidently, one possibility is that these novel reactions are specific for the particular laboratory systems studied and that they play no role in the process of lignin biodegradation in nature. Alternatively, the reaction can be considered as a possible mechanism to open the futile oxidation-reduction cycle (oxidation of the hydroquinones to quinones by LiP or MnP, reduction of the quinones by quinones reductases or other systems). Via the formation of the 4-hydroxycyclohex-2-enones - which are not substrates for LiP- the toxic quinones can be degraded. The further mechanism of degradation of these unique hydroxycyclohex-2-enones is not known as yet. The compounds are rather labile, loss of water under the acidic conditions will lead to re-aromatisation, e.g. 2-hydroxy-4-methoxybenzyl alcohol or the corresponding methyl ether might be formed. Alternatively, it can be envisaged that the further degradation proceeds via rupture of the susceptible keto-enol system. Also, one can rationalize the observations as a possible detoxifying mechanism. It should be noted that the reduction of the hydroquinones is only observed with the hydroxymethyl, or methoxymethyl substituted (hydro)quinones, these being specific degradation products of veratryl alcohol and its methyl ether. No analogous reduction product of 2-methoxyhydroquinone has been detected so far. Most notably, apart from being a degradation product of veratryl alcohol and its methyl ether, 2-methoxy-*p*-quinone is a ubiquitous metabolite in all kinds of lignin substructure degradation pathways. Indeed, we have postulated that rapid fungal uptake and further degradation of 2-methoxyquinones and related quinones is one possible mechanism to shift the polymerization-depolymerization 'equilibrium' towards degradation with formation of CO_2 (Schoemaker, 1990). Somewhere in this process aromatic ring cleavage has to occur. We are still actively pursuing further degradation products derived from quinone intermediates.

CONCLUSION

In recent years, tremendous progress has been made in the field of lignin biodegradation. However, the mechanism of depolymerization and the catabolic pathway of lignin degradation are still poorly understood. From the results described in this paper a number of interesting research topics emerge. First, the redox cycle of LiP should be investigated in more detail, also with substrates other than veratryl alcohol. Second, the influence of pH, the role of active oxygen species and the effects of Mn(II) have been largely neglected so far. Especially Mn(II) seems to play a central role in lignin biodegradation. Addition of Mn(II) influences the distribution of products of the enzymatic reaction (Schmidt et al, 1989).

Manganese regulates the expression of Mn(II)-peroxidase (Brown et al, 1990; Bonnarme and Jeffries, 1990), an enzyme that is present in ligninolytic cultures of a number of white rot fungi, which seems to play a major role in the degradation of phenolic compounds (Wariishi et al, 1989ab, Gold et al, 1989). Mn(III) can, provided that it is properly chelated, mediate oxidation of substrates that are poorly accessible to the large enzymes (Cui and Dolphin, 1990).

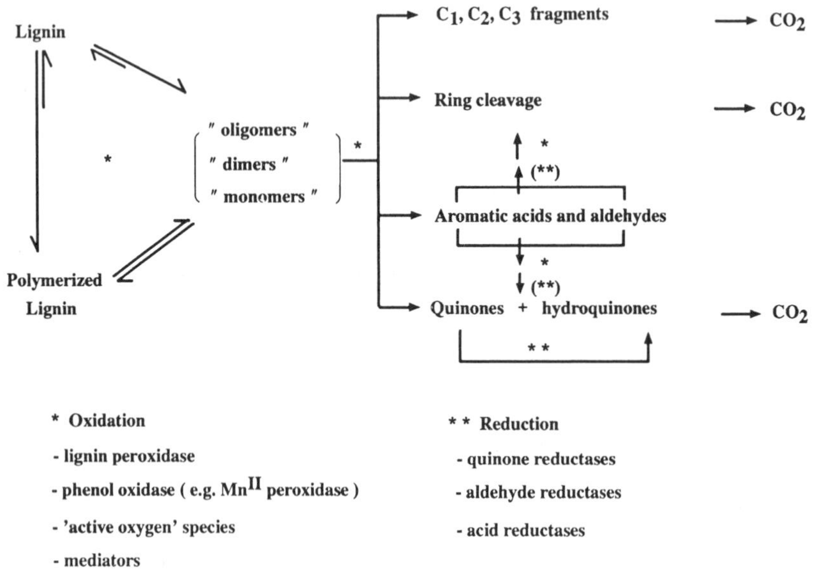

Scheme 9

Third, in this paper one example of formation of a carboxylic acid derivative is described. Sofar, no enzymatic oxidations of alcohols or aldehydes have been described, despite the fact that carboxylic acids are major products isolated from fungal degraded wood. Fourth, the oxidation of phenols is still poorly understood. In view of the fact that phenols tend to polymerize under oxidative conditions, the mechanisms which prevent polymerization should be addressed. As one possible mechanism we have proposed that fungal uptake of small degradation products like quinones and muconic ester derivatives will shift a putative polymerization- depolymerization 'equilibrium' towards degradation as depicted in scheme 9. Obviously, other mechanisms will also be operative. In this context the role of all sorts of carbohydrates (cellulose, hemicelluloses, fungal cell wall polysaccharides etc.) should be investigated further (Leisola and Garcia, 1989).
Fifth, in scheme 9 the role of reductive enzymes is depicted. In addition to aromatic aldehyde and acid reductases present in ligninolytic cultures of white-rot fungi, quinone reductases and possibly 'hydroquinone reductases' should be investigated.
Sixth, it has been suggested that spatial separation and charge-transfer are important factors in the depolymerization

process (Harvey et al, 1989a; Harvey and Palmer, 1990). Finally, a unique conversion of quinones and hydroquinones to the corresponding 4-hydroxy-cyclohex-2-enones has been discovered. This unprecendented reaction in itself merits further investigation.

In conclusion, in view of the extreme complexity of the systems involved, much research remains to be done in this challenging area, before the potential technological applications of lignin biodegradation can become economically feasible.

ACKNOWLEDGEMENTS

Mr. W. Kortenoeven is gratefully acknowledged for his valuable assistance in preparing the manuscript.

REFERENCES

Ander, P., Hatakka, A., Eriksson, K-E., 1980, Vanillic acid metabolism by the white-rot fungus *Sporotrichum pulverulentum*, Arch. Microbiol. 125: 189-202

Ander, P,. Eriksson, K-E., Yu, H-S., 1983, Vanillic acid metabolism by *Sporotrichum pulverulentum*: evidence for demethoxylation before ring-cleavage, Arch. Microbiol. 136: 1-6

Bonnarme, P., Jeffries, T.W., 1990, Mn(II) regulation of lignin peroxidases and manganese-dependent peroxidases from lignin-degrading white-rot fungi, Appl. Environm. Microbiol. 56: 210-217

Brown, J.A., Glenn, J.K., Gold, M.H., 1990, Manganese regulates expression of manganese peroxidase by *Phanerochaete chrysosporium*, J. Bacteriology, 172: 3125-3130

Buswell, J.A., Odier, E., 1987, Lignin biodegradation, CRC Critical Rev.Biotechnol. 6:1-60.

Cai, D., Tien, M., 1989, On the reactions of lignin peroxidase compound III (isozyme H8), Biochem. Biophys. Res. Comm. 162:464-469

Cai, D., Tien, M., 1990, Characterization of the oxycomplex of lignin peroxidase from *Phanerochaete chrysosporium*. Equilibrium and kinetic studies, Biochemistry 29:2085-2091

Cui, F., Dolphin, D., 1990, The role of manganese in model systems related to lignin biodegradation, Holzforschung 44: 279-283

Farrell, R.L., Murtagh, K.E., Tien, M., Mozuch, M.D., Kirk, T.K., 1989, Physical and enzymatic properties of lignin peroxidase isoenzymes from *Phanerochaete chrysosporium*, Enzyme Microb.Technol. 11:322-328

Glenn, J.K., Morgan, M.A., Mayfield, M.B., Kuwahara, M., Gold, M.H., 1983, An extracellular H_2O_2-requiring enzyme preparation involved in lignin biodegradation by the white rot basidiomycete *Phanerochaete chrysosporium*, Biochem. Biophys. Res. Comm. 114:1077-1083

Glumoff, T., Harvey, P.J., Molinari, S., Frank, G., Palmer, J.M., Smit, J.D.G., Leisola, M.S.A., 1990, Lignin peroxidase from *Phanerochaete chrysosporium*. Molecular and kinetic characterization of isozymes, Eur. J. Biochem. 187: 515-520

Glumoff, T,, Winterhalter, K.H., Smit, J.D.G., 1989, Monoclinic crystals of lignin peroxidase, FEBS Letters 257:59-62
Gold, M.H., Wariishi, H., Valli, K., 1989, Extracellular peroxidases involved in lignin degradation by the white-rot basidiomycete *Phanerochaete chrysosporium*, in: Biocatalysis in Agricultural Biotechnology. Whitaker, J.R., Sonnet, P.E. (eds) ACS Symposium series No. 389, Washington. pp 127-140.
Haemmerli, S.D., Leisola, M.S.A., Fiechter, A., 1986, Polymerisation of lignins by ligninases from *Phanerochaete chrysosporium*, FEMS Microbiology Letters, 35:33-36
Haemmerli, S.D., Schoemaker, H.E., Schmidt, H.W.H., Leisola, M.S.A., 1987, Oxidation of veratryl alcohol by the lignin peroxidase of *Phanerochaete chrysosporium*: involvement of activated oxygen, FEBS Letters, 220:149-154
Harvey, P.J., Schoemaker, H.E., Bowen, R.M., Palmer, J.M., 1985, Single-electron transfer processes and the reaction mechanism of enzymic degradation of lignin, FEBS Letters, 183:13-16
Harvey, P.J., Schoemaker, H.E., Palmer, J.M., 1986, Veratryl alcohol as a mediator and the role of radical cations in lignin biodegradation by *Phanerochaete chrysosporium*, FEBS Letters, 195:242-246
Harvey, P.J., Gilardi, G.F., Palmer, J.M., 1989a, Importance of charge transfer reactions in lignin degradation, in: Enzyme systems for lignocellulose degradation, Coughlan, M.P., ed., Elsevier, London. pp 111-120
Harvey, P.J., Palmer, J.M., Schoemaker, H.E., Dekker, H.L., Wever, R., 1989b, Pre-steady-state kinetic study on the formation of compound I and II of ligninase, Biochim. Biophys. Acta, 994:59-63
Harvey, P.J., Palmer, J.M., 1990, Oxidation of phenolic compounds by ligninase, J. Biotechnol., 13:169-179
Hattori, T., Shimada, M., Umezawa, T., Higuchi., Leisola, M.S.A., Fiechter, A., 1988, New mechanism for oxygenative ring cleavage of 3,4-dimethoxybenzyl alcohol catalyzed by the ligninase model, Agric. Biol. Chem., 52: 879-880
Higuchi, T., 1990, Lignin biochemistry: biosynthesis and biodegradation, Wood Sci. Technol., 24:23-63
Kern, H.W., Haider, K., Pool, W., DeLeeuw, J.W., Ernst, L., 1989, Comparison of the action of *Phanerochaete chrysosporium* and its extracellular enzymes (lignin peroxidases) on lignin preparations, Holzforschung 43: 375-384
Kersten, P., Kirk, T.K., 1987, Involvement of a new enzyme,glyoxal oxidase, in extracellullar H_2O_2 production by *Phanerochaete chrysosporium*, J. Bacteriology, 169:2195-2201
Kirk, T.K., Farrell, R.L., 1987, Enzymatic "combustion": the microbial degradation of lignin, Ann. Rev. Microbiol. 41:465-505
Kuila, D., Tien, M., Fee, J.A., Ondrias, M.R., 1985, Resonance raman spectra of extracellular ligninase: Evidence for a heme active site similar to those of peroxidases, Biochemistry, 24:3394-3397
Leisola, M.S.A., Kozulic, B., Meussdoerffer, F., Fiechter, A., 1987, Homology among multiple extracellular peroxidases from *Phanerochaete chrysosporium*, J. Biol. Chem. 262:419-424

Leisola, M.S.A., Haemmerli, S.D., Waldner, R., Schoemaker, H.E., Schmidt, H.W.H., Fiechter, A., 1988, Metabolism of a lignin model compound, 3,4-dimethoxybenzyl alcohol by *Phanerochaete chrysosporium*, Cellulose Chem. Technol. 22:267-277

Leisola, M.S.A., Garcia, S., 1989, The mechanism of lignin degradation, in: Enzyme systems for lignocellulose degradation, Coughlan, M.P., ed., Elsevier, London, pp 89-99

Muheim, A., Waldner, R., Sanglard, D., Reiser, J., Schoemaker, H.E., Leisola, M.S.A., 1990/1991, Purification and properties of an aryl-aldehyde reductase from the white-rot fungus *Phanerochaete chrysosporium*, Eur. J. Biochem., in press

Odier, E., Mozuch, M.D., Kalyanaraman, B., Kirk, T.K., 1988, Ligninase-mediated phenoxy radical formation and polymerization unaffected by cellobiose:quinone oxidoreductase, Biochimie, 70:847-852.

Palmer, J.M., Harvey P.J., Schoemaker H.E., 1987, The role of peroxidases, radical cations and oxygen in the degradation of lignin, Phil. Trans. R. Soc. Lond. **A**, 321, 495-505

Renganathan, V., Gold, M.H., 1986, Spectral characterization of the oxidized states of lignin peroxidase, an extracellular heme enzyme from the white rot basidiomycete *Phanerochaete chrysosporium*, Biochemistry, 25:1626-1631

Schmidt, H.W.H., Haemmerli, S.D., Schoemaker, H.E., Leisola M.S.A., 1989, Oxidative degradation of 3,4-dimethoxybenzyl alcohol and its methyl ether by the lignin peroxidase of *Phanerochaete chrysosporium*, Biochemistry, 28:1776-1783

Schoemaker, H.E., Harvey, P.J., Bowen, R.M., Palmer, J.M., 1985, On the mechanism of enzymatic lignin breakdown, FEBS Letters, 183:7-12

Schoemaker, H.E., Meijer, E.M., Leisola, M.S.A., Haemmerli, S.D., Waldner, R., Sanglard, D., Schmidt, H.W.H., 1989, Oxidation and reduction in lignin biodegradation, in: Biogenesis and biodegradation of plant cell polymers, Lewis, N.G., Paice, M.G., eds., ACS Symposium series Vol 399, Washington. pp 454-471

Schoemaker, H.E., Leisola, M.S.A., 1990, Degradation of lignin by *Phanerochaete chrysosporium*, J. Biotechnol. 13:101-109

Schoemaker, H.E., 1990, On the chemistry of lignin biodegradation, Recl. Trav. Chim. Pays-Bas 109:255-272

Shimada, M.,Hattori, T., Umezawa T., Higuchi, T., Uzura, K., 1987, Regiospecific oxygenations during ring cleavage of a secondary metabolite, 3,4-dimethoxybenzyl alcohol catalyzed by lignin peroxidase, FEBS Letters, 221, 327-331 (1987)

Tien, M., Kirk, T.K., 1983, Lignin-degrading enzyme from the Hymenocyte *Phanerochaete chrysosporium* Burds., Science, 221:661-663

Tien, M., Kirk, T.K., Bull, C., Fee, J.A., 1986, Steady-state and transient-state kinetic studies on the oxidation of 3,4-dimethoxybenzyl alcohol catalyzed by the ligninase of *Phanerochaete chrysosporium* Burds., J. Biol. Chem. 261: 1687-1693

Tien, M., 1987, Properties of ligninase from *Phanerochaete chrysosporium* and their applications, CRC Critical Rev. Microbiol., 15:141-168.

Tuor, U., Haemmerli, S.D., Schoemaker, H.E., Schmidt, H.W.H., Leisola, M.S.A., 1990, On the metabolism of

3,4-dimethoxybenzyl alcohol and its methyl ether by *Phanerochaete chrysosporium*, in: Biotechnology in pulp and paper manufacture, Kirk, T.K., Chang, H.M., eds., in press
Umezawa, T., 1988, Mechanisms for chemical reactions involved in lignin biodegradation by *Phanerochaete chrysosporium*, Wood Research 75:21-79
Wariishi, H., Gold, M.H., 1989, Lignin peroxidase compound III, formation, inactivation and conversion to the native enzyme, FEBS Letters, 243:165-168.
Wariishi, H., Gold, M.H., 1990, Lignin peroxidase compound III, J. Biol. Chem. 265:2070-2077
Wariishi, H., Dunford, H.B., MacDonald, I.D., Gold, M.H., 1989a, Manganese peroxidase from the lignin-degrading basidiomycete *Phanerochaete chrysosporium*, J. Biol. Chem., 264:3335-3340.
Wariishi, H., Valli, K., Gold, M.H., 1989b, Oxidative cleavage of a phenolic diaryl propane lignin model dimer by manganese peroxidase from *Phanerochaete chrysosporium*, Biochemistry, 28:6017-6023
Wariishi, H., Marquez, L., Dunford, H.B., Gold, M.H., 1990, Lignin peroxidase compounds II and III, J. Biol. Chem. 265: 11137-11142

STUDIES ON PLANT TISSUE CULTURE - SYNTHESIS AND BIOSYNTHESIS OF BIOLOGICALLY ACTIVE SUBSTANCES

James P. Kutney

Department of Chemistry
University of British Columbia
2036 Main Mall
Vancouver, B.C., Canada V6T 1Y6

INTRODUCTION

The plant kingdom has provided literally thousands of natural products with widely diverse chemical structures and a vast array of biological activities many of which have seen subsequent application in the pharmaceutical, agrochemical and related industries. Extensive efforts have been expended toward structure elucidation of such compounds and subsequent laboratory syntheses. Due to their structural complexity, these compounds are generally amenable only through multi-step syntheses and/or direct isolation from the living plant. Both processes are often fraught with difficulties. Multi-step syntheses although providing obvious challenges in development of synthetic strategies and training for young synthetic chemists, are rarely useful in any practical production of such natural products often required in large scale for their use as drugs, etc. In a similar manner, isolation of the target compounds by extraction from the plant can be associated with various problems (Figure 1).

We felt that appropriate solutions to at least some of these problems could be addressed by the use of plant cell cultures in combination with chemistry. The advantages of plant cell cultures over the living plant, in terms of natural product (secondary metabolites) production are obvious (Figure 2). Successful development of cell lines from plant tissue allows, hopefully, a superior route to target compounds particularly when such studies are coupled with basic research directed toward the understanding of biosynthetic pathways of the compounds under investigation. Clearly knowledge of such pathways readily derived from studies with enzymes, more easily isolated from cell cultures than plants, can, in turn, allow more efficient routes for development in a synthetic chemical laboratory. Several examples of this strategy are provided from our extensive studies with cell cultures derived from Catharanthus roseus. A brief mention of studies with cell cultures of the Chinese herbal plant, Tripterygium wilfordii, will be mentioned at the end of this discussion.

Bioorganic Chemistry in Healthcare and Technology, Edited by U.K. Pandit and F.C. Alderweireldt, Plenum Press, New York, 1991

1. Active agent often in minute amounts.
2. Separation from other co-occurring substances difficult and costly.
3. Seasonal variation.
4. Plants grow in inaccessible regions.

Figure 1. Medicinal Agents from Plants- Problems.

STUDIES WITH CELL CULTURES OF *C. ROSEUS* - THE VINBLASTINE FAMILY

Detailed accounts of our studies in this area involving development of cell lines, alkaloid production, isolation of enzymes from cell cultures, immobilization of such derived enzymes and biotransformation of synthetic precursors with cell cultures and/or enzymes form the subject of approximately 20 publications but several recent reviews[1-3] summarize these studies for the interested reader. The present discussion will emphasize only those aspects which illustrate the interplay between cell culture methodology and synthetic chemistry and to present the most recent, as yet, unpublished results.

Biotransformation and Biosynthetic Studies: Enzymatic Reactions versus Laboratory Syntheses

In earlier studies[4] and *prior* to the development of our cell culture program, we had completed a partial *chemical* synthesis of the bisindole family (Figure 3). The coupling of catharanthine N-oxide (2) with vindoline (3) under the influence of trifluoroacetic anhydride affords the highly unstable dihydropyridinium intermediate (4) which, without isolation, is then reduced to anhydrovinblastine (5). This process, also studied independently by others[5], afforded an attractive route to the bisindole series but *not* to the clinical drugs vinblastine (6, R = Me, Figure 5) and vincristine (6, R = CHO, Figure 5). However, as our cell culture program developed, approximately four years later, it was logical to investigate whether the cell cultures of *C. roseus*, the plant from which catharanthine (1) and vindoline (3) as well as the bisindole alkaloids were isolated, were capable of performing the above-noted coupling reaction but under *enzymatic* conditions. The questions whether 1 and 3 were the *biosynthetic* building units for the bisindole family and whether there was a relationship between the mechanistic pathways for the "chemical" and "biological" processes leading to the bisindole system, became of considerable relevance to our future studies. With a stable cell line on hand, the preparation of a crude

1. Growth conditions are laboratory controlled, therefore, reproducible.
2. Growth parameters (media, pH, temp. etc.) optimized to provide agent in yields higher than in plant.
3. Cloning provides selected cell lines for optimum production of desired agent.
4. Excellent media for biosynthetic studies.
5. Isolation and studies of enzymes.

Figure 2. Plant Cell Cultures- Advantages over Living Plants.

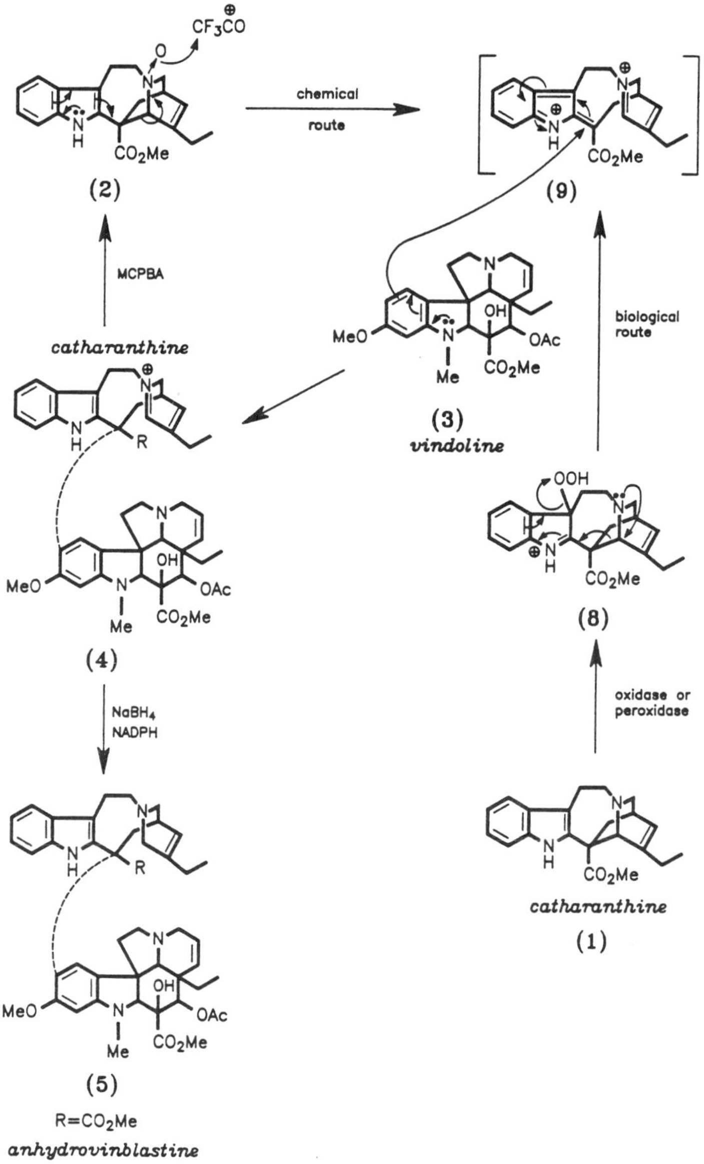

Figure 3. Mechanism for coupling of catharanthine and vindoline. Chemical versus biological route.

enzyme mixture was developed, as summarized in Figure 4, and then utilized in a biosynthetic experiment with [Ar-^{2}H]-catharanthine and ^{14}C-labelled vindoline (label at acetate carbonyl carbon). The study revealed that 1 and 3 were indeed coupled under the influence of the enzyme mixture to the same intermediate 4 obtained several years earlier in our synthetic studies. Subsequent refinement of these conditions (see later) allowed the coupling of the precursor alkaloids 1 and 3 to 4 in high yields (up to 90%) and in short time periods (15-30 min, for example). Longer incubation times revealed that 4 is converted, in part, to anhydrovinblastine (5) through a reductive process, suggestive of NADPH-like enzyme systems. When catharanthine N-oxide (2), the important intermediate in the chemical coupling reaction was utilized as precursor in the enzyme-catalyzed process, no coupling was observed.

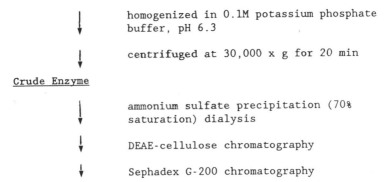

C. roseus Tissue Culture

↓ homogenized in 0.1M potassium phosphate buffer, pH 6.3

↓ centrifuged at 30,000 x g for 20 min

Crude Enzyme

↓ ammonium sulfate precipitation (70% saturation) dialysis

↓ DEAE-cellulose chromatography

↓ Sephadex G-200 chromatography

Partially Purified Enzyme

Figure 4. Isolation of enzymes from cell culture of C. roseus.

Clearly the enzymatic process involved in an activation of the catharanthine unit, as required for the subsequent reaction with vindoline (3), did not involve an N-oxide. The most plausible mode of activation, well known for cleavage of indole systems, for example, tryptophan to kynurenine by peroxidases, is conversion of **1** to the hydroperoxyindolenine (**8**) which then fragments to the same highly reactive intermediate (**9**) obtained from the Polonovski-type fragmentation in the chemical process.

Further studies with the crude enzyme system (Figure 4) were then conducted with anhydrovinblastine (**5**) as precursor and very interesting results were obtained. The alkaloids leurosine (**10**), catharine (**11**) and the clinical drugs vinblastine (**6**, R = Me) and vincristine (**6**, R = CHO), normally isolated from C. roseus plants, were obtained. The major product was leurosine (**10**) being formed initially with short incubation times followed by catharine (**11**), vinblastine and vincristine. In a separate experiment, leurosine (**10**) is enzymatically transformed to catharine (**11**) thereby establishing its role as a biosynthetic precursor to **11**. Clearly the crude enzyme preparation obtained from the cell culture consisted of a multi-enzyme complex capable of selective epoxidation of **5** to **10**, cleavage of **10** to **11**, and finally overall "hydration" of the olefinic linkage to the clinical drugs (**6**).

The enzymatic conversions summarized in Figure 5 were immediately reminiscent of our earlier chemical studies concerned with selective functionalization of **5** as summarized in Figure 6. This interesting comparison revealed that t-butylhydroperoxide, the chemical "equivalent" of peroxidases, was capable of similar conversions. Indeed, subsequent assay of the crude enzyme preparation established that significant levels of peroxidase enzymes are present. Additional information concerning the course of the biosynthetic pathway leading to the various bisindole alkaloids found in C. roseus plants and/or biotransformation of the precursors catharanthine (1), vindoline (3) and anhydrovinblastine (**5**) was obtained from a large series of experiments involving different ages of cells and enzymes derived therefrom. An overall summary is provided in Figure 7. With incubation times for varying time

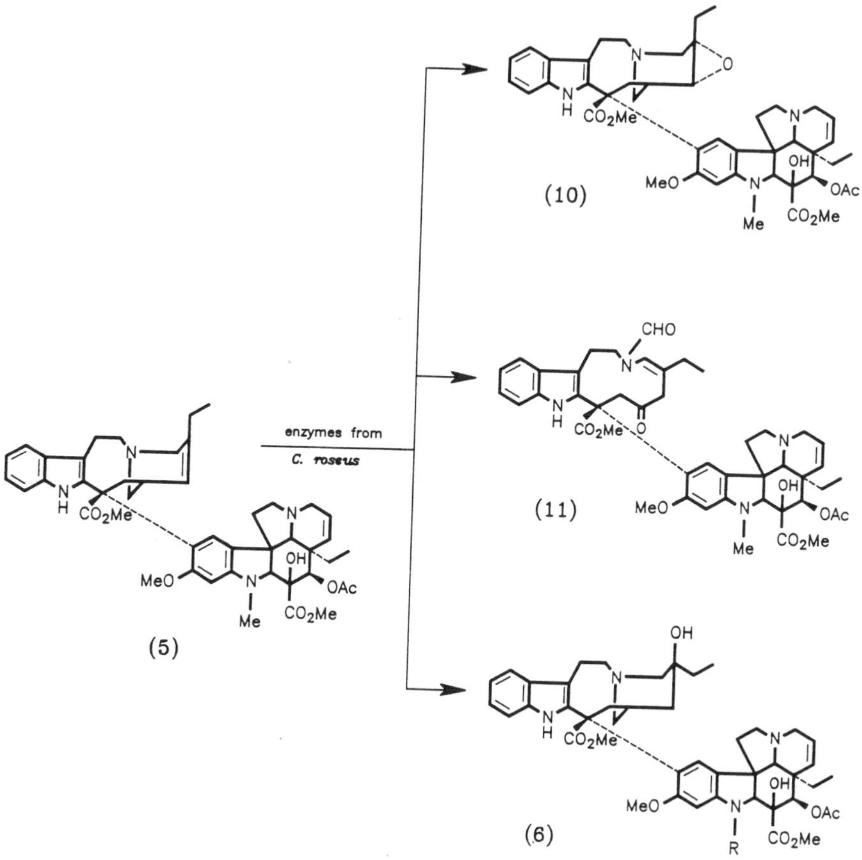

Figure 5. Enzyme-catalysed synthesis of leurosine (10), catharine (11), and vinblastine (6; R = Me), employing cell-free extracts from C. roseus.

periods (15 min to 15 h), anhydrovinblastine (5) via the dihydropyridinium intermediate 4, transforms as noted above to 10, 11, 6 (R = Me) and finally into the known alkaloids vinamidine (12, R = H), and hydroxyvinamidine (12, R = OH), also found in C. roseus plants. In separate experiments, with vinblastine (6, R = Me) and vincristine (6, R = CHO) as precursors, the enzyme preparation as obtained above, yields vinamidine and hydroxyvinamidine as major products. Here again, we have a close parallel between these enzymes-catalyzed conversions and earlier chemical studies in which it was shown that 5, its reduction product deoxyleurosidine (14) and leurosine (10) are converted to 12 (R = H) and 12 (R = OH) by means of $KMnO_4$ (Figure 8).

Enzyme Immobilization

Utilizing enzymes isolated from the plant cell cultures (Figure 4), it was possible to immobilize enzymes employing the technique of affinity gel chromatography[1,6] and it was shown that with the enzyme system, immobilized on 2',5'-ADP Sepharose 4B, a 90% yield of coupling of catharanthine (1) and vindoline (3) to bisindole alkaloids can be achieved. Details of these studies are published elsewhere[1,6].

Biomimetic Studies - Biosynthetic Information Provides Route to Efficient Synthesis of Vinblastine

The above discussion has indicated that the iminium intermediate 4

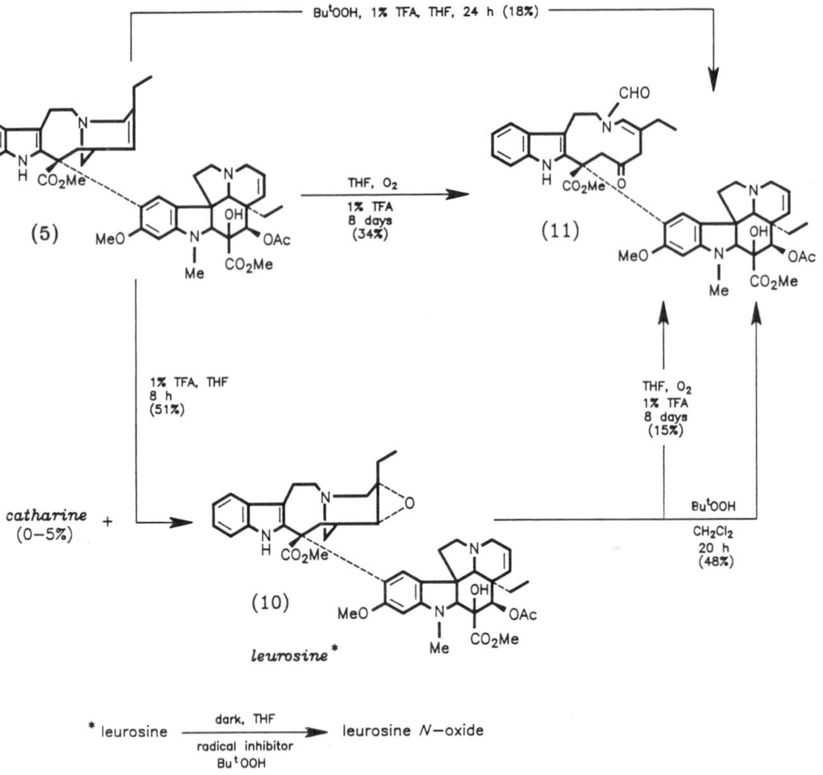

Figure 6. Synthesis of leurosine (10) and catharine (11) from anhydrovinblastine (5).

(Figure 7) plays a pivotal role in the biosynthesis of the various bisindole alkaloids and its enzymatic conversion to the clinical drugs vinblastine and vincristine became of central importance in our program. In consideration of a possible pathway to **6** (R = Me) and recognizing the possible bioconversion of **4** to **6** (R = Me) with NADH-type enzyme systems, as noted above, it was plausible to consider an initial regio-specific 1,4-reduction to enamine **15** (Figure 9), oxidation of **15** to **17** and finally reduction to **6**. Laboratory experiments to evaluate this hypothesis were now developed. Employing the enzyme systems discussed above for the coupling of **1** and **3** to **4**, it was noted that FMN (or FAD) and $MnCl_2$ were effective cofactors in this process. Consequently the conversion of **5** to **4** was undertaken under enzyme-free conditions but with the flavine coenzyme FMN and excellent yields of **4** were obtained[7]. For the subsequent reductive process, **4** → **15**, a Tris-HCl buffer solution of β-NADH (8 equiv.) was added to a methanol-Tris HCl buffer solution of FMN-generated **4** and the reduction allowed to proceed for 4.5 h. The result was an overall conversion (85% yield) of **4** to the desired enamine **15** (1,4-reduction) and **5** (1,2 reduction) in a 4:1 ratio respectively. With **15** in hand, it was possible to demonstrate that in the enzyme-catalyzed coupling of **1** and **3** to **4**, the initial intermediate **4** (15-30 min incubation) is indeed reduced to **15** (major product) and **5** (minor product). Employing HPLC monitoring, **15** was generally observed in a 45 min incubation time and then on further incubation (> 1 h) it was converted, presumably by an oxidative process, to intermediate **17**. The latter is finally reduced to **6**. Support for the presence of **17** in the enzymatic process was again available from

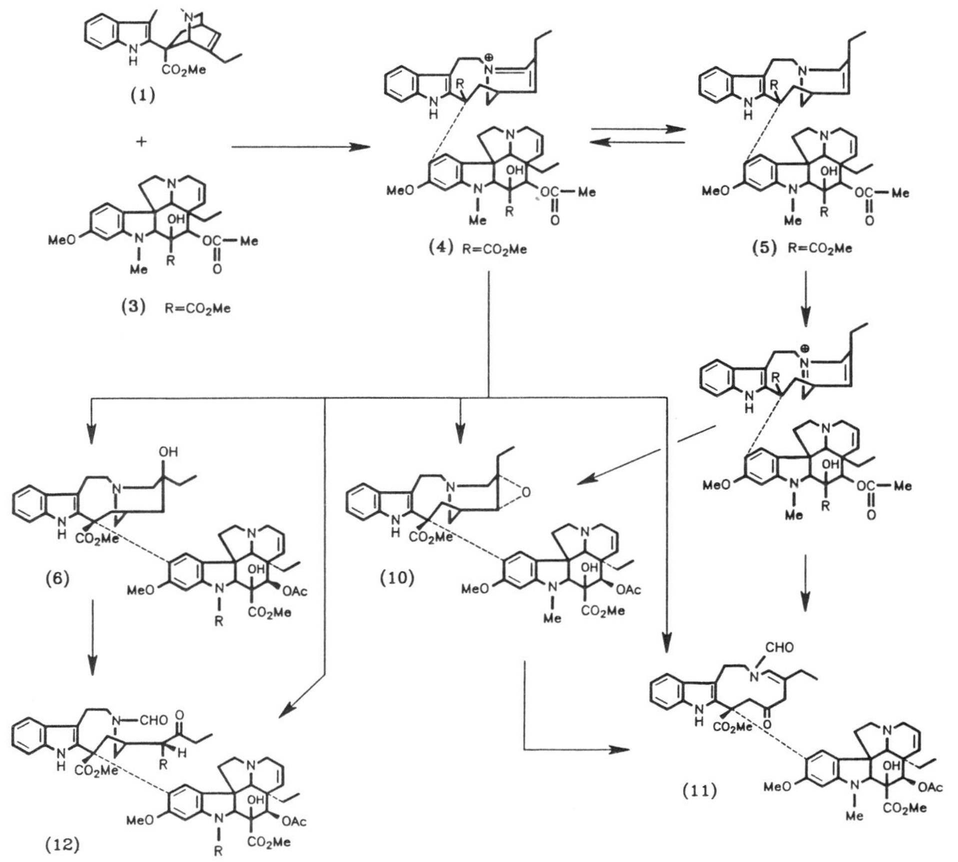

Figure 7. Overall summary of results obtained when catharanthine (1), vindoline (3), and anhydrovinblastine (5) are incubated _either_ with cell cultures of _Catharanthus_ _roseus_ _or_ cell-free extracts derived from such cultures.

chemical experiments. When **15** was reacted with various oxidants (O_2 H_2O_2, $FeCl_3$), it was converted to an intermediate postulated as **17**, identical (HPLC monitoring) with that obtained in the enzyme-catalyzed process. Sodium borohydride reduction of **17** provided vinblastine (**6**, R = Me) and the isomeric alkaloid leurosidine (**19**, Figure 9), the latter being derived from the iminium intermediate **18** (Figure 9). In summary, the combination of synthetic chemistry and enzyme-catalyzed conversions of **5** and/or **1** and **3** established the overall process for the biosynthesis of vinblastine (**6**, R = Me) from the starting alkaloids as shown in Figure 9.

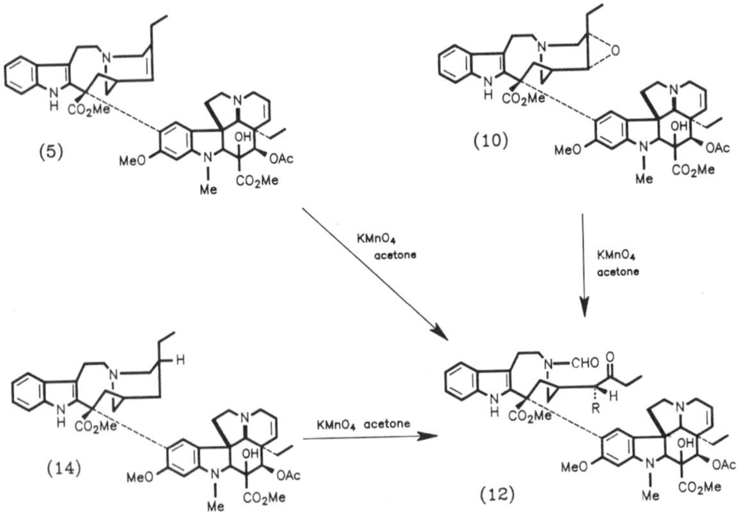

Figure 8. Overall summary of the synthesis of vinamidine (12; R = H) and 3R-hydroxyvinamidine (12; R = OH).

Synthetic Studies - A Highly Efficient and Commercially Important Route to Vinblastine

With the above data in hand, the stage was set for the development of an efficient chemical synthesis of vinblastine and vincristine from the readily available alkaloids catharanthine (**1**) and vindoline (**3**) obtained as major byproducts in the plant extraction of vinblastine and vincristine but, at present, simply discarded.

The first step of the overall process, that is, the coupling of **1** and **3** was already solved through the Polonovski-type fragmentation of **1** and coupling with **3**. Major focus on the regiospecific 1,4-reduction of **4**, accomplished above by NADH enzyme systems, was required. For this purpose, NADH chemical models were considered. A large number of experiments with various 1,4-dihydropyridine systems as reducing agents were performed and the results are summarized in Table 1. Details are published elsewhere[8].

The best reagent in terms of ease of preparation, stability and yield of **15** was the nicotinamide analogue **26**, and the effect of temperature on the course of 1,2- versus 1,4-reduction of **4** was then evaluated. The ratio of 2.2:1 in favor of 15, at the reduction temperature of 20° C (Table 1), could be successively altered to 4.2:1 in favor of **15** and an improved yield (85%) at a lower temperature (-40° C), thereby revealing that the reductive step, 4 → **15**, could now be achieved with the chemical reductant **26**.

Table 1. NADH models for 1,2- versus 1,4-reduction of iminium intermediate (4) at 20° C [a]

Reducing agents	Reduction products [b] 1,4 (15) : 1,2 (5)	Combined yield of 1,2 + 1,4 reduction product (%) [b]
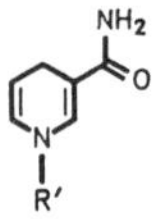		
(20) R' = CH₂Ph	1 : 1	75
(21) = CHPh₂	0.9 : 1	60
(22) = 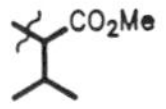	2 : 1	65
(23) = 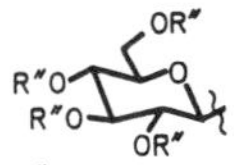 R" = Ac	3 [c] : 1	70
(24) = 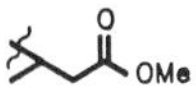	1.1 : 1	65
(25) = 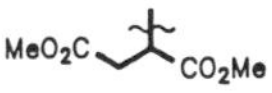	2.3 : 1	70
(26) = 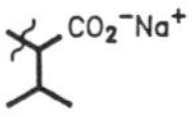	2.2 : 1	70
(27) 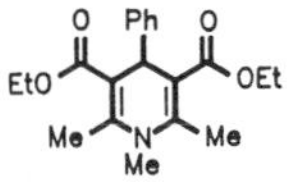	0.4 : 1	60
(28) 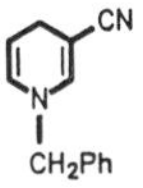	1 : 1	40
(29) 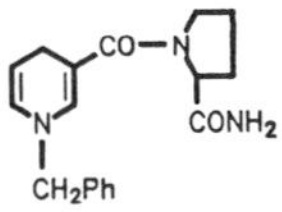	1.1 : 1	60

(a) Typical procedure: 100 mg (4) in MeOH (6 ml) at 20° C to which reducing agents (20-29) (1-6 eq.) in MeOH (6 ml) were added.
(b) HPLC quantitation.
(c) Reduction rate very slow. Prolonged reaction resulted in significant further reduction of (15) to the saturated derivative (14) and (16) (50%).

The third step in the overall process, that is, oxidation of 15 to 17, was achieved most successfully with $FeCl_3$ as oxidant and 17, without isolation, was immediately reduced with sodium borohydride.

The final objective, the direct conversion of 1 and 3, to vinblastine, <u>without isolation</u> of intermediates, that is, a "one-pot" process could be realized. A summary of this process is provided in Figure 10. The overall sequence involving five distinct and separate chemical

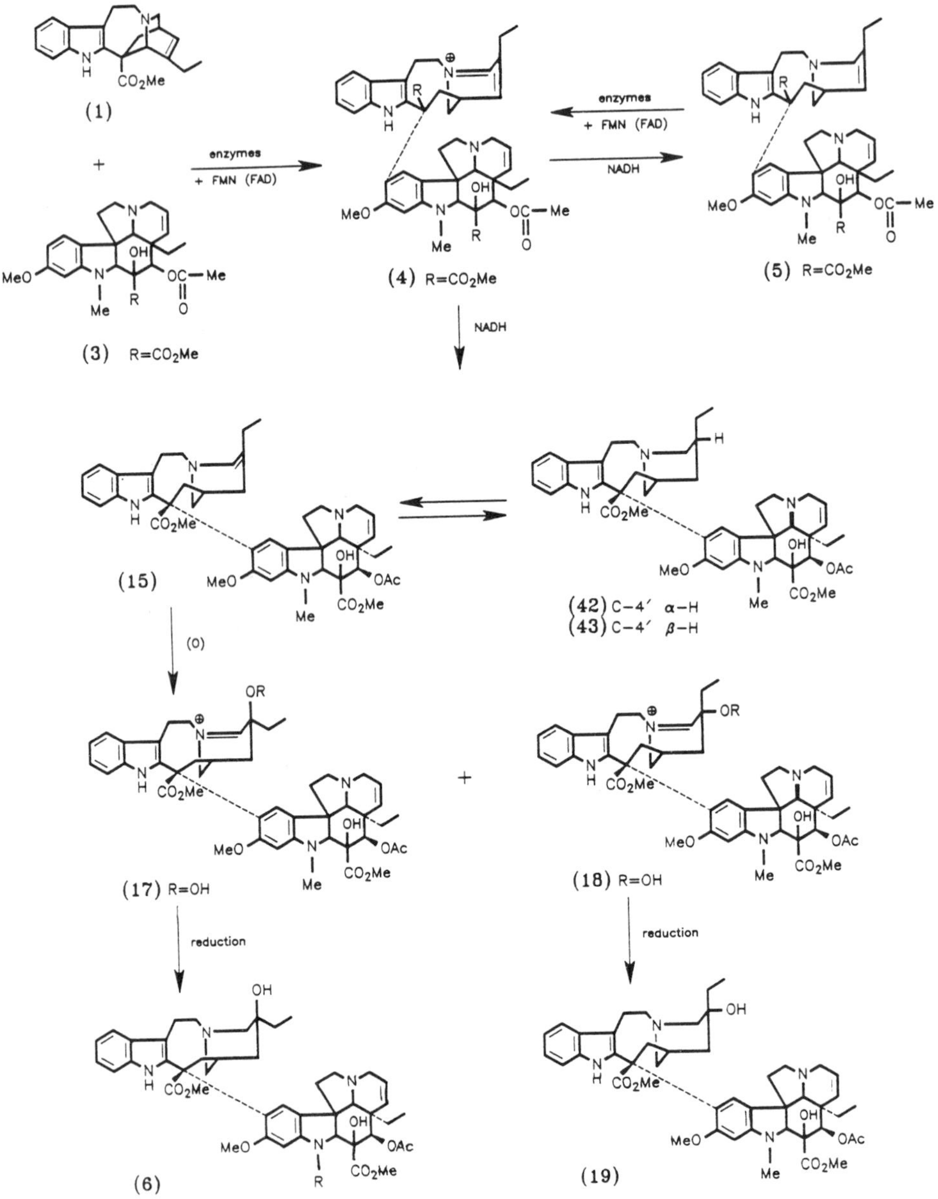

Figure 9. Overall summary of the biosynthetic pathway of vinblastine (6; R = Me) and leurosidine (19) from catharanthine (1) and vindoline (3).

reactions, and providing an overall yield of vinblastine (6, R = Me) in 40% yield, demands that each reaction within the sequence must proceed with yields in excess of 80%. When the two additional bisindole products, anhydrovinblastine (5, 18%) and leurosidine (19, 17%) are considered, it is clear that the majority of reactions proceed in almost quantitative yield.

In conclusion, the above discussion reveals that the combination of plant cell cultures and synthetic chemistry has provided a highly efficient and commercially important route to the clinical drugs from waste by-products presently discarded in the commercial preparation of these clinical drugs.

STUDIES WITH CELL CULTURES OF C. ROSEUS - THE PODOPHYLLOTOXIN FAMILY

The podophyllotoxin family of natural products (30, Figure 11)[9] are a well studied family of compounds from both a chemical and pharmacological focus. Their medicinal properties have been extensively evaluated and current importance relates to the clinical efficacy of etoposide as a clinical agent in cancer chemotherapy. This drug has shown effective use in the treatment of myelocytic leukemia, neuroblastoma, bladder, testicular and small-cell lung cancers. The present process considered for commercial production starts with podophyllotoxin which, in turn, is chemically converted to 4'-demethylepipodophyllotoxin (31) and subsequently converted to etoposide by the sequence shown in Figure 12. As in the case of vinblastine-vincristine production, the basic starting material must be derived from a plant source, for example, Podophyllum peltatum or related species, and the difficulties

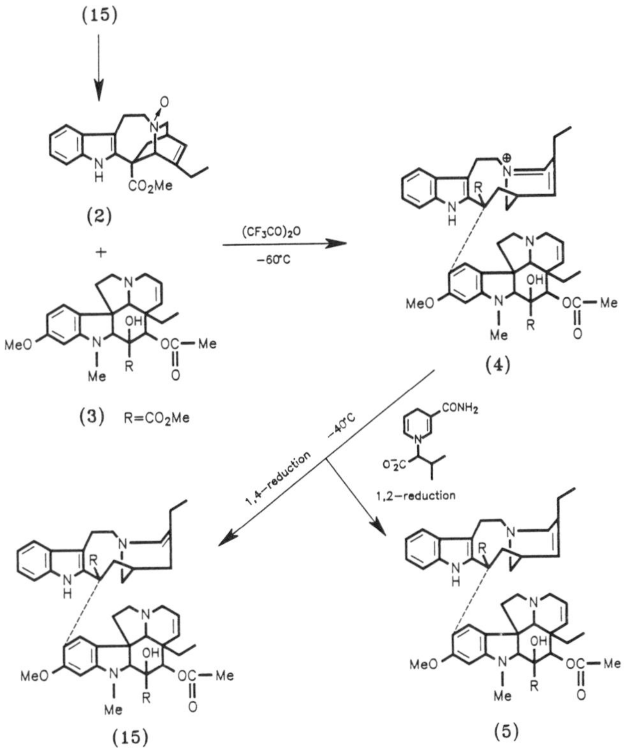

Figure 10. A highly efficient 'one-pot' process for the synthesis of vinblastine (6; R = Me) and leurosidine (19) from catharanthine (1) and vindoline (3).

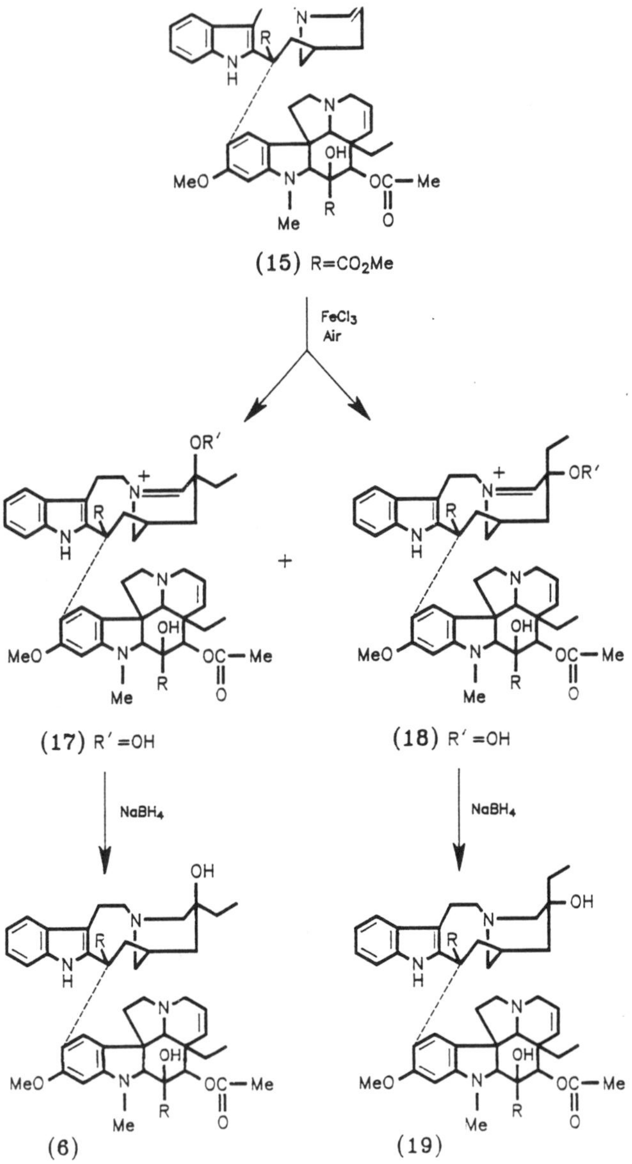

Figure 10 - continued

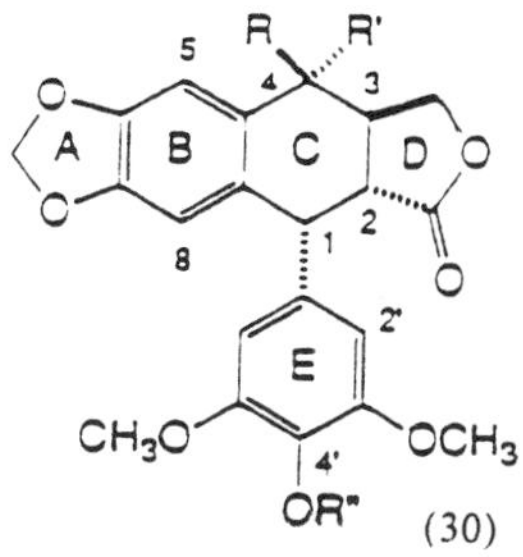

(30)

Podophyllotoxin : R = H; R' = OH; R" = CH_3

Epipodophyllotoxin : R = OH; R' = H; R" = CH_3

Deoxypodophyllotoxin : R = R' = H; R" = CH_3

4'-Demethylpodophyllotoxin : R = H; R' = OH; R" = H

4'-Demethylepipodophyllotoxin : R = OH; R' = H; R" = H

Etoposide : R = 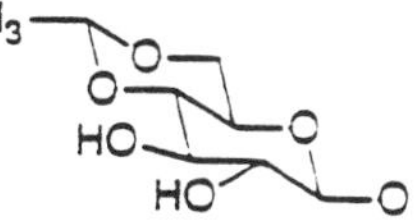 ; R' = H; R" = H

Figure 11. The podophyllotoxin family of compounds.

noted earlier (Figure 1) obviously prevail. One possible alternative to the synthesis of **31** would be to involve a combination of plant cell culture methodology and synthetic chemistry in an overall strategy similar to the above.

Consideration of the biosynthetic pathway generally postulated for the podophyllotoxins (lignans, in general) as shown in Figure 13, involves phenol oxidative coupling, (originally proposed by D.H.R. Barton for phenolic natural products) in the later steps of the pathway (**38** → **39**, Figure 13). Such ring closure reactions are considered to be achieved through peroxidase enzymes present in the plant. Since the enzyme systems isolated from C. roseus cell cultures (Figure 4) revealed significant peroxidase activity, it was of interest to determine whether these enzymes were capable of utilizing "foreign" synthetic precursors similar in structure to **36** and/or **38** (Figure 13) for such ring closure reactions. The synthetic pathway developed for this purpose is outlined in Figure 14.

The readily available aldehydes ($R = R^1 = H$ or $R = H$; R^1 = alkyl) were used for these studies. The sequence, summarized in Figure 14, afforded excellent overall yields of the corresponding precursors of general structure **45** ($R = R^1 = H$; $R = H$; $R^1 = CH_3$ or isopropyl; R, R^1 = methylenedioxy) and a number of these were evaluated in an enzyme-catalyzed process. The results for two of these precursors are presented in Figure 15. The crude enzyme system (CFE), from the C. roseus cell line (coded AC3) studied above, and prepared according to the procedure outlined in Figure 4, was incubated with the synthetic precursor **46**, the latter being available via catalytic desulfurization of **42**, in the presence of H_2O_2 as cofactor, to afford an essentially

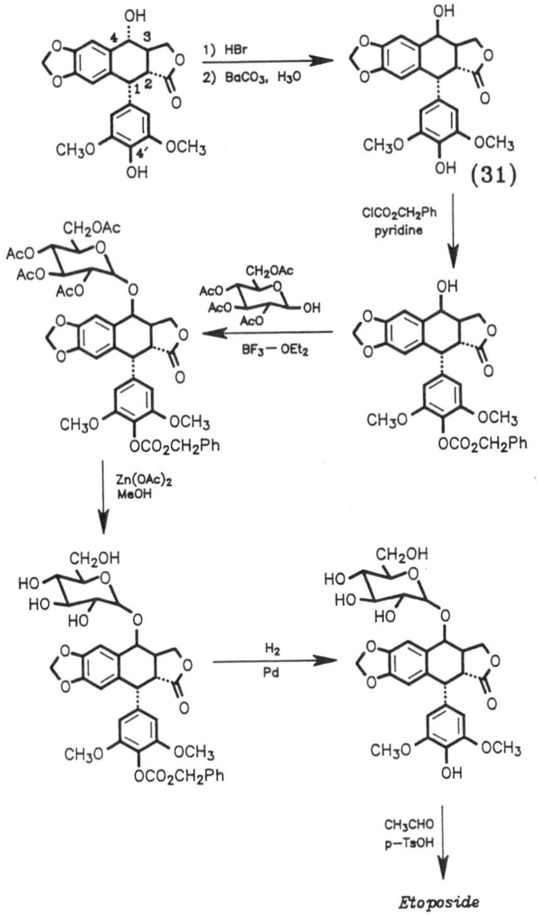

Figure 12. Commercial production of etoposide from podophyllotoxin.

quantitative yield of the desired ring-closed product 47. Extensive studies were conducted as to age of culture from which the enzymes are isolated, pH dependency, incubation time, etc. The conditions shown in Figure 15 are optimum for the cyclization of 46.

In a similar manner, the precursor 48 is cyclized in excellent yield to 49. We have already developed the selective removal of the isopropyl group in 49 and converted the corresponding phenolic compound to the methylenedioxy system thereby completing the interrelationship between this compound and intermediate 31 (Figure 12), the latter being employed in the commercial production of etoposide.

In summary, it is clear from the latter study that there is considerable versatility in plant cell culture-derived enzymes in terms of their utilization as "reagents" in organic synthesis. The peroxi-

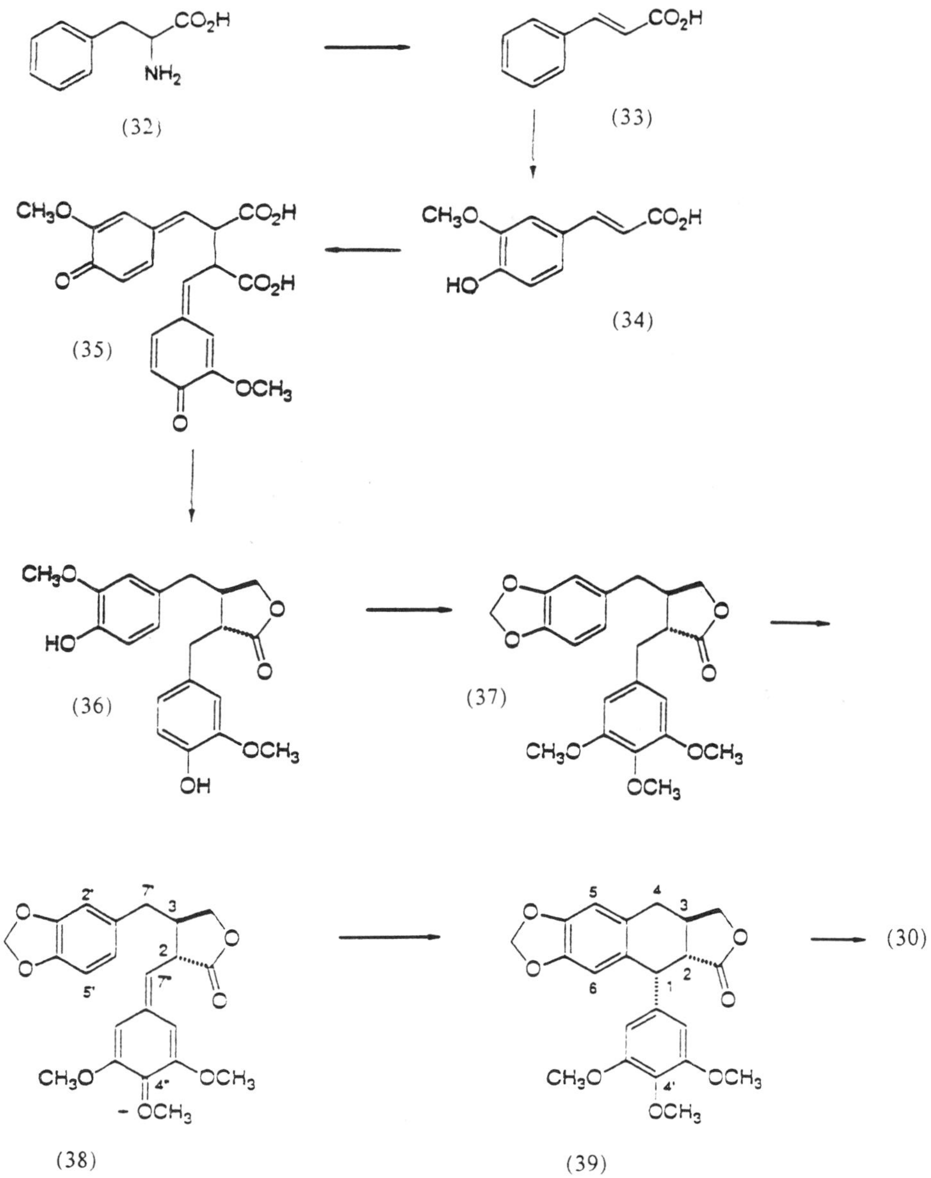

Figure 13. Proposed biosynthetic pathway of podophyllotoxin.

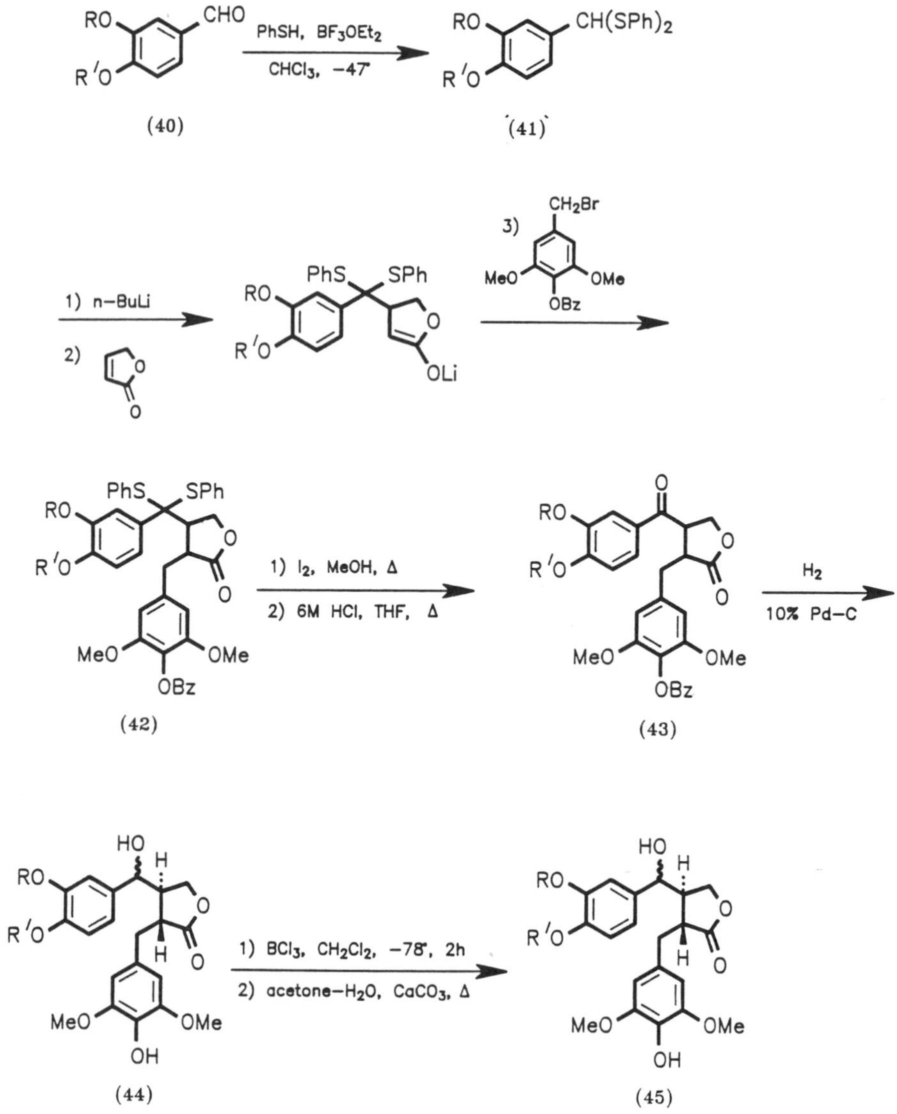

Figure 14. Synthesis of a 4'-demethylepipodophyllotoxin precursor.

dases in C. roseus cultures originally employed and developed for indole alkaloid production are now of considerable interest in synthetic routes to the clinical anti-cancer drug etoposide. Again, the combination of plant cell culture methodology and synthetic chemistry affords an attractive route to this commercially important compound.

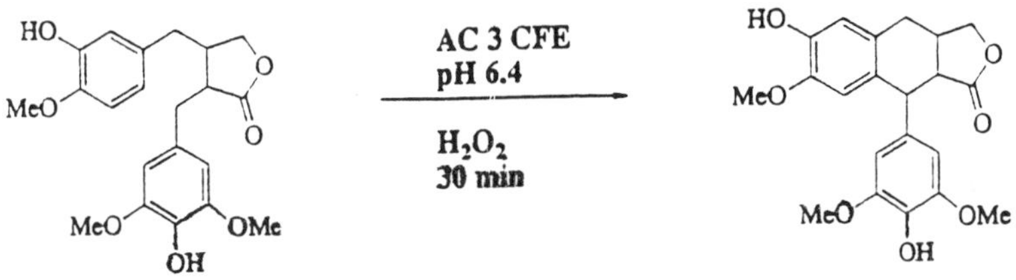

(46) (47)

Conditions: Peroxidase activity in CFE: 1.71 units mL^{-1}
Protein concentration: 1.1 - 1.6 mg mL^{-1}
Buffer: 0.02M phosphate

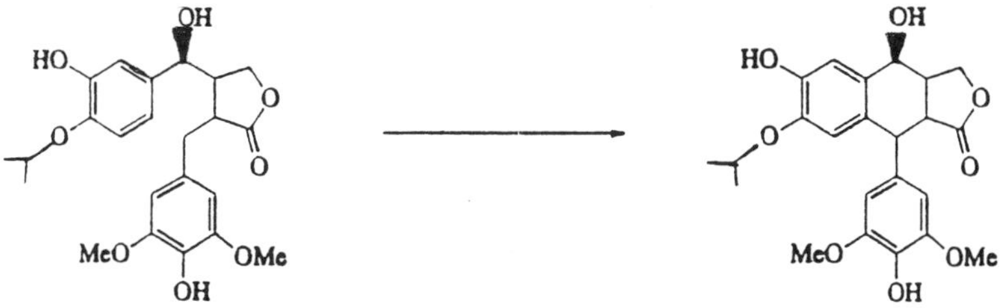

(48) (49)

Figure 15. Formation of carbon-carbon bonds with enzymes from Catharanthus roseus cell free extracts (CFE).

STUDIES WITH CELL CULTURES OF T. WILFORDII

Tripterygium wilfordii Hook f is a perennial twining vine of the family Celastraceae, which is cultivated in many parts of southern China such as Zhejiang, Anhui, Jiangxi, Fujian and Guangdong provinces and also in Taiwan. The herb is commonly known in China as Lei Gong Teng (Thunder God vine) or Mang Cao (rank grass). Its use in Chinese traditional medicine dates back many centuries, and it is first mentioned in the Saint Peasant's Scripture of Materia Medica[10], written about two thousand years ago, as being used for the treatment of fever, chills, oedema and carbuncle. Chinese gardeners used the powdered root to protect their crops from chewing insects. Most recently, crude extracts and refined extracts (a so-called multi-glycoside extract, or GTW) have been used increasingly to treat such disorders as rheumatoid arthritis, ankylosing spondilitis and a variety of dermatological disorders.[11-13]

In 1972, Morris Kupchan and co-workers first isolated the novel diterpenoid triepoxides, tripdiolide (50) and triptolide (51), from the roots of Tripterygium wilfordii.[14] These were the first reported natural products containing the 18(4 → 3)abeo-abietane skeleton and the first recognised diterpenoid triepoxides. Tripdiolide (50) and triptolide (51) were shown to have significant antileukaemic activity and these data generated a great deal of interest in these compounds.

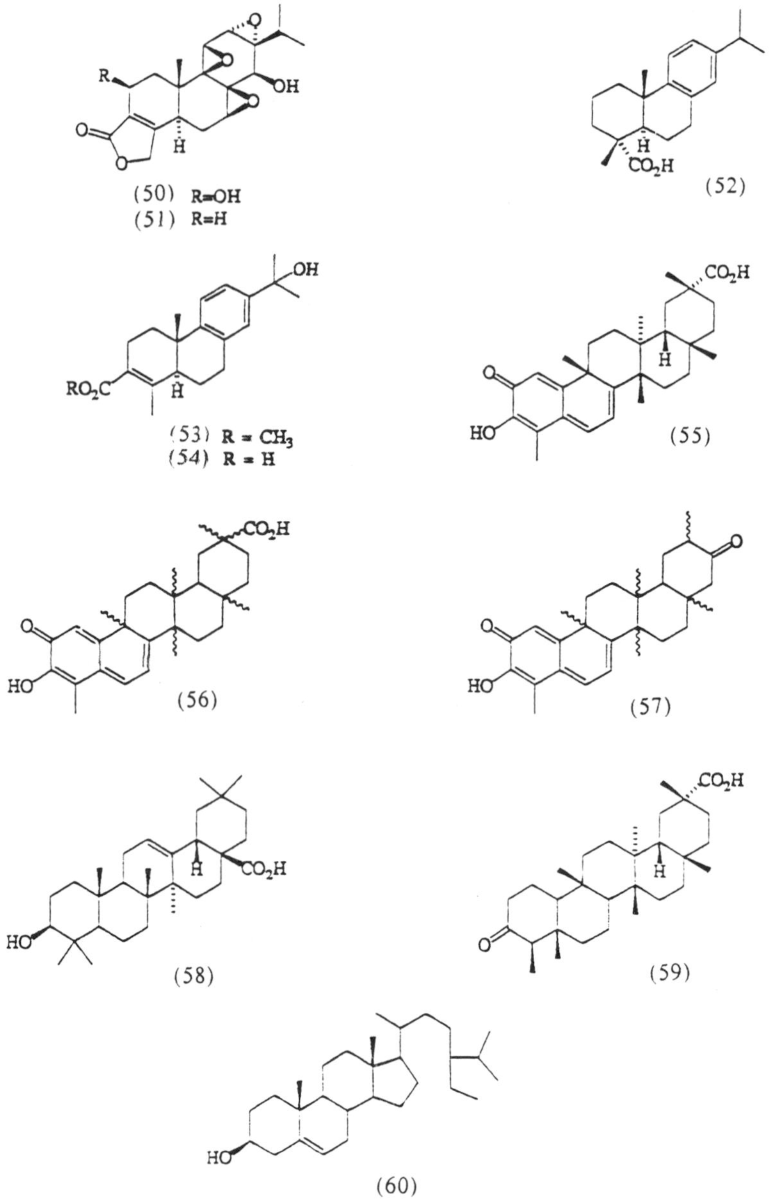

Figure 16. Natural products isolated from Tripterygium wilfordii cultures.

Since then, much research has been carried out into the production of the diterpene triepoxides for further biological evaluation. Numerous compounds have been isolated from Tripterygium wilfordii plants and their biological properties investigated (vide infra).

Our own interests in this area were stimulated originally by the cytotoxic properties associated with **50** and **51** and the results of our earlier studies are published.[15-17] However, the more recent data concerning biological activities of potential interest in such areas as rheumatoid arthritis, skin allergies and male contraception encouraged

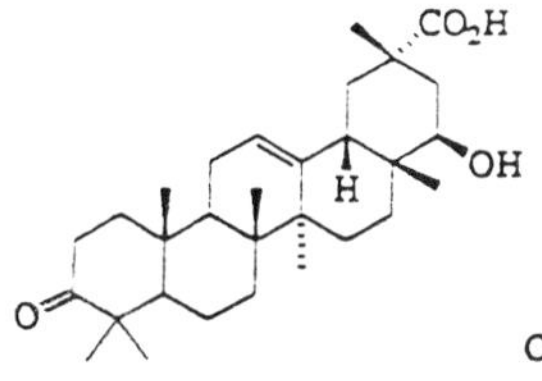

(61)

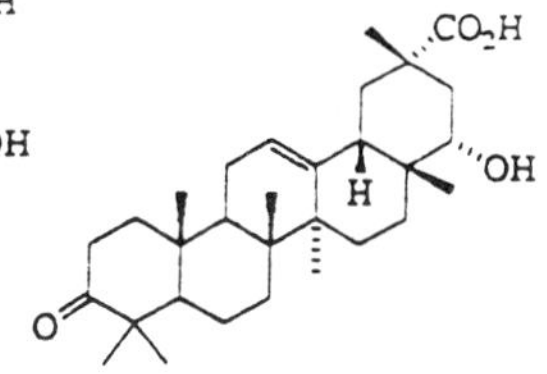

(62)

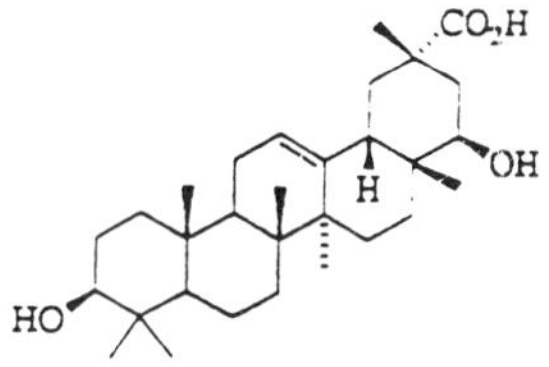

(63)

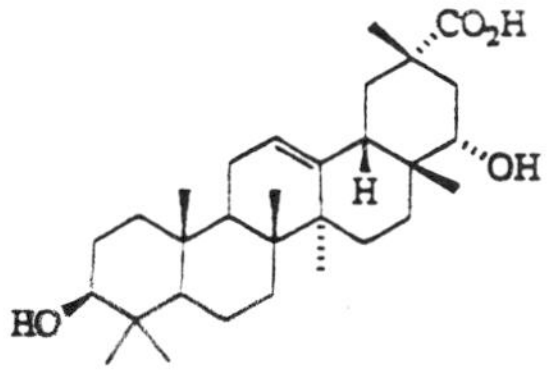

(64)

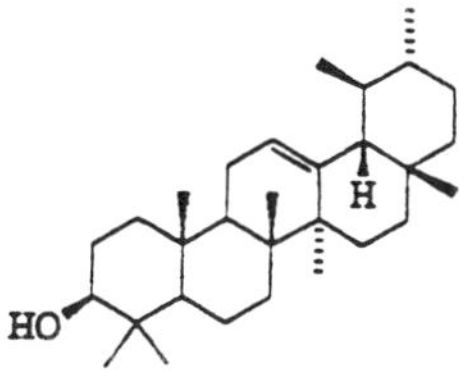

(65)

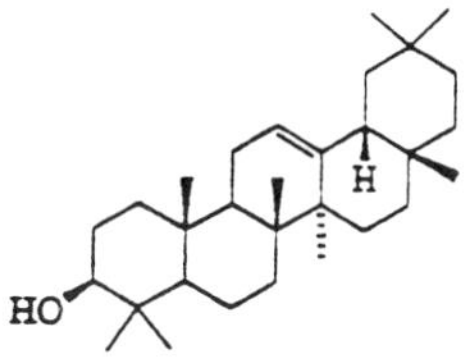

(66)

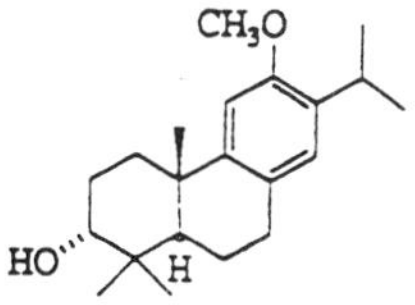

(67)

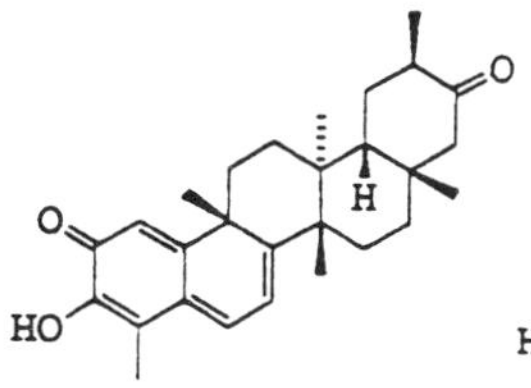

(68)

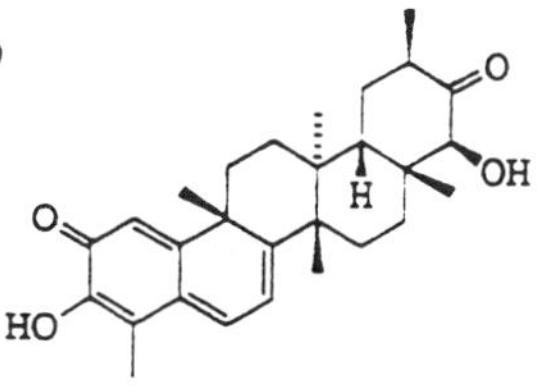

(69)

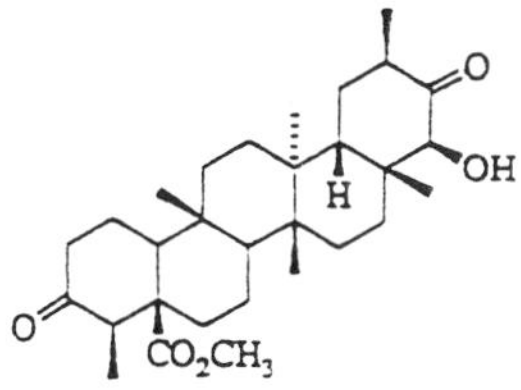

(70)

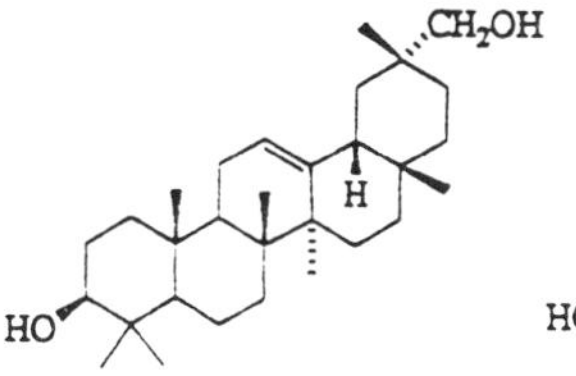

(71)

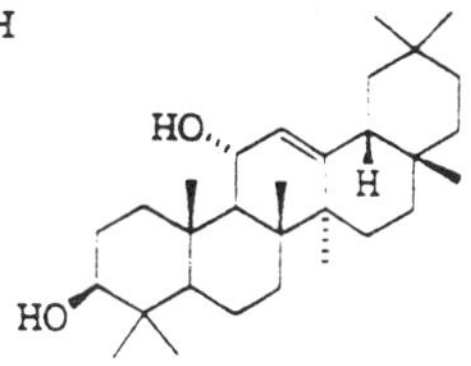

(72)

Figure 16 (continued)

us to expand our studies in this area. The Tripterygium wilfordii cell line (coded as TRP4a) has been grown in large bioreactors and several hundred liters of fermentation broth and cells have been processed. The structures shown in Figure 16 summarize our studies.

CONCLUSION

It is hoped that the above discussion has illustrated the interesting avenues of research that can be derived from such an interdisciplinary program.

Acknowledgement

It is a great pleasure to acknowledge the many enthusiastic and dedicated group of researchers who made this presentation possible.

Studies in the C. roseus area and relating to the vinblastine family include the following: B. Aweryn, J. Balsevich, G. H. Bokelman, B. Botta, C. Boulet, C. Buschi, L. S. L. Choi, M. Gummulka, W. Gustowski, G. M. Hewitt, T. Hibino, T. Honda, I. Itoh, E. Jahngen, A.V. Joshua, P. Kolodziecjczyk, G.C. Lee, N.G. Lewis, P.H. Liao, M. McHugh, T. Matsui, J. Nakano, T. Nikaido, T. Okutani, J. Onodera, I. Perez, A. Ratcliffe, P. Salisbury, T. Sato, S. K. Sleigh, K. L. Stuart, R. Suen, A. Treasurywala, H. Tsukamoto, B.R. Worth and S. Wunderly.

Studies in the C. roseus area and relating to the podophyllotoxin family include the following: S. Ansell, M. Arimoto, S. Gao, G. M. Hewitt, T. Jarvis, R. Milanova, J. Palaty, K. Piotrowska.

Studies with T. wilfordii represent a collaborative program between the author's laboratory and that of Prof. P.M . Townsley, Department of Food Science at this University. The development of the cell culture methods was performed jointly at Food Science by P. M. Townsley, W.T. Chalmers, D. J. Donnelly, K. Nilsson and F. Webster; G. G. Jacoli, Canada Department of Agriculture, Vancouver; and by P. Salisbury and G. M. Hewitt at the Chemistry Department. Analytical methods, isolation and characterization of the metabolites and development of synthetic methodology were performed at the Chemistry Department by M. H. Beale, L. S. L. Choi, E. Chojecka-Koryn, R. Duffin, M. Horiike, H. Jacobs, N. Kawamura, T. Kurihara, R.D. Sindelar, K. L. Stuart, Y. Umezawa, B. Vercek, and B.R. Worth. I am grateful to Prof. G. B. Marini-Bettolo, Universita Cattolica, Rome, for samples of polpunonic acid, methyl polpunonate and tingenone.

I am grateful to Professor U.K. Pandit, University of Amsterdam, for providing generous samples of 1,4-dihydronicotamide analogues and Hantzch ester employed in our studies.

Financial aid was provided by grants from the Natural Sciences and Engineering Research Council of Canada, and contracts from the National Research Council of Canada.

References

1. J. P. Kutney, Nat. Prod. Rep., 7:85 (1990), and references cited therein.
2. J. P. Kutney, Studies in Natural Products Chemistry; Atta-ur-Rahman, Ed., Interscience, Amsterdam, Vol. 2, Structure Elucidation (Part A), (1988), p. 365, and references cited therein.
3. J. P. Kutney, Heterocycles, 25:617 (1987)..
4. J. P. Kutney, T. Hibino, E. Jahngen, T. Okutani, A. Ratcliffe, A. M. Treasurywala, S. Wunderly, Helv. Chim. Acta, 59:2858 (1976).
5. N. Langlois, F. Gueritte, Y. Langlois, P. Potier, J. Am. Chem. Soc., 98:7017 (1976).
6. J. P. Kutney, C. A. Boulet, L. S. L. Choi, W. Gustowski, M. McHugh, J. Nakano, T. Nikaido, H. Tsukamoto, G. M. Hewitt, R. Suen, Heterocycles, 27:621 (1988).
7. J. P. Kutney, L. S. L. Choi, J. Nakano, H. Tsukamoto, Heterocycles, 27:1827 (1988).

8. J. P. Kutney, L. S. L. Choi, J. Nakano, H. Tsukamoto, M. McHugh, C. A. Boulet, Heterocycles, 27:1845 (1988).
9. I. Jardine, Anticancer Agents Based on Natural Product Models; J. M. Cassady, J. D. Douros, Eds., Academic Press, New York (1980), p. 319, and references cited therein.
10. S. Huang, The Saint Peasant's Scripture of Materia Medica; Publishing House for Chinese Medical Classics, Beijing, republ. (1982), p. 309.
11. D. Y. J. Yu, Traditional Chinese Med., 3:125 (1983).
12. J. Guo, S. Yuan, X. Wang, S. Xu, D. Li, Chinese Med. J., 7:405 (1981).
13. W. Xu, J. Zheng, X. Lu, Intl. J. Dermatol., 24:152 (1985).
14. S. M. Kupchan, W. A. Court, R. G. Dailey, C. J. Gilmore, R. F. Bryan, J. Am. Chem. Soc., 94:7194 (1972).
15. J. P. Kutney, M. H. Beale, P. J. Salisbury, R. D. Sindelar, K. L. Stuart, B. R. Worth, P. M. Townsley, W. T. Chalmers, D. J. Donnelly, K. Nilsson, G. G. Jacoli, Heterocycles, 14:1465 (1980).
16. J. P. Kutney, G. M. Hewitt, T. Kurihara, P. J. Salisbury, R. D. Sindelar, K. L. Stuart, P. M. Townsley, W. T. Chalmers, G. G. Jacoli, Can. J. Chem., 59:2677 (1981).
17. J. P. Kutney, L. S. L. Choi, R. Duffin, G. Hewitt, N. Kawamura, T. Kurihara, P. Salisbury, R. Sindelar, K. L. Stuart, P. M. Townsley, W. T. Chalmers, F. Webster, G. G. Jacoli, Planta Med., 48:158 (1983).

NMR STUDIES OF COMPLEXES OF *L. CASEI* DIHYDROFOLATE REDUCTASE WITH ANTIFOLATE DRUGS: MULTIPLE CONFORMATIONS AND CONFORMATIONAL SELECTION OF BOUND ROTATIONAL ISOMERS

J. Feeney

Laboratory of Molecular Structure
National Institute for Medical Research
Mill Hill, London NW7 1AA, U.K.

INTRODUCTION

Dihydrofolate reductase (DHFR) catalyses the reduction of dihydrofolate (and folate with lower efficiency) to tetrahydrofolate using NADPH as a coenzyme[1]. The enzyme is of considerable pharmacological interest being the target for several clinically useful 'antifolate' drugs such as methotrexate (anti-neoplastic) 1, trimethoprim (anti-bacterial) 2 and pyrimethamine (anti-malarial) 3.

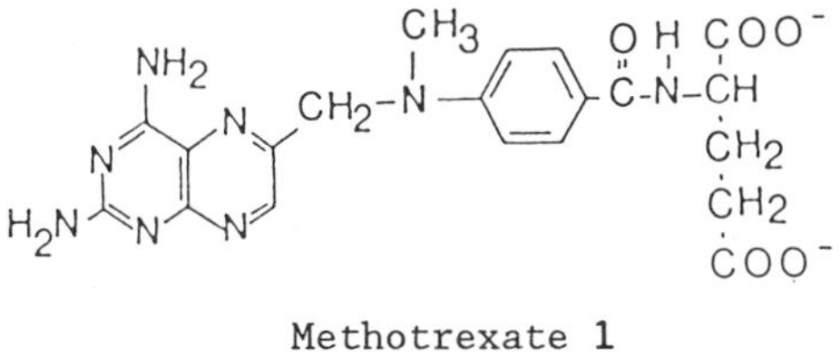

Methotrexate 1

In each case the drug acts by inhibiting this essential enzyme in invasive cells[1-3]. A great deal of effort has gone into structural studies involving the enzyme from various sources and numerous X-ray crystallographic[4-9] and NMR[10-54] studies have been reported for a wide range of inhibitor complexes formed with the enzyme. These studies are aimed at obtaining an improved understanding of the factors which control the specificity of ligand binding. Such information could help in the design of new antifolate drugs. Much of the work has been directed at defining the conformations of the bound inhibitors and in characterising specific interactions between the ligands and the protein. Chemical modifications of the ligands or single-site modifications of the protein by site-directed mutagenesis have been used to investigate important interactions between ligand and protein[13,46-48,51,52]. From available X-ray structural data it is possible to identify sites on the protein with possible binding potential and then to design inhibitors containing groups

Bioorganic Chemistry in Healthcare and Technology, Edited by U.K. Pandit and F.C. Alderweireldt, Plenum Press, New York, 1991

capable of interacting with these additional binding sites. For example, several analogues of trimethoprim, modified such that they could interact with binding sites normally only used in substrate binding, have been prepared and investigated[8,48]. Several tightly binding inhibitors of DHFR were obtained using this approach and some of the modified inhibitors have binding constants three orders of magnitude larger than that of the parent molecule[48]. Such binding studies always need to be accompanied by structural investigations to determine whether the predicted interactions have taken place and also to see if any additional conformational changes accompany the binding. Both X-ray crystallography[8] and NMR spectroscopy[48] have been used for this purpose. While the X-ray studies provide very detailed structural information they do require the samples to be studied as crystals. NMR examines the complexes in the solution state and can provide much useful information about specific interactions (particularly those involving charged residues), conformational states and dynamic processes within the complexes. Because of the rapid developments in methods of assigning resonances in complex protein spectra[55,57] and in methods of interpreting NOE data in structural terms it seems likely that NMR will become increasingly used for providing detailed conformational information in the solution state.

ASSIGNMENT OF LIGAND AND PROTEIN SIGNALS

To obtain any detailed structural information from NMR spectra of protein complexes it is first necessary to make specific resonance assignments for ligand and protein signals. Ligand resonance assignments are relatively easy to achieve either directly by using selective isotopic labelling (^{13}C, ^{15}N, ^{2}H)[11-15,17,19] or indirectly from transfer of saturation or 2D exchange experiments which connect signals in bound and free ligand species[16,18,19,54].

Until recently it was much more difficult to obtain protein resonance assignments for moderately sized proteins. For example our early assignment studies on *L. casei* DHFR relied on correlating X-ray structural data with NOE data from protons in close proximity. This method assumes that the crystal and solution conformations are similar. Since 1989 technical developments have allowed the sequential assignment method pioneered by Wuthrich and coworkers[58] to be applied to proteins of the size of *L. casei* DHFR (M_r 18,300). This method does not rely directly on crystal structure data but obtains the complete resonance assignments by correlating NMR data with sequence information. The various spin systems are first identified in the NMR spectrum and the spin systems of adjacent residues are then connected to each other by using NOE data: in this way a number of continuous stretches of assigned residues are obtained and matched to the known sequence of the protein. The most important NOE connections for making sequential assignments are those between amide NH protons and protons in adjacent residues (NH_i-NH_{i+1}, αCH_{i-1}-NH_i and βCH_{i-1}-NH_i NOEs). This method works very well for small proteins but for medium sized proteins such as DHFR the considerable overlap of cross-peaks in the 2D NOESY and COSY spectra make the method difficult to apply directly. Some simplification can be obtained by examining selectively deuterated proteins. For example by examining a DHFR sample in which all the αCH protons of the 16 valine residues have been replaced by deuterium one could easily detect the removal of the corresponding crosspeaks in the αCH-NH region of the COSY spectrum. This not only allows immediate residue assignment of these valine resonances but also makes the assignments of other, previously overlapped, cross-peaks much simpler to interpret. However by far the most powerful and generally applicable methods for removing ^{1}H chemical shift degeneracy involves examining isotopically labelled proteins (^{13}C or ^{15}N) with various 3D and 4D methods[55-57]. We have prepared a uniformly ^{15}N-labelled DHFR (90% labelled) and subjected it to NOESY.HMQC and HOHAHA.HMQC 3D experiments (HMQC is a heteronuclear multiple quantum coherence experiment which detects protons directly bonded to ^{15}N and characterises them by their appropriate ^{15}N chemical shift). These experiments give the NOESY and HOHAHA crosspeaks involving the relevant NH protons edited in slices according to the ^{15}N chemical shifts of each particular amide nitrogen. For example in the NOESY.HMQC spectra, each ^{15}N edited slice contains only NOE connections involving the NH protons from amide groups where the nitrogens have the particular ^{15}N chemical shift for that slice. Each slice represents a small-subset of the total number of NOE connections and therefore the chance of overlap is much reduced. If two NH protons have the same (or nearly the same) chemical shift then the NOESY.HMQC experiment does not allow detection of the NOE connection between such protons. This problem has been resolved by using a HMQC.NOESY.HMQC 3D experiment[60] which causes the NH protons in the NOESY slices to be characterised by their ^{15}N chemical shifts rather than their ^{1}H shifts: since it is unlikely that NH protons with degenerate ^{1}H chemical shifts will be associated with ^{15}N nuclei with degenerate chemical shifts, the ^{1}H shift degeneracy is removed[60]. By applying these various methods to the DHFR.MTX complex we have been able to obtain backbone resonance assignments for 147 of the 162 residues in DHFR and side-chain assignments for more than 80 residues. An analysis of the data (NOE patterns, intrastrand NOEs and non-exchanging NH protons) revealed that the β-sheet and most of the helical secondary structure seen in the crystal is essentially retained intact in solution. The small deviations noted are discussed elsewhere[59]. It is worth noting that assignments made previously using the crystal structure/NOE data correlation method were all in agreement with those obtained using the sequential assignment method. This indicates that the solution and crystal structures are fairly similar.

Now that we have a large body of securely assigned protein ^{1}H resonances these can be used as reporter groups for detecting NOEs with protons on interacting ligands and for monitoring conformational changes within the protein.

NMR CHARACTERISATION OF PROTEIN-LIGAND COMPLEXES

Information relating to ionisation states[11-15], protein-ligand interactions[11,12,46,48,51,54], conformations[15,19,21], multiple conformations[25-33] and dynamic processes[34,35] within complexes has been obtained from NMR studies. A brief survey of the results obtained for the well-studied complex trimethoprim.DHFR will provide a guide to the type of information available. These studies were facilitated by having available a series of isotopically labelled (^{13}C and ^{15}N) trimethoprim analogues as indicated in Figure 1.

For example, the ^{13}C chemical shift of ^{13}C-2-labelled trimethoprim in its complex with DHFR is characteristic of the shift expected if the adjacent N1 is protonated within the complex[11,14]. This protonation was later confirmed by examining DHFR complexes formed with a trimethoprim analogue labelled with ^{15}N at the N-1 position where the observed ^{15}N chemical shift is again characteristic of a protonated nitrogen[12]. In this case the ^{1}H signal from the proton directly attached to ^{15}N could easily be detected in the ^{1}H spectrum of the complex with DHFR since it features a 90 Hz doublet characteristic of scalar coupling between directly bonded ^{1}H and ^{15}N atoms[12]. In earlier studies, Cocco et al.[43] used ^{13}C NMR measurements to demonstrate that the N1 of methotrexate is also protonated in its complex with DHFR. From the crystal structure studies[6] it is known that the N1 position is near to the conserved Asp 26 residue (*L. casei* numbering). Thus the N1H group forms a charge-charge interaction with the negatively charged carboxylate group of Asp 26.

Several dynamic processes have been characterised within the DHFR.trimethoprim complex[35]. From ^{1}H line-width measurements made on the N1H signal of bound trimethoprim over a range of temperatures it was possible to estimate the rate of breaking and reforming of the hydrogen bond between the N1H proton and the Asp 26 carboxylate group (34 s^{-1} at 298 K)[35]. ^{13}C relaxation measurements on the ^{13}C-[7, 4'-OMe]-trimethoprim analysed in its complex with DHFR suggest that the benzyl ring is librating rapidly ($>10^{10}s^{-1}$) over angles of ± 30°[35]. Finally, a ^{13}C line shape analysis of the ^{13}C-[3'-OMe]-trimethoprim DHFR complex as a function of temperature indicates that the benzyl ring is flipping between two equivalent environments (250 s^{-1} at 298 K)[35]. Such ring flipping about the C7-C1' bond would be sterically impossible if the trimethoprim retained its bound conformation as measured previously: ring flipping can only occur if

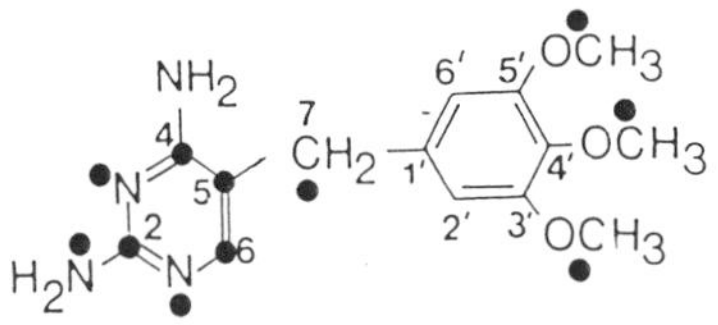

Trimethoprim **2**

Fig. 1. Trimethoprim structure labelled (●) to indicate the sites at which ^{13}C and ^{15}N have been introduced in a series of selectively-labelled compounds.

there is a momentary fluctuation in structure where the torsion angle about C6-C7 bond changes by at least 30° to remove the steric hindrance between the benzyl and pyrimidine ring. This will also be accompanied by a fluctuation in the protein structure. At 298 K the dissociation rate constant for the complex is 2 s^{-1} (measured from transfer of saturation experiments): thus the various dynamic processes mentioned above take place within the lifetime of the complex. Thus the breaking and reforming of the N1H interaction with Asp 26 and the ring flipping with its accompanying fluctuations in conformation within the complex both involve the breaking of interactions between protein and ligand many times within the lifetime of the complex[35].

With regard to conformational studies, while some limited information about the solution conformation of trimethoprim bound to DHFR has already been obtained[15,19,20] (from ring current chemical shift calculations and intramolecular NOE measurements) more detailed conformational studies must await analysis of NOE data between protons on the trimethoprim and recently assigned protons in the protein. NMR has proved to be particularly useful for studying multiple conformational states in protein-ligand complexes in solution. In the ternary complex of DHFR with trimethoprim and $NADP^+$ the presence of two almost equally populated conformations (designated Forms I and II) that give rise to separate NMR spectra have been detected[30-32]. For example, it was observed that all 7 His C2 proton resonances appear as doublets in the spectra of this complex: a line shape analysis on these doublets as a function of temperature allowed the rate of interconversion between Forms I and II to be estimated (e.g. at 304 K the rate is 16 s^{-1}). The two conformations have been characterised by examining isotopically labelled (^{13}C and ^{15}N) ligands in the ternary complex: the labels at each position all give rise to separate signals corresponding to Forms I and II. One can change the populations of the two forms by examining modified analogues of trimethoprim or $NADP^+$. The simplest way of quantitating the two conformational states is to measure the relative intensities of the two sets of ^{31}P signals observed for the pyrophosphate phosphorus nuclei in each form. The main difference between the two forms is in the binding of the nicotinamide ring of $NADP^+$: in Form I this is in a binding pocket within the protein with its glycosidic bond fixed in an anti-conformation while in Form II the conformation of the pyrophosphate moiety is altered such that the nicotinamide ring extends away from the protein in a mixture of syn- and anti-conformations.

Clearly it is important to be aware of any conformational equilibria of this type if one is undertaking structure-activity studies. In such cases, one needs to take into account that two independent structure-activity relationships, one for each bound form, must be considered if a proper understanding of the ligand binding is to be achieved.

CONFORMATIONAL SELECTION OF ROTATIONAL ISOMERS OF PYRIMETHAMINE ANALOGUES BOUND TO DHFR

If a flexible ligand can exist in more than one rotameric state this also provides the possibility of multiple conformations when bound to the protein. For example, pyrimethamine analogues which have a biphenyl like structure can give rise to rotational isomers resulting from hindered rotation about the pyrimidine-phenyl bond[29,54]. Such restricted rotation has been well-characterised in ortho-substituted biphenyls and one would expect ortho-substituted phenyl pyrimidines to behave in a similar manner. We have used NMR spectroscopy to investigate complexes of DHFR with pyrimethamine and several of its analogues[29,54].

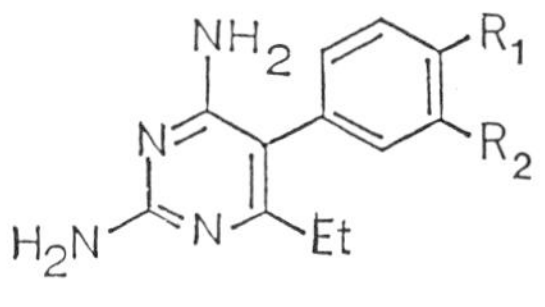

		R_1	R_2
3	Pyrimethamine	Cl	H
4	Fluoropyrimethamine	F	H
5	Fluoronitropyrimethamine	F	NO_2

An analogue such as fluoronitropyrimethamine **5** containing an asymmetrically substituted aromatic ring can exist as mixtures of two rotational isomers (an enantiomeric pair, see Fig 2). One might expect some conformational selection of these rotamers on binding to DHFR. The ^{19}F NMR spectrum of the fluoronitropyrimethamine.DHFR complex reported by Tendler and coworkers[19] illustrates clearly that two different conformational states exist since two separate ^{19}F signals are detected for the bound ligand. Tendler et al.[29] reported that the populations of the two forms (designated A and B) are different, the A/B intensity ratio being 0.6/0.4 in the binary complex. Addition of $NADP^+$ to this complex caused the preference for binding to be reversed with the A/B ratio in the ternary complex being 0.3/0.7. These results in themselves do not provide direct evidence that the two observed conformations are related to different rotameric states involving the pyrimidine-phenyl bond although they are strongly suggestive of this. In order to characterise fully these binary and ternary complexes it was necessary to examine the different bound forms using ^1H NMR studies of the complexes. All three pyrimethamine analogues (**3**-**5**) bind tightly to *L. casei* DHFR ($Ka > 5 \times 10^5\ M^{-1}$) and the exchange between the bound and free species is slow on the NMR chemical shift time scale resulting in separate NMR spectra for the bound and free species. Birdsall et al.[54] have used 2D exchange ^1H NMR spectroscopy to connect the signals from bound species with their corresponding signals in the free ligand for complexes involving all the ligands (**3**-**5**). In the complexes where there is a symmetrically substituted phenyl ring, such as pyrimethamine **3** and fluoropyrimethamine **4** four signals are observed for the four non-equivalent aromatic protons. This clearly indicates the presence of hindered rotation about the phenyl-pyrimidine bond with the phenyl ring taking up a fixed position within its binding site such that the four phenyl protons are now in separate shielding environments on the protein. Had there been rapid interconversion between the different rotamers then only a single averaged signal would have been detected for the pairs of protons on opposite sides of the ring (H2′, H6′ and H3′, H5′) as a result of the ring flipping. In the case of the other symmetrically substituted pyrimethamine analogue, fluoropyrimethamine, again 4 non-equivalent aromatic proton signals were detected and these have very similar shielding contributions resulting from the binding. This indicates that the phenyl ring is binding in a similar environment in the two complexes. DHFR complexes containing pyrimethamine analogues with asymmetrically substituted phenyl rings such as fluoronitropyrimethamine **5** showed two complete sets of signals (Form A and Form B) for the phenyl protons in the bound ligands. By comparing the ^1H chemical shifts of these bound protons with those in the pyrimethamine.DHFR complex it was possible to show that the phenyl ring protons in the two Forms A and B are experiencing the same protein environment in each of the two forms as that experienced for the corresponding protons in the pyrimethamine.DHFR complex. This can only be true if Forms A and B correspond to two rotational isomers which result from a ~180° rotation about the pyrimidine-phenyl bond with the 2,4-diamino pyrimidine ring being bound similarly in the two forms.

Baker et al.[9] have examined the crystal structure of a complex of pyrimethamine with *E. coli* DHFR and found that the 2,4-diaminopyrimidine ring binds in essentially the same binding site as that occupied by the corresponding part of the methotrexate ring it its complex with *E. coli* DHFR (Bolin et al.[6]) and has its phenyl ring orientated at approximately 90° to the plane of the pyrimidine ring. Based on this information, Birdsall et al.,[54] built a model of *L. casei* DHFR.pyrimethamine complex and noted that a pair of aromatic ring protons on one side of the phenyl ring of bound pyrimethamine are oriented towards Phe 30. Two of the phenyl protons of bound pyrimethamine have substantial upfield shielding contributions which would be consistent with these being on the side of the ring oriented towards Phe 30 (assuming that Phe 30 has a similar conformation in the DHFR complex formed with pyrimethamine and methotrexate). Subsequent NOE measurements[54] confirmed these assignments by showing connections between an ortho proton on the phenyl ring of bound pyrimethamine and the H4 and H3,5 aromatic protons of Phe 30.

MULTIPLE CONFORMATIONS AND ROTATIONAL ISOMERISM IN TERNARY COMPLEXES OF DHFR WITH PYRIMETHAMINE AND NADP+

It was mentioned earlier that in the complex DHFR.trimethoprim NADP+ two different conformations (Forms I and II) have been detected[30,31]. The different conformations have the nicotinamide ring of NADP+ bound differently in the two forms. Recently Birdsall et al.[54] have found that similar conformations are present in ternary complexes of DHFR with pyrimethamine analogues and NADP+. The two forms can be characterised quantitatively by measuring the relative intensities of the two different sets of pyrophosphate ^{31}P signals corresponding to Forms I and II. For the complex of fluoronitropyrimethamine.NADP+.DHFR such measurements indicate 65% Form I and 35% ± 10% Form II. For this complex it was also possible to measure the populations of the rotational isomeric bound Forms A and B (30% and 70% ± 10% respectively). These results raise the intriguing possibility that the two types of conformational states could be strongly correlated with only forms IB and IIA being populated. This would be

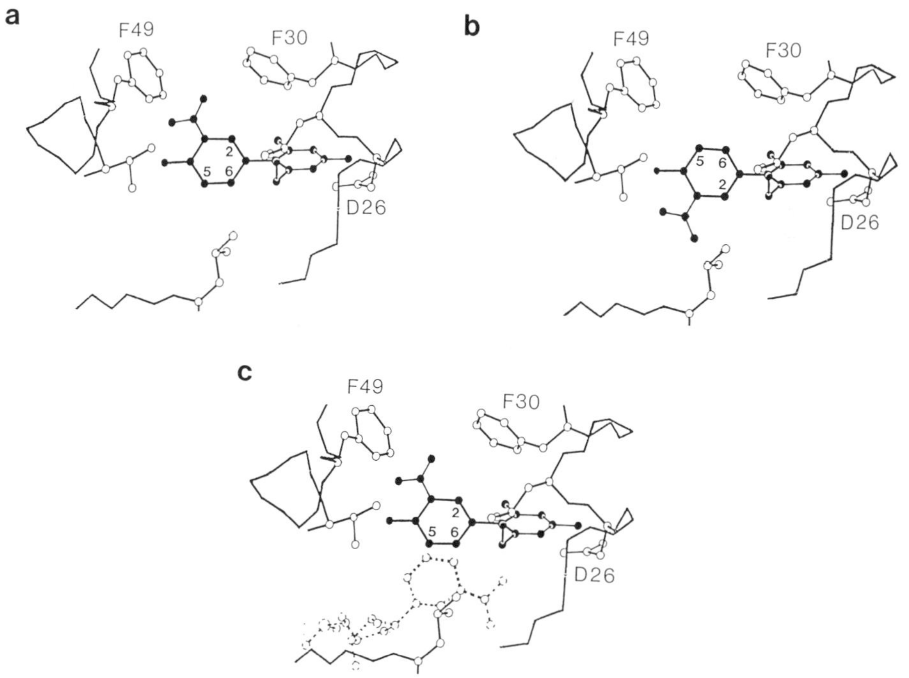

Fig. 2. Model of the binding site in the fluoronitropyrimethamine.DHFR complex for (a) Form A, (b) Form B. (c) Model of the complex indicating the position of the NADP+ nicotinamide ring binding site as a broken-line structure. (Modified from Birdsall et al., 1990 Biochemistry, in press).

consistent with a model with the NO_2 substituent oriented towards the vacant site for the nicotinamide ring binding in Form IIA and oriented away from the nicotinamide ring binding site in Form IB (Fig. 2). However one cannot exclude the alternative explanation that all four forms exist but that none of the detected NMR signals is affected by more than one of the different conformational states.

The conformational preference for binding Form A in the binary complex is fluoronitropyrimethamine with DHFR can also be considered in terms of the model of the complex shown in Fig. 2. In Form A the NO_2 substituent would be directed towards the vacant nicotinamide ring binding site and this could assist in favourable binding. Addition of $NADP^+$ to form the ternary complex which reverses the preference for Form A would be consistent with the unfavourable steric interaction between the bulky NO_2 group and the nicotinamide ring of bound $NADP^+$ in Form I of the complex.

CONCLUSION

The availability of numerous assigned ligand and protein resonances in the NMR spectra of complexes of DHFR with antifolate drugs allows detailed studies of interactions and conformational states within the complexes. It is clear that the protein-ligand complexes in solution can exist as mixtures of conformational states. NMR provides a unique method for studying multiconformational equilibria in a quantitative manner. Clearly, structure-activity studies must take into account such multiple conformations if a proper understanding of the binding is to be achieved.

ACKNOWLEDGEMENTS

The NMR measurements were made at the MRC Biomedical NMR Centre at Mill Hill. The work described resulted from collaborations with several coworkers including Berry Birdsall, Gordon Roberts, Mark Carr, Saul Tendler, Roger Griffin, Tom Frenkiel, Chris Bauer, Gill Ostler and John McCormick.

REFERENCES

1. R. L. Blakley, The Biochemistry of Folic Acid and Related Pteridines, Elsevier/North Holland, Amsterdam (1979).
2. D. R. Seeger, D. B. Cosulich, J. M. Smith, and M. E. Hultquist, Analogues of pteroylglutamic acid III. 4-Amino derivatives, J. Am. Chem. Soc. 71:1753 (1949).
3. B. Roth and C. C. Cheng, Selective inhibitors of bacterial dihydrofolate reductase: structure activity relationships, Prog. Med. Chem. 19:1 (1982).
4. D. A. Matthews, R. A. Alden, J. T. Bolin, D. J. Filman, S. T. Freer, R. Hamlin, W. G. J. Hol, R. L. Kisliuk, E. J. Pastore, L. T. Plante, N. Xuong, and J. Kraut, Dihydrofolate reductase from Lactobacillus casei X-ray structure of the enzyme-methotrexate-NADPH complex, J. Biol. Chem. 253:6946 (1978).
5. D. A. Matthews, R. A. Alden, S. T. Freer, N. Xuong, and J. Kraut, Dihydrofolate reductase from L. casei: Stereochemistry of NADPH binding, J. Biol. Chem. 254:4144 (1979).
6. J. T. Bolin, D. J. Filman, D. A. Matthews, R. C. Hamlin, and J. Kraut, Crystal structures of E. coli and L. casei dihydrofolate reductase refined to 1.7 A resolution. I. General features and binding of methotrexate, J. Biol. Chem. 257:13650 (1982).

7. D. A. Matthews, J. T. Bolin, J. M. Burridge, D. J. Filman, K. W. Volz, B. T. Kaufman, C. R. Beddell, J. N. Champness, D. K. Stammers, and J. Kraut, Refined crystal structures of Escherichia coli and chicken liver dihydrofolate reductase containing bound trimethoprim, J. Biol. Chem. 260:381 (1985).
8. L. F. Kuyper, B. Roth, D. P. Baccanari, R. Ferone, C. R. Beddell, J. N. Champness, D. K. Stammers, J. G. Dann, F. E. A. Norrington, D. J. Baker, and P. J. Goodford, Receptor based design of DHFR inhibitors: comparison of crystallographically determined enzyme binding with enzyme affinity in a series of carboxy-substituted trimethoprim analogues, J. Med. Chem. 25:1122 (1982).
9. D. J. Baker, C. R. Beddell, J. N. Champness, P. J. Goodford, F. E. A. Norrington, B. Roth, and D. K. Stammers, in: Chemistry and Biology of Pteridines, J. D. Blair, ed., W. de Gruyter, Berlin (1982).
10. J. Feeney, NMR studies of interactions with ligand with dihydrofolate reductase, Biochem. Pharmacol. 40:141 (1990).
11. H. T. A. Cheung, M. S. Searle, J. Feeney, B. Birdsall, G. C. K. Roberts, I. Kompis, and S. J. Hammond, Trimethoprim binding to L. casei dihydrofolate reductase: A ^{13}C NMR study using selectively ^{13}C-enriched trimethoprim, Biochemistry. 25:1925 (1986).
12. A. W. Bevan, G. C. K. Roberts, J. Feeney, and I. Kuyper, ^{1}H and ^{15}N NMR studies of protonation and hydrogen-bonding in the binding of trimethoprim to dihydrofolate reductase, Eur. Biophys. J. 11:211 (1985).
13. R. E. London, E. E. Howell, M. S. Warren, J. Kraut, and R. L. Blakely, NMR study of the state of protonation of inhibitors bound to mutant DHFR lacking the active site carboxyl, Biochemistry. 25:7229 (1986).
14. G. C. K. Roberts, J. Feeney, A. S. V. Burgen, and S. Daluge, The charge state of trimethoprim bound to L. casei dihydrofolate reductase, FEBS Letters 131:85 (1981).
15. J. Cayley, J. P. Albrand, J. Feeney, G. C. K. Roberts, E. A. Piper, and A. S. V. Burgen, NMR studies of the binding of trimethoprim to dihydrofolate reductase, Biochemistry. 18:3886 (1979).
16. E. I. Hyde, B. Birdsall, G. C. K. Roberts, J. Feeney, and A. S. V. Burgen, ^{1}H NMR saturation transfer studies of coenzyme binding to Lactobacillus casei dihydrofolate reductase, Biochemistry. 19:3738 (1980).
17. J. L. Way, B. Birdsall, J. Feeney, G. C. K. Roberts, and A. S. V. Burgen, An NMR study of NADP binding to L. casei dihydrofolate reductase, Biochemistry. 14:3470 (1975).
18. E. I. Hyde, B. Birdsall, G. C. K. Roberts, J. Feeney, and A. S. V. Burgen, ^{31}P NMR studies of the binding of oxidised coenzymes to Lactobacillus casei dihydrofolate reductase, Biochemistry. 19:3746 (1980).
19. B. Birdsall, G. C. K. Roberts, J. Feeney, J. G. Dann, and A. S. V. Burgen, Trimethoprim binding to bacterial and mammalian dihydrofolate reductase: a comparison by proton and carbon-13 NMR, Biochemistry. 22:5597 (1983).
20. J. P. Albrand, B. Birdsall, J. Feeney, G. C. K. Roberts, and A. S. V. Burgen, The use of transferred nuclear Overhauser effects in the study of the conformations of small molecules bound to proteins, Internat. J. Biol. Macromolecules 1:37 (1979).
21. J. Feeney, B. Birdsall, G. C. K. Roberts, and A. S. V. Burgen, The use of transferred nuclear Overhauser effect measurements to compare the binding of coenzyme analogues to dihydrofolate reductase, Biochemistry. 22:628 (1983).
22. J. Feeney, B. Birdsall, G. C. K. Roberts, and A. S. V. Burgen, ^{31}P NMR studies of NADPH and NADP+ binding to L. casei dihydrofolate reductase, Nature. 257:564 (1975).

23. P. J. Cayley, J. Feeney, and B. J. Kimber, ^{31}P NMR studies of complexes of NADPH and NADP+ with E. coli dihydrofolate reductase, Internat. J. Biol. Macromolecules 2:251 (1980).
24. B. Birdsall, G. C. K. Roberts, J. Feeney, and A. S. V. Burgen, ^{31}P NMR studies of the binding of adenosine-2'-phosphate to L. casei dihydrofolate reductase, FEBS Letters 80:313 (1977).
25. B. Birdsall, A. Gronenborn, E. I. Hyde, G. M. Clore, G. C. K. Roberts, J. Feeney, and A. S. V. Burgen, ^{1}H, ^{13}C and ^{31}P NMR studies of dihydrofolate reductase-NADP+-folate complex: characterisation of three co-existing conformational states, Biochemistry. 21:5831 (1982).
26. B. Birdsall, A. Gronenborn, G. M. Clore, G. C. K. Roberts, J. Feeney, and A. S. V. Burgen, 1C NMR evidence for three slowly interconverting conformations of the dihydrofolate reductase-NADP+-folate complex, Biochem. Biophys. Res. Commun. 101:1139 (1981).
27. B. Birdsall, J. De Graw, J. Feeney, S. J. Hammond, M. S. Searle, G. C. K. Roberts, W. T. Colwell, and J. Crase, ^{15}N and ^{1}H NMR evidence for multiple conformations of the complex of dihydrofolate reductase with its substrate, folate, FEBS Letters 217:106 (1987).
28. B. Birdsall, J. Feeney, S. J. B. Tendler, S. J. Hammond, and G. C. K. Roberts, Dihydrofolate reductase: Multiple conformations and alternative modes of substrate binding, Biochemistry. 28:2297 (1989).
29. S. J. B. Tendler, R. J. Griffin, B. Birdsall, M. F. G. Stevens, G. C. K. Roberts, and J. Feeney, Direct F-19 NMR observation of the conformational selection of optically active rotamers of the antifolate compound fluoronitroropyrimethamine bound to the enzyme dihydrofolate reductase, FEBS Letters 240:201 (1988).
30. A. Gronenborn, B. Birdsall, E. I. Hyde, G. C. K. Roberts, J. Feeney, and A. S. V. Burgen, Direct observation by NMR of two co-existing conformations of an enzyme-ligand complex in solution, Nature. 290:273 (1981).
31. A. Gronenborn, B. Birdsall, E. I. Hyde, G. C. K. Roberts, J. Feeney, and A. S. V. Burgen, ^{1}H and ^{31}P NMR characterisation of two conformations of the trimethoprim-NADP+-dihydrofolate reductase complex, Molecular Pharm. 20:145 (1981).
32. B. Birdsall, A. W. Bevan, C. Pascual, G. C. K. Roberts, J. Feeney, A. Gronenborn, and G. M. Clore, Multinuclear NMR characterisation of two coexisting conformational states of Lactobacillus casei dihydrofolate reductase trimethoprim-NADP+ complex, Biochemistry. 23:4733 (1984).
33. R. E. London, G. P. Groff, and R. L. Blakley, ^{13}C NMR evidence for the slow exchange of tryptophans in dihydrofolate reductase between stable conformations, Biochem. Biophys. Res. Commun. 86:779 (1979).
34. G. M. Clore, A. M. Gronenborn, B. Birdsall, J. Feeney, and G. C. K. Roberts, NMR studies of 3',5'-difluoromethotrexate binding to Lactobacillus casei dihydrofolate reductase. Molecular motion and coenzyme induced conformational change, Biochem. J. 217:659 (1984).
35. M. S. Searle, M. J. Forster, B. Birdsall, G. C. K. Roberts, J. Feeney, H. T. A. Cheung, I. Kompis, and A. J. Geddes, The dynamics of trimethoprim bound to dihydrofolate reductase, Proc. Natl. Acad. Sci. (USA) 85:3787 (1988).
36. J. Feeney, G. C. K. Roberts, B. Birdsall, D. V. Griffiths, R. W. King, P. Scudder, and A. S. V. Burgen, ^{1}H Nuclear magnetic resonance studies of the tyrosine residues of selectively deuterated Lactobacillus casei dihydrofolate reductase, Proc. Roy. Soc. Lond.B. 196:267 (1977).
37. B. Birdsall, J. Feeney, D. V. Griffiths, S. Hammond, B. J. Kimber, R. W. King, G. C. K. Roberts, and M. S. Searle, The combined use of selective deuteration and doube resonance experiments in assigning the ^{1}H resonances of valine and tyrosine residues of dihydrofolate reductase, FEBS Letters 175:364 (1984).

38. M. S. Searle, S. J. Hammond, B. Birdsall, G. C. K. Roberts, J. Feeney, R. W. King, and D. V. Griffiths, Identification of the ^{1}H resonances of valine and leucine residues in dihydrofolate reductase by using a combination of selective deuteration and two-dimensional correlation spectroscopy, FEBS Letters 194:165 (1986).
39. J. Feeney, B. Birdsall, J. Akiboye, S. J. B. Tendler, J. Jimenez-Barbero, G. Ostler, J. R. P. Arnold, G. C. K. Roberts, A. Kuhn, and K. Roth, Optimising selective deuteration of proteins for 2D ^{1}H NMR detection and assignment studies: application to the residues of L. casei dihydrofolate reductase, FEBS Letters 248:57 (1989).
40. B. J. Kimber, D. V. Griffiths, B. Birdsall, R. W. King, P. Scudder, J. Feeney, G. C. K. Roberts, and A. S. V. Burgen, 19F Nuclear magnetic resonance studies of ligand binding to 3-fluorotyrosine- and 6-fluorotryptophan-containing dihydrofolate reductase from Lactobacillus casei, Biochemistry. 16:3492 (1977).
41. B. J. Kimber, J. Feeney, G. C. K. Roberts, B. Birdsall, A. S. V. Burgen, and B. D. Sykes, Proximity of two tryptophan residues in dihydrofolate reductase determined by 19F NMR, Nature (Lond.) 271:184 (1978).
42. S. J. Hammond, B. Birdsall, M. S. Searle, G. C. K. Roberts, and J. Feeney, Dihydrofolate reductase: ^{1}H resonance assignments and coenzyme-induced conformational changes, J. Mol. Biol. 188:81 (1986).
43. L. Cocco, J. P. Groff, J. Temple C, J. A. Montgomery, R. E. London, N. S. Matwiyoff, and R. L. Blakley, ^{13}C Nuclear magnetic resonance study of protonation of methotrexate and aminopterin bound to dihydrofolate reductase, Biochemistry. 20:3926 (1981).
44. P. Wyeth, A. Gronenborn, B. Birdsall, G. C. K. Roberts, J. Feeney, and A. S. V. Burgen, The histidine residues of L. casei dihydrofolate reductase: paramagnetic relaxation and deuterium exchange studies and partial assignments, Biochemistry. 19:2608 (1980).
45. A. Gronenborn, B. Birdsall, E. I. Hyde, G. C. K. Roberts, J. Feeney, and A. S. V. Burgen, The effects of coenzyme binding on the histidine residues of L. casei dihydrofolate reductase, Biochemistry. 20:1717 (1981).
46. D. J. Antonjuk, B. Birdsall, A. S. V. Burgen, H. T. A. Cheung, G. M. Clore, J. Feeney, A. Gronenborn, G. C. K. Roberts, and W. Tran, A ^{1}H NMR study of the role of the glutamate moiety in the binding of methotrexate to dihydrofolate reductase, Brit. J. Pharmacol. 81:309 (1984).
47. S. J. Hammond, B. Birdsall, J. Feeney, M. S. Searle, G. C. K. Roberts, and H. T. A. Cheung, Structural comparisons of complexes of methotrexate analogues with L. casei dihydrofolate reductase by 2D ^{1}H NMR at 500 MHz, Biochemistry. 26:8585 (1987).
48. B. Birdsall, J. Feeney, C. Pascual, G. C. K. Roberts, I. Kompis, R. L. Then, K. Muller, and A. Kroehn, A ^{1}H study of the interactions and conformations of rationally designed brodimoprim analogues in complexes with Lactobacillus casei dihydrofolate reductase, J. Med. Chem. 23:1672 (1984).
49. B. Birdsall, E. I. Hyde, A. S. V. Burgen, G. C. K. Roberts, and J. Feeney, Negative cooperativity between folinic acid and coenzyme in their binding to L. casei dihydrofolate reductase, Biochemistry. 20:7186 (1981).
50. J. Feeney, B. Birdsall, J. P. Albrand, G. C. K. Roberts, A. S. V. Burgen, P. A. Charlton, and D. W. Young, A ^{1}H NMR study of the complexes of the diastereoisomers of folinic acid with dihydrofolate reductase, Biochemistry. 20:1837 (1981).
51. M. A. Jimenez, J. R. P. Arnold, J. Andrews, J. A. Thomas, G. C. K. Roberts, B. Birdsall, and J. Feeney, Dihydrofolate reductase: control of the mode of substrate binding by aspartate 26, Protein Engineering 2:627 (1989).

52. B. Birdsall, J. Andrews, G. Ostler, S. J. B. Tendler, J. Feeney, M. S. Searle, G. C. K. Roberts, R. W. Davies, and H. T. A. Cheung, NMR studies of differences in the conformations and dynamics of ligand complexes formed with mutant dihydrofolate reductases, Biochemistry. 28:1353 (1989).
53. B. Birdsall, J. R. P. Arnold, J. Jimenez Barbero, T. A. Frenkiel, C. J. Bauer, S. J. B. Tendler, M. D. Carr, J. A. Thomas, G. C. K. Roberts, and J. Feeney, The ^{1}H NMR assignments of the aromatic resonances in complexes of Lactobacillus casei dihydrofolate reductase and the origins of their chemical shifts, Eur. J. Biochem. : (in press).
54. B. Birdsall, S. J. B. Tendler, J. R. P. Arnold, J. Feeney, R. J. Griffin, M. D. Carr, J. A. Thomas, G. C. K. Roberts, and M. F. G. Stevens, NMR studies of multiple conformations in complexes of L. casei dihydrofolate reductase with analogues of pyrimethamine, Biochemistry. : (in press).
55. D. Marion, P. C. Driscoll, L. E. Kay, P. T. Wingfield, A. Bax, A. M. Gronenborn, and G. M. Clore, Overcoming the overlap problem in the assignment of ^{1}H NMR spectra of larger proteins by use of three-dimensional heteronuclear ^{1}H-^{15}N Hartmann-Hahn-multiple quantum coherence and nuclear Overhauser-multiple quantum coherence spectrosopy: Application to Interleukin 1β, Biochemistry. 28:6150 (1989).
56. E. R. P. Zuiderweg and S. W. Fesik, Heteronuclear three-dimensional NMR spectroscopy of the inflammatory protein C5a, Biochemistry. 28:2387 (1989).
57. L. E. Kay, G. M. Clore, A. Bax, and A. M. Gronenborn, Four-dimensional heteronuclear triple-resonance NMR spectroscopy of Interleukin-1β in solution, Science 249:411 (1990).
58. K. Wuthrich, NMR of Proteins and Nucleic Acids, John Wiley and Sons Inc., New York (1986).
59. M. D. Carr, B. Birdsall, J. Jimenez-Barbero, V. I. Polshakov, C. J. Bauer, T. A. Frenkiel, G. C. K. Roberts, and J. Feeney, Dihydrofolate reductase: Sequential assignments and secondary structure in solution, (submitted for publication).
60. T. Frenkiel, C. Bauer, M. D. Carr, B. Birdsall, and J. Feeney, HMQC-NOESY-HMQC: A three-dimensional NMR experiment which allows detection of nuclear Overhauser effects between protons with overlapping signals, J. Magn. Reson. (in press).

STRUCTURE-FUNCTION RELATIONSHIP OF GASTRIN HORMONES: A RATIONAL APPROACH TO DRUG DESIGN

Evaristo Peggion, Maria Teresa Foffani, and Stefano Mammi

Biopolymer Research Center, Department of Organic Chemistry
University of Padova, Via Italy

INTRODUCTION

The gastrins are a family of gastro-intestinal peptide hormones performing a wide range of biological functions in the digestive process. The sequences of the most common forms of gastrins are the following:

*p*Glu-Leu-Gly-Pro-Gln-Gly-His-Pro-Ser-Leu-Val-Ala-Asp-Pro-Ser-Lys-Lys-Gln-Gly-Pro-Trp-Leu-(Glu)$_5$-Ala-Tyr-Gly-Trp-Met-Asp-Phe-NH$_2$ (big gastrin);

*p*Glu-Gly-Pro-Trp-Leu-(Glu)$_5$-Ala-Tyr-Gly-Trp-Met-Asp-Phe-NH$_2$ (little gastrin);

H-Trp-Leu-(Glu)$_5$-Ala-Tyr-Gly-Trp-Met-Asp-Phe-NH$_2$ (minigastrin).

They exist with a free tyrosine side chain (form I), or with a O-sulfate tyrosyl side-chain (form II). The most abundant form of the hormone is little gastrin.

From extensive structure-function relationship studies performed in several laboratories it is presently well established that the *message* sequence responsible for the biological activity of gastrins rests on the C-terminal tetrapeptide amide -Trp-Met-Asp-Phe-NH$_2$. More than 600 analogs and derivatives of this sequence have been synthesized and their biological activity investigated in detail.[1] From these studies the requirements in terms of chemical structure essential for hormonal activity are known in detail. However, the C-terminal sequence alone exhibits only 10 % of the biological potency of little gastrin. Thus, the N-terminal portion of the molecule plays a role in optimizing the hormonal message.

In general, the determination of the conformational features required to perform a biological function of small, linear peptides is a very difficult task. In fact, linear peptides do not show a strong conformational preference (like enzymes and proteins), they do not crystallize and, in solution, they exist as a population of rapidly interconverting conformers. All these features make the determination of the conformational aspects responsible for biological activity more difficult.

In aqueous solution all natural gastrin peptides and their synthetic analogs, either active or inactive, exhibit CD patterns in the far UV typical of a random conformation.[2] In the absorption region of the aromatic chromophores there are very

Bioorganic Chemistry in Healthcare and Technology, Edited by U.K. Pandit and F.C. Alderweireldt, Plenum Press, New York, 1991

weak signals indicating the absence of any significant constrain of the aromatic side chains. These results led to the hypothesis that, if a "bioactive structure" does exist, it must be formed in the lipophilic environment at the receptor binding site. In the attempt to contribute to a better understanding of structure-function relationship gastrin hormones , and ultimately to determine the bioactive structure of gastrins, a systematic investigation was undertaken by our group on a series of gastrin fragments of increasing chain length. In this paper the most recent results will be briefly reviewed.

RESULTS AND DISCUSSION

The following peptides were synthesized and studied:

*p*Glu-Gly-Pro-Trp-Leu-$(Glu)_5$-Ala-Tyr-Gly-Trp-Nle-Asp-Phe-NH_2 (Nle[15]-little gastrin);
H-Leu-$(Glu)_5$-Ala-Tyr-Gly-Trp-Nle-Asp-Phe-NH_2 (des-Trp[1],Nle[12]-minigastrin);
*p*Glu-$(Glu)_4$-Ala-Tyr-Gly-Trp-Nle-Asp-Phe-NH_2 (gastrin dodecapeptide);
*p*Glu-$(Glu)_3$-Ala-Tyr-Gly-Trp-Nle-Asp-Phe-NH_2 (gastrin undecapeptide);
*p*Glu-$(Glu)_2$-Ala-Tyr-Gly-Trp-Nle-Asp-Phe-NH_2 (gastrin decapeptide);
*p*Glu-Glu-Ala-Tyr-Gly-Trp-Nle-Asp-Phe-NH_2 (gastrin nonapeptide);
*p*Glu-Ala-Tyr-Gly-Trp-Nle-Asp-Phe-NH_2 (gastrin octapeptide).

In all these products methionine was replaced by norleucine, which offers the advantage of preserving full biological activity, while preventing inactivation via oxidation of the thioether group.[1] The peptides were first characterized by CD spectroscopy in trifluoroethanol (TFE).[3] From CD results in the near and far UV (figure 1) we note first that the structure of the shortest fragment gastrin octapeptide is not random. There is in fact a double minimum at 215 nm and 200 nm and a positive band at 190 nm. This pattern is completely different from the one observed in water, where the structure is completely random. The extent of structural order increases upon chain elongation. This is shown by the strong increase of the intensity of the two negative CD bands at 216 and 207 nm and also by the enhancement of the positive band at 192 nm. The increase of the amount of ordered conformation is not linearly proportional to the length of the peptide chain (Figure 2). Apparently, the larger extent of conformational change takes place upon elongation of the molecule from the undecapeptide to the dodecapetide. Parallel biological tests on all fragments revealed that the potency of their biological action increases from 34 to 87% of that of little gastrin upon elongation of the chain from the undecapeptide to the dodecapeptide. Most interestingly, the change in conformation and the increase of biological potency match the same profile (figure 2). This observation led to the hypothesis that the conformation assumed by gastrin hormones in TFE is of biological relevance. This hypothesis was further supported by the observation that both little gastrin and minigastrin in a membrane-like environment (such as aqueous solutions containing containing detergent micelles) adopt a conformation very similar to that observed in TFE.[4,5]

On the basis of the CD properties of minigastrin, little gastrin and the shorter fragments, a working hypothesis on the peptide conformation in TFE was made. The chiroptical properties of little gastrin and minigastrin led Fromageot and coworkers to suggest the presence of an α-helical segment at the N-terminus, comprising the sequence Leu-$(Glu)_5$Ala.[6] In a subsequent work we suggested that the helical segment should not involve more than 4-5 residues and the possibility of a β-bend was not ruled out.[3] The chiroptical properties of the fragments with 3 and 4 Glu residues in the sequence proved to be extremely useful to clarify the conformational features of mini and little gastrin. Both these fragments exhibit the characteristic double minimum CD pattern with the same shape as that of the longer hormones (Figure 1). Quite surprisingly, the CD spectra of these fragments are insensitive to the extent of ionization of the glutamic acid side chains. It follows that these residues cannot be comprised into an α-helical segment, since addition of KOH should cause the collapse of this structure. Since the rest of the sequence is hardly compatible with an α-helix we concluded that the observed CD spectrum is not that of an α-helix. Actually, CD spectra very similar to those of the undeca and dodecapeptide have been reported for well documented examples of type I and type III β-bends. We therefore proposed that in the structure of these two fragments there is a bend. According to the analysis of Chou-Fasman,[7] the chain reversal should be located in the central sequence Ala-Tyr-Gly-Trp.

The increase of ordered structure when the chain is elongated from the dodecapeptide to des-Trp[1],Nle[12]-minigastrin suggests the onset of an helical segment at the N-terminus. In agreement with this interpretation, ionization of the Glu side-chains causes a decrease of the intensity of the CD bands and a spectrum is obtained which is very close to that of the dodecapeptide. The sequence H-Leu-$(Glu)_5$-Ala-Tyr-OH alone is not able to fold into the α-helical structure in TFE. Therefore an interaction of the N-terminus, possibly with the C-terminus should provide the stabilization energy allowing the proper folding of the N-terminus.

Finally,, when the peptide chain is elongated to little gastrin, there is an additional increment of ordered structure. We interpreted these data by assuming that the length of the N-terminal helical segment is further increased, possibly starting from Trp[1].

With these results we started the characterization of minigastrin by proton NMR. Our goal was to verify the proposed U-shaped conformational model of minigatrin and little gastrin with an helical segment at the N-terminus.

The achievement of reliable NMR data in TFE is complicated by the strong tendency to aggregation of gastrin peptides at concentration levels of the order of 10^{-5}molar or higher. Aggregation is easily revealed by typical concentration-dependent and temperature-dependent CD patterns. For instance, in the near UV there is an extraordinarily intense CD signal probably due to stacking interactions among indole chromophores. These features disappear in the presence of 10% water. In such a system solubility increases substantially and it was possible to reach mM concentrations without intermolecular association. The assignment of all proton resonances in TFE containing 10% water proved to be extremely difficult because of the large OH peak and the methylene quartet of the solvent. Thus we chose to assign the spectrum in water and follow the shift of resonance peaks by solvent titration from 100% water to 90% TFE.[8] Homonuclear 2D NMR experiments were performed in the usual way using standard procedures. From the 2D Hartmann-Hahn spectrum (figure 3) the proton resonaces in water were nearly fully assigned. The following ambiguities remained: 1) only 4 of the 5 Glu-NH peaks were visible; 2) the 4 amide protons of the Glu residues could not be distinguished from one another. The amide proton of Glu[3] is expected to have a very broad resonance peak, as frequently observed in a peptide when the adjacent N-teminal group is a free amine. Thus, the very broad NH peak at 9.2 ppm in 90% TFE (see below) was assigned to this residue. The other ambiguities were clarified by synthesizing the following minigastrin analogs containing α-deuterated Glu residues at positions 6, 7, and 5 and 7, respectively:

H-Leu-Glu-Glu-Glu-*Glu-Glu-Ala-Tyr-Gly-Trp-Nle-Asp-Phe-NH_2
H-Leu-Glu-Glu-Glu-Glu-*Glu-Ala-Tyr-Gly-Trp-Nle-Asp-Phe-NH_2
H-Leu-Glu-Glu-*Glu-Glu-*Glu-Ala-Tyr-Gly-Trp-Nle-Asp-Phe-NH_2
(* denotes α-deuteration sites)

From the collapse of amide doublets into singlets, the assignment of Glu[5]-NH, Glu[6]-NH, and Glu[7]-NH was unambiguously achieved.[9] During the water-TFE titration experiments we were able to follow all amide resonances except though the intersection of Tyr, Asp, and Phe at approximately 20% water (figure 4). These ambiguities were clarified by a Hartmann-Hahn experiment in 94% TFE with a multiple peak saturation of the methylene quartet. Numerous connectivities were found which confirmed our previous assignments (figure 5). Furthermore, we were able to distinguish the amide peaks of Asp, Tyr, and Phe.[8] The assigned one-dimensional spectrum of the minigastrin analog in 90% TFE is shown in figure 6. Because of the particular nature of the system, we have been unable to obtain reliable results from NOESY experiments in 90% TFE, which would have allowed a more complete description of the peptide conformation. However, some conclusions can be drawn from the temperature coefficients of amide protons (Table 1). There are two sets of low temperature coefficients at the two ends of the molecule. The presence of an helical segment at the N-terminus implies that the amide protons of Glu residues in position 6 and 7 should be hydrogen bonded to the carbonyl groups of Glu residues in position 3 and 4, respectively. The results of Table 1 indicate that Glu^6 NH only is clearly solvent shielded, and therefore the helical segment should end with this residue. Surprisingly, Glu-NH in position 5 also exhibits a very low and positive temperature coefficient. This point remains an open question and will be discussed below. From table 1 it appears that the amide protons of the C-terminal portion of the molecule are also solvent shielded. According to our conformational hypothesis the Trp-NH should be hydrogen bonded to Ala-CO to form a β-bend. Again the results of table 1 are consistent with this hypothesis. The new features emerging from the NMR data are the low temperature coefficients of the amide protons at the C-terminus. A conformation compatible with the H-bonds suggested by NMR data and also consistent with the conformational preference of the sequence starting from Ala^8 and with the CD results comprises a series of concatenated type I or III β-bends.[8] This corresponds to a short segment of 3_{10}-helix at the C-terminus starting from Ala^8.

Taken together, both CD and NMR results are consistent with the hypothesis of a U-shaped conformation comprising helical segments at both ends, separated by a flexible region in the central sequence. In this structure the Gly residue in position 10 appears to play a key role in determining the conformational preference of the entire molecule. A simple sequence analysis according to the Chou-Fasman method[7] shows that replacement of Gly with an helix forming residue changes completely the conformational preference of the molecule, extending the α-helix thoughout the entire molecule. If the U-shaped structure represents the bioactive conformation of the gastrins, this sequence modification should affect substantially the biological potency of the hormone. We therefore synthesized the Ala^{10}-analog of minigastrin and studied in detail its conformational and biological properties.[10]

The CD results in the far UV (data not shown) indicate that there is an increase of the helical content with respect to that observed in minigastrin in TFE. The CD spectrum becomes almost identical to that of little gastrin which should contain at the N-terminus an helical segment longer than minigastrin. In the near UV the CD properties reveal that the environment of the aromatic chromophores in the two minigastrin analogs is different. Proton NMR studies were carried out in the usual way using standard 2D techniques. The Ala^{10}-analog is more soluble than des-Trp^1,Nle^{12}-minigastrin, and in 90% TFE it was possible to reach a concentration of the order of 5 mM without aggregation. The resonace assignment was accomplished with a 2D homonuclear Hartmann-Hahn experiment and by using the assignment of the Gly^{10} analog as a reference.[10] The main features of the TOCSY spectrum are reported in Figure 7 and the chemical shifts of the two analogs are compared in Figure 8. The resonance positions of the amide protons of the two analogs are all very different, with the exception of Trp-NH, the C-terminal cis-amide, and all but one of the Glu NH's. The Nle NH is the most different one, with a downfield shift of 0.29 ppm. This residue also shows considerable differences in the resonances of its aliphatic protons. All these findings indicate that the structural differences between Ala^{10}-minigastrin and Gly^{10}-mingastrin, visible in the CD spectrum, are mainly located in the C-terminal portion of the molecule. The temperature coefficients of the amide protons of the two analogs are compared in Table 1. With the exception of Nle NH and Tyr NH all temperature coefficients of the Ala^{10}-analog range from -1.6 to -4.3 ppb/K and are of the same

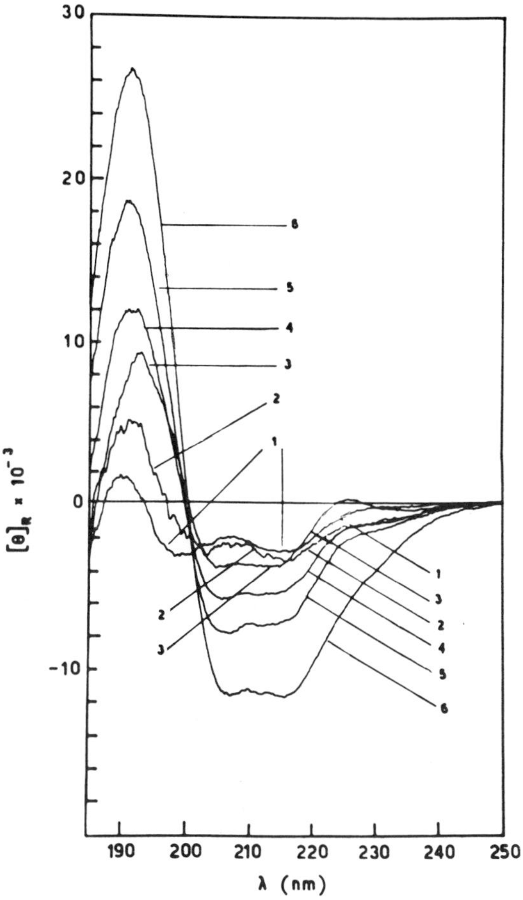

Figure 1. Far-uv CD spectra of gastrin fragments (see text) from octapeptide to tridecapeptide in 98% TFE; increasing numbers correspond to increasing chain length.

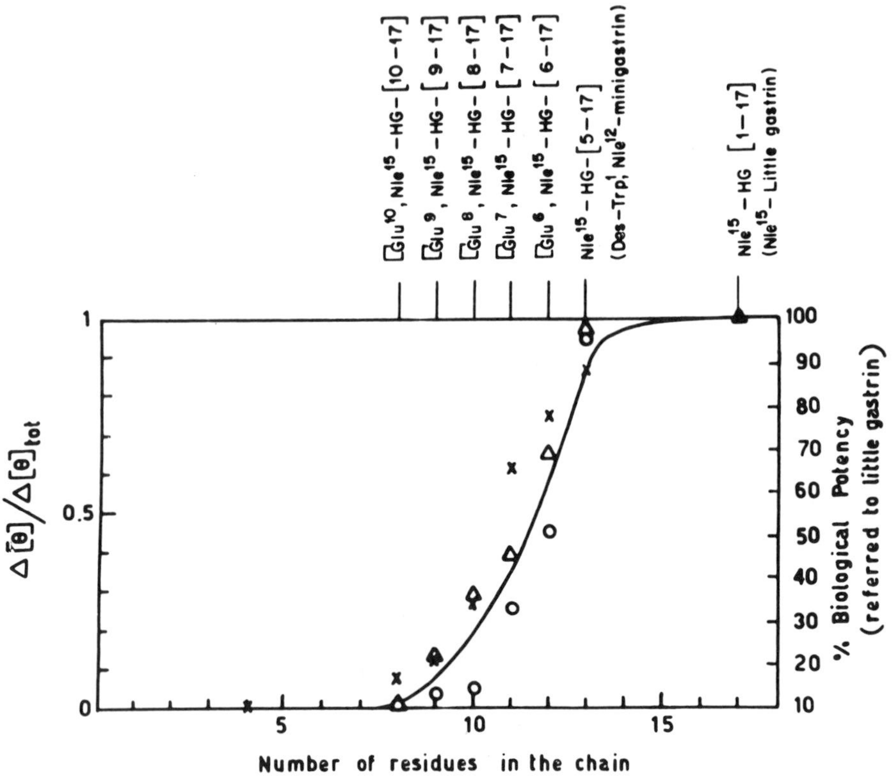

Figure 2. Relative variation of molar ellipticity at 216 nm (circles) and 192 nm (triangles) and of biological potency (crosses) of gastrin fragments as a function of chain length.

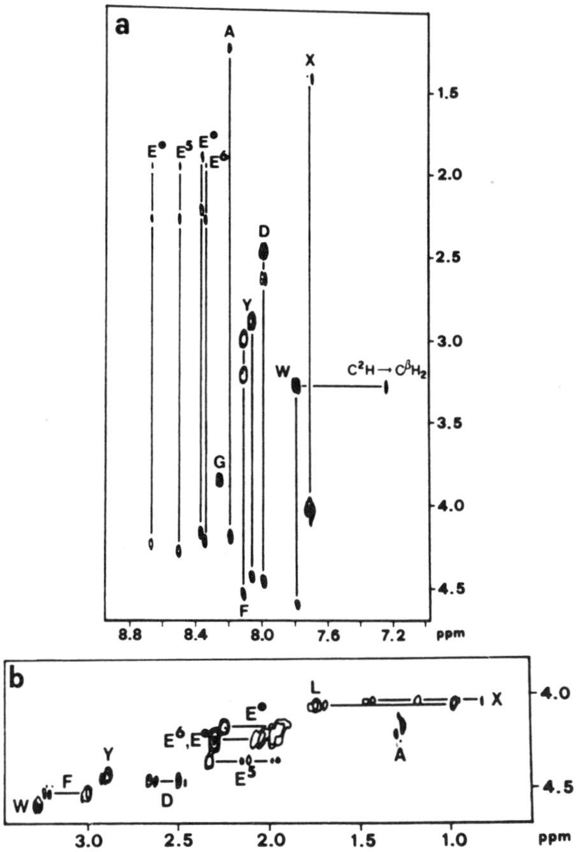

Figure 3. Portions of a two-dimensional TOCSY spectrum of 4.43 mM *des*-Trp[1],Nle[12]-minigastrin in H_2O (10% D_2O) showing the NH/aromatic to aliphatic region (a) and the $C^{\alpha}H$ to aliphatic region (b).

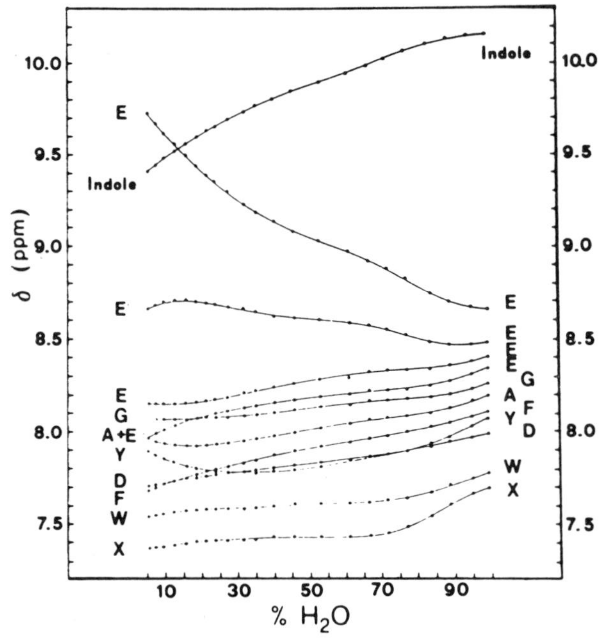

Figure 4. Solvent dependence of the chemical shift of amide proton resonances of *des*-Trp[1],Nle[12]-minigastrin in water-trifluoroethanol mixtures.

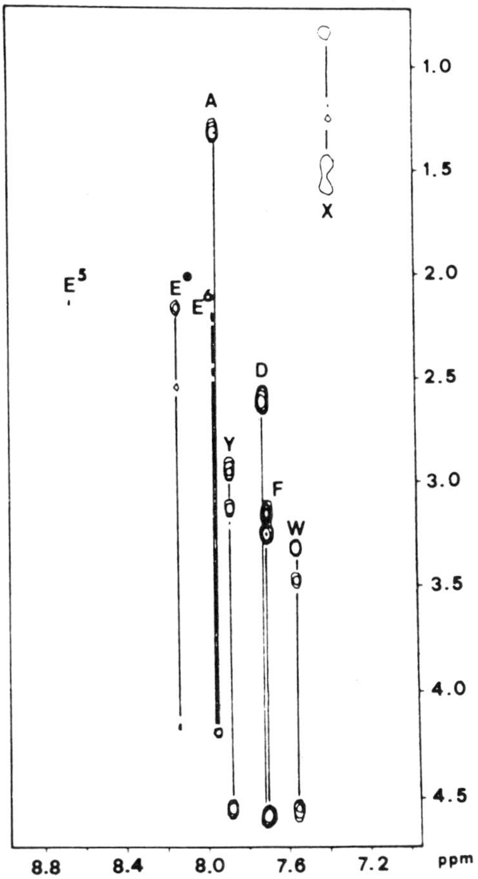

Figure 5. Portion of a two-dimensional TOCSY spectrum of ~4.0 mM *des*-Trp[1],Nle[12]-minigastrin in 94% TFE/6% D_2O showing the NH/aromatic to aliphatic region.

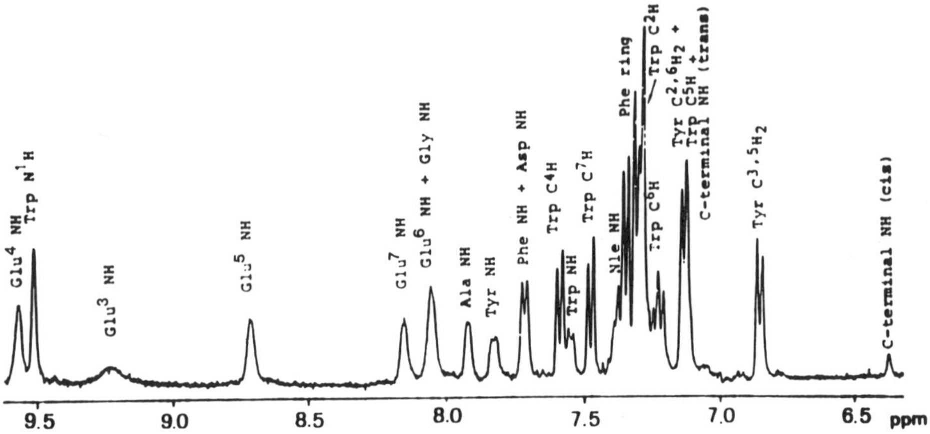

Figure 6. Assigned spectrum of *des*-Trp[1],Nle[12]-minigastrin in TFE (10% D_2O).

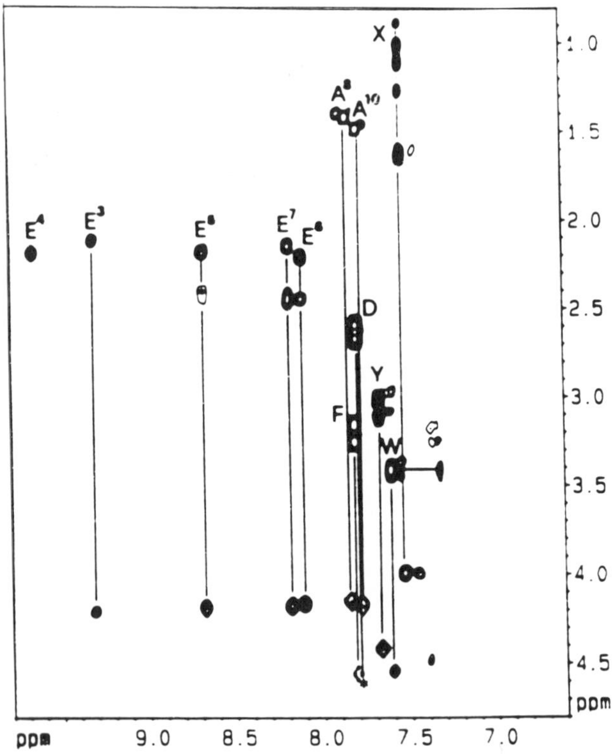

Figure 7. Portion of a two-dimensional TOCSY spectrum of 4.9 mM *des*-Trp[1],Ala[10],Nle[12]-minigastrin in 90% TFE-d_3/10% H_2O showing the NH/aromatic to aliphatic region.

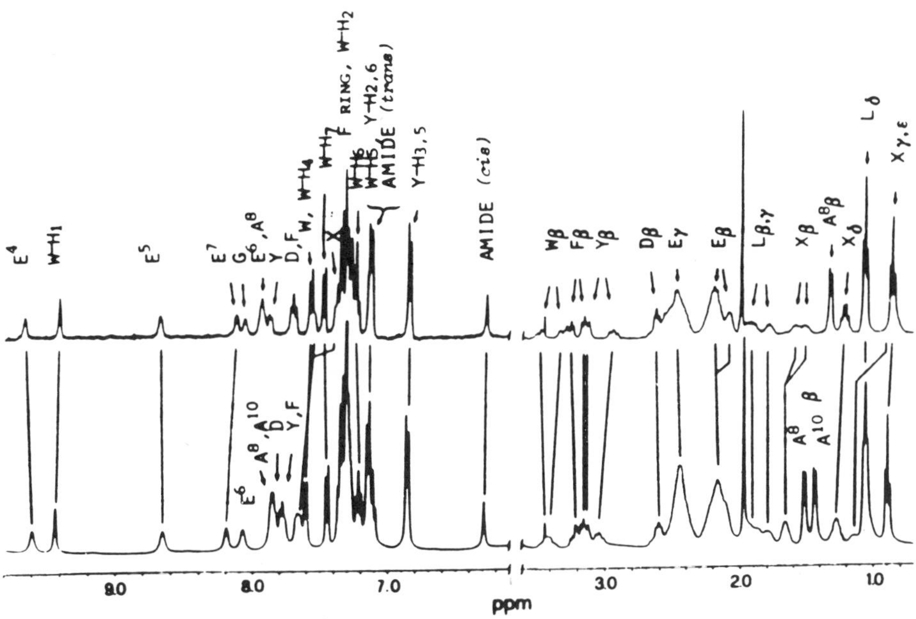

Figure 8. Comparison of the monodimensional spectra of *des*-Trp^{1},Nle^{12}-minigastrin (top) and *des*-Trp^{1},Ala^{10},Nle^{12}-minigastrin (bottom) in 90% TFE/10% D_2O.

Table I. Position at T = 296 K (ppm) and Temperature Coefficients (Δppb/K) of Amide Resonances in 90% TFE[a,b]

	des-Trp^{1},Nle^{12}-minigastrin		des-Trp^{1},Ala^{10},Nle^{12}-minigastrin	
Glu^{4}	9.660	−8.9 ± 1.4	9.660	−6.7 ± 0.2
Glu^{5}	8.681	+0.9 ± 1.3	8.659	+0.8 ± 0.4
Glu^{6}	7.934	−0.7 ± 0.8	8.071	−1.6 ± 0.4
Glu^{7}	8.126	−4.5 ± 0.9	8.196	−3.7 ± 0.2
Ala^{8}	7.934	−4.8 ± 0.9	7.858	−4.3 ± 0.6
Tyr	7.879	−7.1 ± 0.7	7.767	−6.3 ± 0.6
Gly	8.063	−5.3 ± 0.7		
Ala^{10}			7.858	−4.3 ± 0.6
Trp	7.651	−2.3 ± 0.6	7.668	−4.2 ± 0.2
Nle	7.373	−1.6 ± 0.8	7.659	−5.2 ± 0.7
Asp	7.710	−1.8 ± 0.9	7.844	−4.0 ± 0.6
Phe	7.689	−3.0 ± 0.7	7.787	−3.1 ± 0.4
amide (cis)	6.269	−4.4 ± 1.4	6.286	−4.3 ± 0.2

order of magniture as those reported by Ribeiro et al.[11] for oligoglutamates in TFE containing about 30% α-helix. These results again suggest that in this analog there is an extension of the N-terminal helical segment toward the C-terminus. This conclusion is also supported by the results of a ROESY experiment showing cross-peaks between Tyr $C^{\alpha}H$ and Nle C^{β} protons,[10] and such interaction is diagnostic for a helical structure. Connectivities of the type i - (i+3) are also found between the methyl protons of Ala [10] and the gamma protons of Glu. The Tyr ring protons are also close in space to Nle side chains. These data show that both Tyr and Ala[10] are part of a helical segment which extends at least to Nle.

The presence of a clearly shielded amide proton corresponding to the Glu[5] residue in both minigastrin and in its Ala[10] analog is an unsolved problem. An hydrogen bond with a proton acceptor within the C-terminal sequence, brought about by an U-shaped structure is ruled out by the fact that the Ala[10] analog does not assume a hairpin conformation. Model studies suggest the possibility of hydrogen bonding of this proton to the side chain carboxylate moiety of a Glu residue, but we do not have direct experimental evidence for this.

In conclusion, CD and NMR results indicate that in the Ala[10]-analog of minigastrin there is an extension of the helical segment from the N-terminus to the C-terminus. The replacement of Gly with Ala not only changes completely the conformational preference of the molecule but also affects substantially its biological potency. In vivo biological tests (gastric acid secretion in rats) showed that the potency of this gastrin analog is only 10% of that of des-Trp[1],Nle[12]-minigastrin.[10] This residual activity corresponds, within the limit of error of the assay system, to that of the C-terminal tetrapeptide fragment. The impact of the change of conformation of the Ala[10] analog on the hormonal potency provides further support to the hypothesis of the U-shaped conformation as the optimal bioactive state of the hormone. When the preference for this conformation is reduced, only the intrinsic potency of the C-terminal tetrapeptide amide is present. On the other hand, we have also shown that replacement of the $-(Glu)_5-$ sequence with $-(Asp)_5-$ sequence prevents the formation of the α-helix at the N-terminus and also reduces to a large extent the biological potency of the hormone.[12]

REFERENCES

1. J. S. Morley, H. J. Tracy, and R. A. Gregory, Structure-function relationships in the active C-terminal tetrapeptide sequence of gastrin, Nature 207:1356 (1965).
2. E. Peggion, E. Jäger, S. Knof, L. Moroder, and E. Wünsch, Conformational aspects of gastrin-related peptides: a circular dichroism study, Biopolymers 20:633 (1981).
3. E. Peggion, M. T. Foffani, E. Wünsch, L. Moroder, G. Borin, M. Goodman, and S. Mammi, Conformational properties of gastrin fragments of increasing chain length, Biopolymers 24:647 (1985).
4. C. C. Wu, A. Hachimori, and J. T. Yang, Lipid-induced ordered conformation of some peptide hormones and bioactive oligopeptides: predominance of helix over β form, Biochemistry 21:4556 (1982).
5. S. Mammi, N. J. Mammi, M. T. Foffani, E. Peggion, L. Moroder, and E. Wünsch, Conformation of human little gastrin and minigastrin analogs in surfactant solution, Biopolymers 26:S1 (1987).
6. E. Abillon, P. Pham Van Chuong, and P. Fromageot, Conformational calculations on gastrin C-terminal tetrapeptide, Int. J. Peptide Protein Res. 17:480 (1981).
7. P. Y. Chou, and G. D. Fasman, Empirical prediction of protein conformation, Ann. Rev. Biochem 47:251 (1978).

8. S. Mammi, N. J. Mammi, and E. Peggion, Conformational studies of human *des*-Trp[1],Nle[12]-minigastrin in water-trifluoroethanol mixtures by ^{1}H NMR and circular dichroism, Biochemistry 27:1374 (1988).
9. K. Amonraksa, M. Simonetti, S. DaRin-Fioretto, S. Mammi, and E. Peggion, Synthesis and ^{1}H NMR studies of α-deuterated analogues of *des*-Trp[1],Nle[12]-human minigastrin, Biopolymers 30:000 (1990) (in press).
10. S. Mammi, M. T. Foffani, E. Peggion, J. C. Galleyrand, J. P. Bali, M. Simonetti, W. Göring, L. Moroder, and E. Wünsch, Conformational and biological properties of the Ala[10] analogue of human *des*-Trp[1],Nle[12]-minigastrin, Biochemistry 28:7182 (1989).
11. A. A. Ribeiro, R. Saltman, and M. Goodman, Conformational analyses of polyoxyethylene-bound homo-oligo-L-glutamates in a helix-supporting environment, trifluoroethanol, Biopolymers 24:2469 (1985).
12. E. Wünsch, R. Scharf, E. Peggion, M. T. Foffani, and J. P. Bali, Conformational properties of gastrin peptides: synthesis, biological activity, and preliminary CD characterization of the $(Asp)_5$ analog of *des*-Trp[1],Leu[12]-minigastrin, Biopolymers 25:229 (1986).

INDIRECT MODELLING OF DRUG-RECEPTOR INTERACTIONS

Uli Hacksell, Anette M. Johansson, Anders Karlén, and Charlotta Mellin

Department of Organic Pharmaceutical Chemistry, Box 574, Uppsala Biomedical Centre, Uppsala University, S-751 23 Uppsala, Sweden

INTRODUCTION

The rapid progress in the area of molecular biology has provided us with the primary structures of a large variety of receptor proteins, many of which belong to the G-protein coupled receptor super family (e.g., dopamine (DA) D_1 and D_2 receptors, serotonin (5-HT) 5-HT_{1A}, 5-HT_{1C} and 5-HT_2 receptors, and muscarinic m_1-m_5 receptors).[1-4] These receptor proteins are believed to form seven transmembrane helixes which associate to generate a hydrophilic pore (or groove). The binding site for the ligand is thought to be located within this hydrophilic environment. However, experimental evidence on the 3D-structures of various G-protein coupled receptors is still lacking. Therefore, it is not possible to design ligands for these receptors based on receptor/ligand complementarity. Instead, one has to rely on indirect design methods. These may be based on (a) classical QSAR-techniques, which efficiently describe lipophilic and electronic properties, (b) methods related to the active analogue approach, which in a qualitative sense, takes into account topology and certain other molecular features, and (c) the more recent 3D-QSAR method,[5] which may provide quantitative information on the relation between biological activity and steric as well as electronic properties. In the present review, we will highlight the use of the active analogue approach in studies of structure-activity relationships. Examples will be given from the two areas: centrally active DA D_2 and 5-HT_{1A}-receptor agonists. In both areas, it has been essential to have access to information on the stereoselectivity of the receptor/ligand interaction.

Bioorganic Chemistry in Healthcare and Technology, Edited by U.K. Pandit and F.C. Alderweireldt, Plenum Press, New York, 1991

INDIRECT MODELS FOR RECEPTOR/LIGAND INTERACTIONS

Traditionally, models involving three attachment points between drugs and receptors have been used to rationalize biological stereoselectivity. According to such models the more potent enantiomer (the eutomer) can develop three bonds to the receptor surface whereas the less potent enantiomer (the distomer) only is able to form two bonds. This is a consequence of the idea that the drug has to adopt one particular orientation in relation to the receptor site. If we consider less abstract models, such as possible structures of receptor sites, it becomes apparent that the distomer also may be able to develop three intermolecular bonds to the receptor provided that it approaches the receptor site in a different manner or in a different conformation. In addition, a three point attachment is not at all neccessary to explain stereoselectivity. In fact, three interactions, repulsive or attractive, are sufficient. It should be noted, however, that numerous contacts between the receptor and the ligand are possible when docking e.g. a substrate with the active site of an enzyme. An additional factor to consider is that a mutual structural rearrangement of receptor and ligand may occur during complex formation. Thus, not only stereochemistry but also conformational mobility has to be considered when evaluating possible modes of binding of drugs.

Most qualitative models used to rationalize structure-activity relationships have been based on the pharmacophore concept. A pharmacophore may be defined as the 3D-arrangement of functional groups/atoms that is required in order for a drug to produce a particular response. Frequently, the pharmacophore consists of three pharmacophore points/ structural elements. These pharmacophore points are choosen based on their ability to participate in important intermolecular interactions. Thus, charged groups, hydrogen acceptors/donators, polar groups and aromatic rings are frequently selected as potential pharmacophore points. The positions and properties of the pharmacophore points are normally deduced from structure-activity relationship (SAR) studies of series of analogues.

The possibility to take into account both a pharmacophore and steric factors has been discussed by Marshall in the context of the active analogue approach.[6] According to this model, active analogues have a common volume which is part of the receptor (enzyme) excluded volume, that is, an area in space which is not occupied by the receptor. On the other hand, an inactive analogue which is able to produce a pharmacophore by adopting a conformation of accessible energy, should, by definition, produce steric bulk which is part of the receptor (enzyme) essential volume, that is, an area in space which is occupied by the receptor. The receptor-excluded volume is defined as the combined van der Waals volume produced by the active analogues when superimposed on the pharmacophore. Parts of the receptor-essential volume may be defined by superimposing inactive analogues onto the pharmacophore and identifying common excess volumes. The active analogue approach is most readily adopted on series of drugs with limited conformational mobility. However, also compounds which possess a considerable flexibility may be used provided that conformational energetics are taken into account.

A MODEL FOR DA D_2-RECEPTOR AGONISTS

We have a long standing interest in compounds being able to activate DA D_2-receptors, because selective dopaminergic agents may offer interesting therapeutic alternatives to present drugs in the treatment of dysfunctions in the central nervous system (CNS). Our synthetic efforts in this area have been focussed on phenolic 2-aminotetralins and phenylpiperidine derivatives (for a review, see ref 7). The potent and selective dopaminergic actions of 5-hydroxy-2-(dipropylamino)-tetralin (5-OH DPAT) provided us with a lead compound for SAR studies. To map the steric requirements of the D_2 receptor we prepared a number of stereochemically well-defined C1-, C2-, and C3-methyl substituted analogues of 5-OH DPAT.[8-13] The introduction of methyl groups in the non-aromatic ring of a 2-aminotetralin affects the conformational energetics.[14,15] Consequently, this set of compounds made it possible to study steric as well as conformational factors of importance for D_2-receptor activation.

It is noteworthy that a pronounced stereoselectivity was observed in the series of methyl substituted 5-OH DPAT derivatives (Table 1). This appears to be a general feature of dopaminergic 2-aminotetralin derivatives, e.g. (*S*)-5-OH DPAT is a potent DA-receptor agonist whereas the *R* enantiomer has been classified as a weakly potent antagonist.[16] Similarly, (1*S*,2*R*)-UH-242 is an antagonist whereas (1*R*,2*S*)-UH-242 is an agonist at DA D_2-receptors. This trend simplified the SAR-analysis considerably.

Since the 2-aminotetralin derivatives are conformationally flexible, the conformational preferences of the compounds were studied by a combination of theoretical (molecular mechanics calculations) and experimental (NMR-spectroscopy and X-ray crystallography) methods.[12,14,15] In general, an excellent agreement was obtained between the calculations and the experiments. It was observed, however, that several of the molecular geometries identified by X-ray crystallography correspond to conformations of fairly high relative steric energies. This is not a contradiction since packing forces in the crystal may be quite large. However, our

Table 1. Dopaminergic potencies and conformational energies of some phenolic 2-(dipropylamino)derivatives.

Code	Structure	Dopaminergic potency	Relative steric energy of the pharmacophore conformation (kcal/mol)	Ref.
(*S*)-5-OH-DPAT	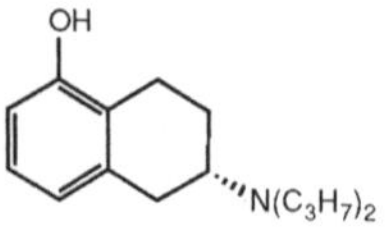	very potent agonist	0.5	15
(*R*)-5-OH-DPAT	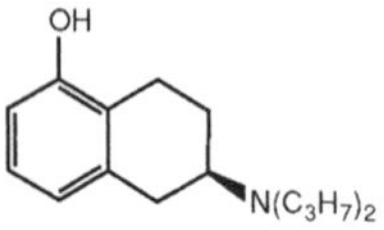	weakly potent antagonist	>2.5	16
(1*R*,2*S*)-UH-242	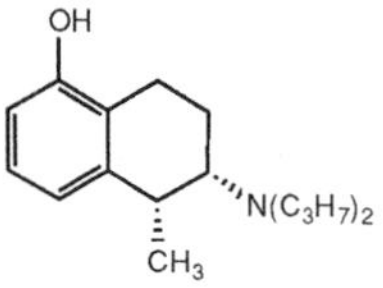	potent agonist	0	11
(1*S*,2*R*)-UH-242	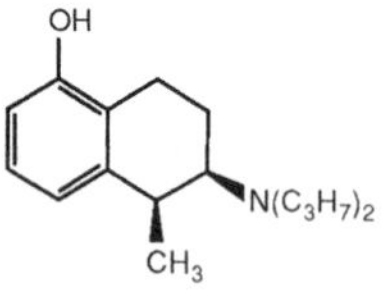	weakly potent antagonist	>2.5	11
(1*S*,2*S*)-AJ-116	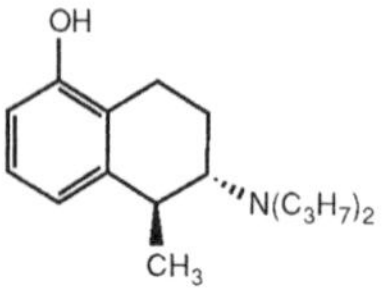	weakly potent agonist	2.4	11
(1*R*,2*R*)-AJ-116	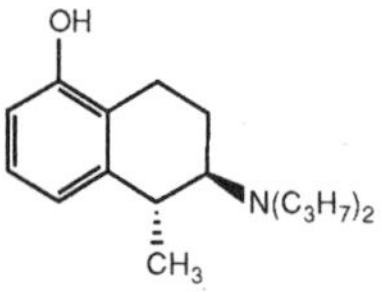	inactive	>2.5	11
(2*R*,3*S*)-AJ-166	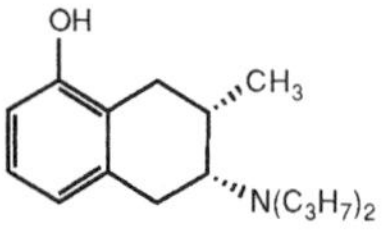	very potent agonist	0	12
(2*S*,3*R*)-AJ-166	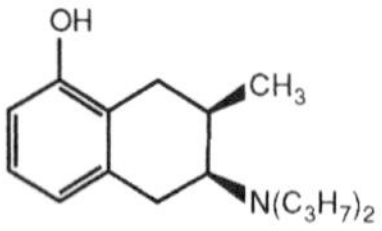	inactive	>2.5	12
(2*R*,3*R*)-AJ-164	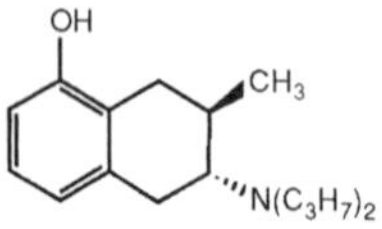	weakly potent agonist	>2.5	12
(2*S*,3*S*)-AJ-164	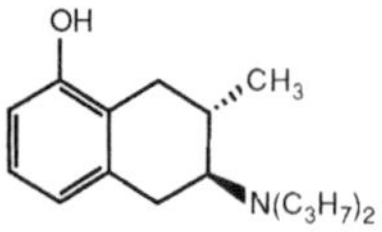	weakly potent agonist	>2.5	12

results emphasize the problem associated with modelling studies of flexible compounds in which only X-ray derived geometries are taken into account.

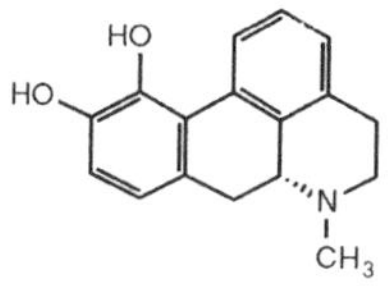

(*R*)-APO

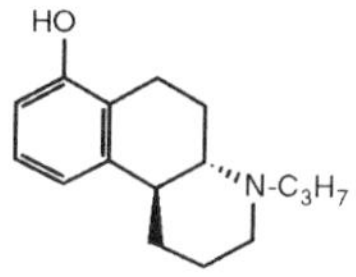

(4a*S*,10b*S*)-7-OH-OHB[f]Q

In order to define a pharmacophore for DA D_2-receptor agonists, we selected two potent, stereoselective and conformationally restricted agonists, (6a*R*)-apomorphine [(6a*R*)-APO] and (4a*S*,10b*S*)-7-hydroxy-4-propyl-1,2,3,4,5,6,10b-octahydrobenzo[**f**]quinoline [(4a*S*,10b*S*)-7-OH-OHB[**f**]Q],[17] as starting points. The conformational space for each of the compounds was investigated (using molecular mechanics and semi-empirical calculations) and low-energy conformations were identified. When computer-generated superpositions of these conformations were analyzed we observed that the minimum-energy conformations produced the best superposition of the fitted putative pharmacophore elements (the phenolic hydroxyl group positioned *meta* to the phenethylamine moiety, the aromatic ring, and the nitrogen and its lone pair; under physiological conditions the nitrogen lone pair corresponds to the N^+-H structural element. We believe that a reinforced electrostatic interaction between the protonated amino group and a carboxylate ion at the receptor constitutes the primary interaction between the ligand and its receptor). In the fitted conformations, the 2-aminotetralin moieties of the two structures adopt half chairs with pseudoequatorial amino substituents (Fig 1). Thus, this superposition was used to define a putative pharmacophore for DA D_2-receptor agonists.[15]

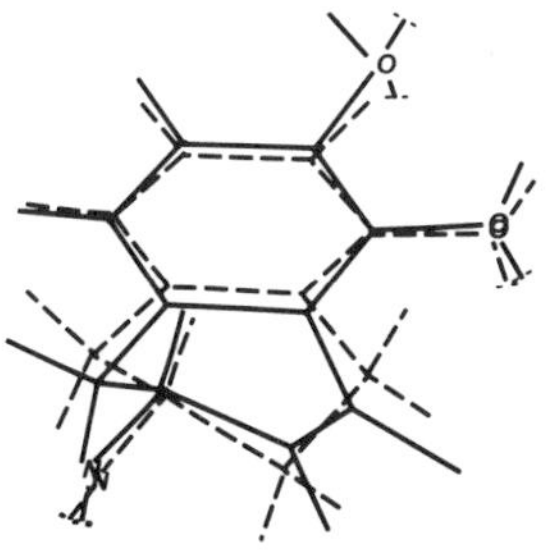

Fig. 1. Computer-generated best fit of the 2-aminotetralin moieties of (6a*R*)-APO (solid lines) and (4a*S*,10b*S*)-7-OH-OHB[f]Q. The compounds were fitted in their lowest energy conformations. The superposition defines a pharmacophore for DA D_2-receptor agonists (see also ref 15).

The relevance of the putative pharmacophore was then tested by applying it to the set of methyl-substituted analogues of 5-OH DPAT derivatives with the gratifying result that a good correlation was obtained between the ability of the compounds to adopt a pharmacophore conformation and their ability to activate DA D_2 receptors. Thus, the pharmacophore that was generated from the two tri- and tetracyclic analogues appears to be relevant.

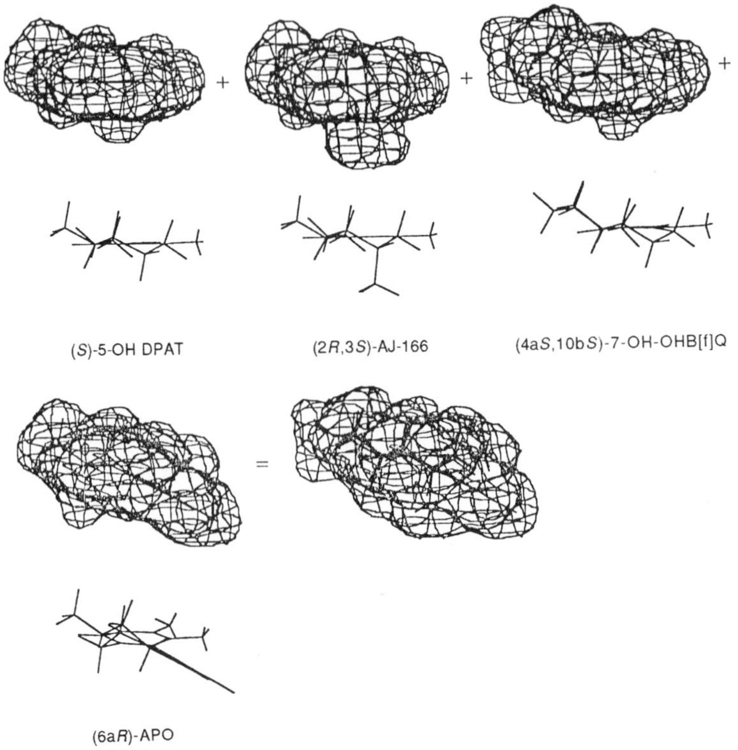

Fig. 2. A partial DA D_2-receptor excluded volume formed by addition of the van der Waals volumes of potent agonists in their pharmacophore conformations (see also refs 14, 15, and 19).

By superimposing the van der Waals volumes of the potent agonists (6a*R*)-APO, (4a*S*,10b*S*)-7-OH-OHB[f]Q, (*S*)-5-OH DPAT, (1*R*,2*S*)-UH-242, and (2*R*,3*S*)-AJ-166, we were able to generate a partial DA D_2-receptor excluded volume (Fig 2).[12] Compounds which are able to adopt pharmacophore conformations without producing excess volume (van der Waals volume outside that of the receptor-excluded volume) should, by definition be able to activate DA D_2 receptors. Based on this model it may be possible to design novel dopaminergic agonists with an atomic framework different from that in e.g. the 2-aminotetralins. In addition, compounds may be assigned as inactive without informations from biological tests. Today, we would neither have synthesized nor tested UH-199[18] since only conformations with relative steric energies above 6 kcal/mol fit to the pharmacophore. In addition, the *R*-enantiomer produces large excess volume when compared with the DA D_2-receptor excluded volume.[19]

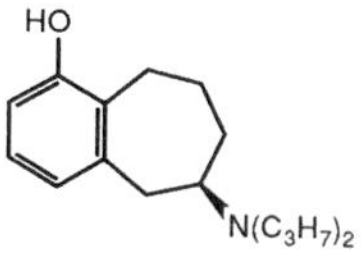

(*R*)-UH-199

A MODEL FOR $5\text{-}HT_{1A}$-RECEPTOR AGONISTS

8-Hydroxy-2-(dipropylamino)tetralin, 8-OH DPAT, the C8-hydroxylated regioisomer of the DA-receptor agonist 5-OH DPAT, is a potent $5\text{-}HT_{1A}$-receptor agonist.[21] Such derivatives have been the subject of an intense interest during the recent years at least in part due to the fact that the anxiolytic agent buspirone binds with high affinity to $5\text{-}HT_{1A}$-receptors.[21] In order to obtain an understanding of factors of importance for $5\text{-}HT_{1A}$-receptor activation we have synthesized and tested a number of derivatives of 8-OH DPAT (Figure 3).[22-29] In contrast to the dopaminergic 2-aminotetralin derivatives discussed above, the 8-OH DPAT derived analogues do not consistently interact in a stereoselective fashion with $5\text{-}HT_{1A}$-receptors. This indicates that the $5\text{-}HT_{1A}$-receptor/ligand interaction may be fundamentally different from that between DA D_2-receptors and its ligands. In addition, the different stereoselectivities made it much more difficult to deduce a model for $5\text{-}HT_{1A}$-receptor agonists than for DA D_2-receptor agonists.[30]

The 8-OH DPAT derivatives presented in Fig 3 constitute a good set for model development since they possess different conformational characteristics, different steric properties and a good spread in pharmacological potencies. In order to describe the observed structure-activity relationships within the series, we first divided the compounds into three subsets: (**A**) potent agonists (with K_i-values 32.3 nM), (**B**) moderately potent agonists (K_i-values between 38 and 394 nM), and (**C**) compounds which lack agonist properties. In addition, the potent ligand *d*-LSD was added to group A. Second, putative pharmacophore elements were selected: the hydroxyl group, the aromatic ring, the nitrogen, the nitrogen lone pair of electrons (the N^+-H bond), and a dummy atom located 2.6 Å from the nitrogen and aligned with the N^+-H vector (the dummy atom was supposed to mimic a carboxylate ion at the receptor). Third, various putative pharmacophores were evaluated as follows: the compounds in subset **a** were fitted to the

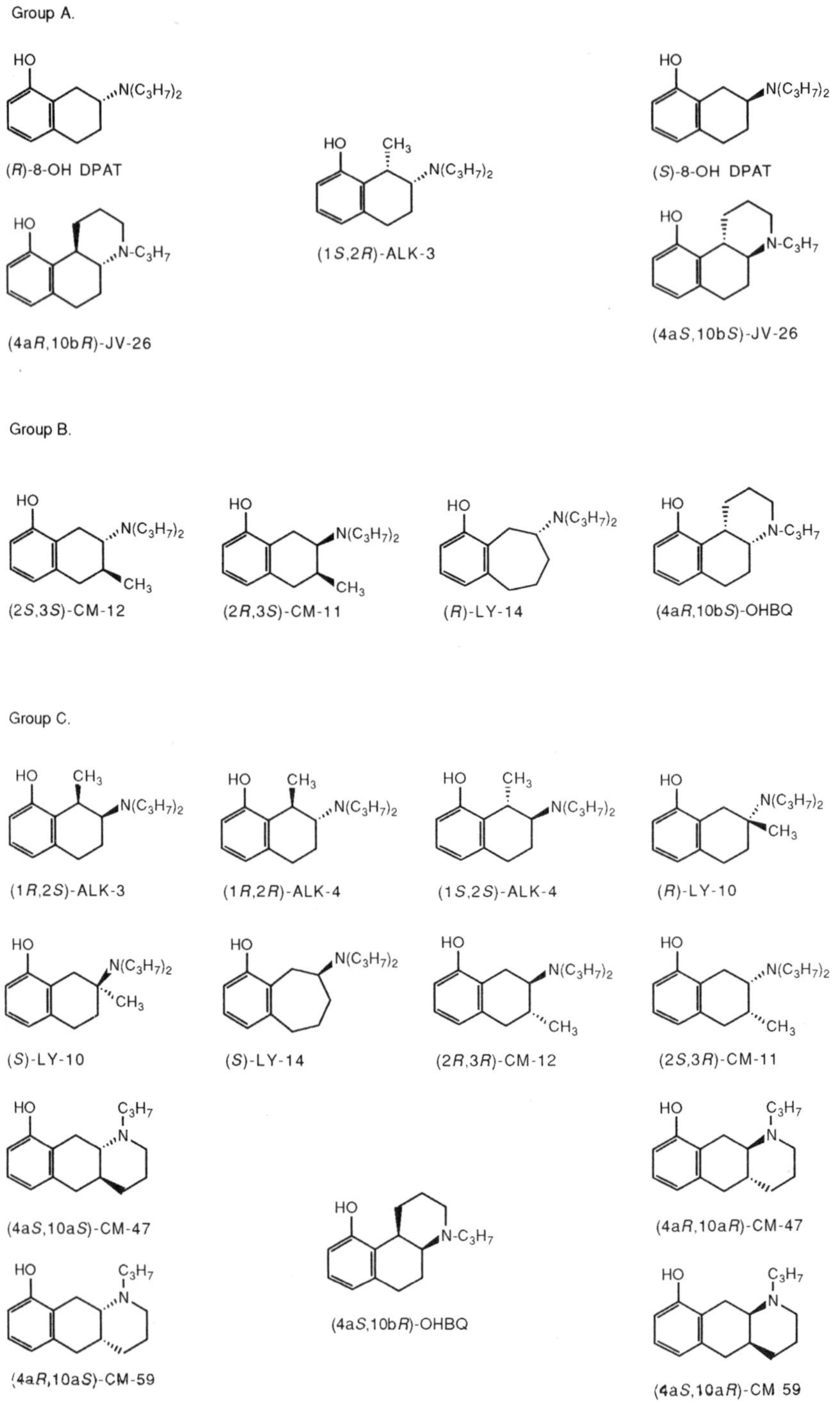

Fig. 3. Structures of the compounds used to develop a model for $5\text{-}HT_{1A}$-receptor agonists. The three groups consist of potent agonists (A), moderatly potent agonists (B) and derivaties that lack ability to stimulate $5\text{-}HT_{1A}$-receptors (C). The compounds are described in refs 20, 22-29, and 32.

pharmacophore and a receptor-excluded volume was generated by adding their pharmacophore conformations. Since (1*S*,2*R*)-ALK-3, *d*-LSD and the enantiomers of JV-26 possess a limited flexibility, few conformations had to be considered in the fitting procedure. In order for a pharmacophore hypothesis to be valid the following criteria had to be fullfilled: it should be able to rationalize the difference between groups **A** - **C**, that is, all compounds in set **A** should be able to fit to the pharmacophore in low-energy conformations and none of the compounds in subsets **B** or **C** should be able to fit to the pharmacophore (in an energetically acceptable conformation) without generating excess volume. Similarly, the compounds in subset **C** should be unable to fit to the pharmacophore without producing volume in excess to that generated by subsets **A** and **B**.[29]

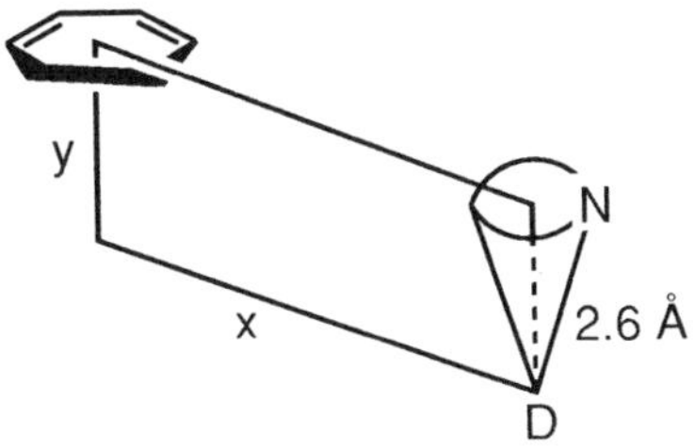

Fig. 4. A flexible pharmacophore for 5-HT_{1A}-receptor agonists. D is the dummy atom which corresponds to a carboxylate ion at the receptor. x and y represent key distances. Allowed positions for the (protonated) nitrogen are defined by the top of the cone.

By applying these rules it was possible to reject all but one of the putative pharmacophores. The procedure may be examplified by the pharmacophore consisting of a hydroxyl group, an aromatic ring and a nitrogen. This pharmacophore could be excluded since the inactive enantiomer of ALK-3 did not produce excess volume. The only pharmacophore that fulfills all the criteria stated above consists of an aromatic binding site and a dummy atom (Fig 4).[29] In addition, the orientation of the N-dummy atom vector was restricted to that defined by the potent agonists (in the model describing potent agonists) or by the potent and moderately potent agonists (in the model distinguishing between all agonists and "inactives"). Thus, in contrast to the DA D_2-agonist pharmacophore, the 5-HT_{1A}-agonist pharmacophore does not include a hydrogen donating hydroxyl group. Some illustrations of the model are given in Fig 5.

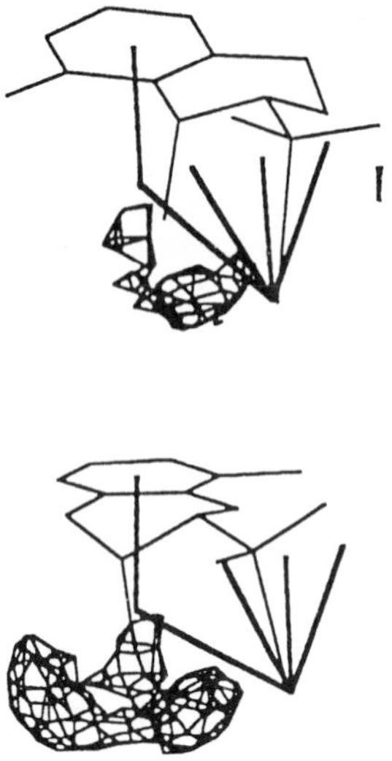

Fig. 5. Excess volumes formed by comparison of the van der Waals volumes of the inactive (1*R*,2*S*)-ALK-3 (top) and (2*S*,3*R*)-CM-11 (bottom) with the partial 5-HT_{1A}-receptor excluded volume. The thick lines represent elements of the pharmacophore model.

The deduced model for 5-HT_{1A}-receptor agonists may turn out to be useful in the design of novel agonists. However, (*S*)-UH-301, which only differs from the potent agonist (*S*)-8-OH DPAT by a C5-fluoro substituent, binds to 5-HT_{1A}-receptors but appears to lack stimulating ability.[31] Based on a steric analysis, the above 3D-model classifies (*S*)-UH-301 as an agonist although it is an antagonist. Apparently, the intrinsic activity of 8-OH DPAT may be altered by changes in the electronic properties of the aromatic ring. Consequently, the model has to be extended since it does not take into account electronic properties.

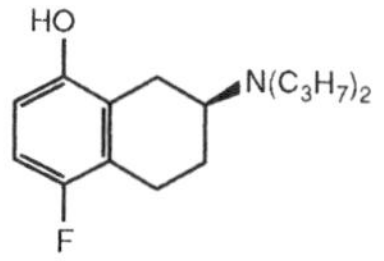

(*S*)-UH-301

CONCLUSION

The two applications presented herein demonstrate some of the weaknesses and strenghts of the active analogue approach. It is apparent that the method allows for a strict definition of pharmacophores and receptor/enzyme excluded volumes even in cases where sets of flexible analogues are used. However, the method does not deal explicitly with electronic properties and will not provide good results in cases where electrostatic ligand/receptor interactions are important. In such cases it is likely that 3D-QSAR methods, based on a pharmacophore generated by use of the active analogue approach, will produce more useful models.

ACKNOWLEDGEMENTS

The excellent contributions from our co-workers, whose names are cited in the references, are gratefully acknowledged. We also thank the Swedish Natural Science Council and the Swedish National Board for Technical Development for financial support and Dr. Uno Svensson for valuable help with the preparation of this manuscript.

REFERENCES

1. R.A. Dixon, C.D. Strader, and I.S. Sigal, Structure and function of G-protein coupled receptors. Annu. Rep. Med. Chem. 23:221-233 (1988).
2. J.C. Venter, C.M. Fraser, A.R. Kerlavage and M.A. Buck, Molecular biology of adrenergic and muscarinic cholinergic receptors. A Perspective. Biochem. Pharmacol. 38:1197-1208 (1989).
3. E.C. Hulme, N.J.M. Birdsall, and N.J. Buckley, Muscarinic receptor subtypes. Annu. Rev. Toxicol. 30:633-673 (1990).
4. P. Hartig, Molecular biology of 5-HT receptors. Trends Pharmacol. Sci. 10:64-69 (1989).
5. R.D. Cramer, III, D.E. Patterson, and J.D. Bunce, Comparative molecular field analysis (CoMFA). 1. Effect of shape on binding of steroids to carrier proteins. J. Am. Chem. Soc. 110:5959-5967 (1988).
6. J.R. Sufrin, D.A. Dunn, and G.R. Marshall, Steric mapping of the L-methionine binding site of ATP:L-methionine S-adenosyltransferase. Mol. Pharmacol. 19:307-313 (1981).
7. U. Hacksell, A.M. Johansson, A. Karlén, K. Svensson, and C.J. Grol, Stereochemical parameters are of importance for dopamine D_2-receptor activation, in: "Chirality and Biological Activity", H. Frank, B. Holmstedt and B. Testa, eds., Alan R. Liss, N.Y. (1990).
8. U. Hacksell, A.M. Johansson, L.-E. Arvidsson, J.L.G. Nilsson, S. Hjorth, A. Carlsson, H. Wikström, D. Sanchez, and P. Lindberg, C_1-Methylated 5-hydroxy-2-(dipropylamino)tetralins - Central dopamine-receptor stimulating activity. J. Med. Chem. 27:1003-1007 (1984).
9. U. Hacksell, L.-E. Arvidsson, A.M. Johansson, J.L.G. Nilsson, D. Sanchez, B. Andersson, P. Lindberg, H. Wikström, S. Hjorth, K. Svensson, and A. Carlsson, 5-Hydroxy-2-methyl-2-(di-n-propylamino)tetralin: Synthesis and central pharmacological effects. Acta Pharm. Suec. 22:65-74 (1985).
10. A.M. Johansson, L.-E. Arvidsson, U. Hacksell, J.L.G. Nilsson, K. Svensson, S. Hjorth, D. Clark, A. Carlsson, D. Sanchez, B. Andersson, and H. Wikström, Novel dopamine-receptor agonists and antagonists with preferential action on autoreceptors. J. Med. Chem. 28:1049-1053 (1985).
11. A.M. Johansson, L.-E. Arvidsson, U. Hacksell, J.L.G. Nilsson, K. Svensson, and A. Carlsson, Resolved cis- and trans-2-amino-5-methoxy-1-methyltetralins: Central dopamine receptor agonists and antagonists. J. Med. Chem. 30:602-611 (1987).

12. A.M. Johansson, J.L.G. Nilsson, A. Karlén, U. Hacksell, K. Svensson, A. Carlsson, L. Kenne, and S. Sundell, C3-Methylated 5-hydroxy-2-(dipropylamino)tetralins: Conformational and steric parameters of importance for central dopamine receptor activation. J. Med. Chem. 30:1135-1144 (1987).
13. A. Krotowska, G. Nordvall, C. Mellin, A.M. Johansson, U. Hacksell, and C.J. Grol, Dopamine D_2-receptor affinities of resolved C1-dimethylated 2-aminotetralins. Acta Pharm. Suec. 24:145-152 (1987).
14. A. Karlén, A.M. Johansson, L. Kenne, L.-E. Arvidsson, and U. Hacksell, Conformational analysis of the dopamine-receptor agonist 5-hydroxy-2-(di-n-propylamino)tetralin and its C(2)-methyl substituted derivative. J. Med. Chem. 29:917-924 (1986).
15. A.M. Johansson, A. Karlén, C.J. Grol, S. Sundell, L. Kenne, and U. Hacksell, Dopaminergic 2-aminotetralins. Affinities for dopamine D_2 receptors, molecular structures and conformational preferences. Mol. Pharmacol. 30:258-269 (1986).
16. A. Karlsson, L. Björk, C. Pettersson, N.-E. Andén, and U. Hacksell, (R)- And (S)-5-hydroxy-2-(dipropylamino)-tetralin (5-OH DPAT): Assessment of optical purities and dopaminergic activities. Chirality 2:90-95 (1990).
17. H. Wikström, B. Andersson, D. Sanchez, P. Lindberg, L.-E. Arvidsson, A.M. Johansson, J.L.G. Nilsson, K. Svensson, S. Hjorth, and A. Carlsson, Resolved monophenolic 2-aminotetralins and 1,2,3,4,5,5,10b-octahydrobenzo[f]quinolines: Structural and stereochemical considerations for centrally acting pre- and postsynaptic dopamine-receptor agonists. J. Med. Chem. 28:215-225 (1985).
18. U. Hacksell, L.-E. Arvidsson, U. Svensson, J.L.G. Nilsson, H. Wikström, P. Lindberg, D. Sanchez, S. Hjorth, A. Carlsson, and L. Paalzow, Monophenolic 2-(dipropylamino)indanes and related compounds: Central dopamine-receptor stimulating activity. J. Med. Chem. 24: 429-434 (1981).
19. A. Karlén, A. Helander, L. Kenne, and U. Hacksell, Topography and conformational preferences of 6,7,8,9-tetrahydro-1-hydroxy-N,N-5H-6-benzocycloheptenylamine. A rationale for the dopaminergic inactivity. J. Med. Chem. 32:765-774 (1989).
20. L.-E. Arvidsson, U. Hacksell, J.L.G. Nilsson, S. Hjorth, A. Carlsson, D. Sanchez, P. Lindberg, and H. Wikström, 8-Hydroxy-2-di-n-propylaminotetralin, a new centrally acting 5-HT-receptor agonist. J. Med. Chem. 24:921-923 (1981).
21. J. Traber and T. Glaser, 5-HT_{1A}-Related anxiolytics. Trends Pharmacol. Sci. 8:432-437 (1987).
22. L.-E. Arvidsson, U. Hacksell, A. Johansson, J.L.G. Nilsson, P. Lindberg, D. Sanchez, H. Wikström, K. Svensson, S. Hjorth, and A. Carlsson, 8-Hydroxy-2-(alkylamino)tetralins and related compounds as central 5-hydroxytryptamine receptor agonists. J. Med. Chem. 27:45-51 (1984).
23. C. Mellin, Y. Liu, L. Björk, N.-E. Andén, and U. Hacksell, A C2-methylated derivative of the 5-hydroxytryptamine-receptor agonist 8-hydroxy-2-(dipropylamino)tetralin (8-OH DPAT). Acta Pharm. Suec. 24:153-160 (1987).

24. L.-E. Arvidsson, A.M. Johansson, U. Hacksell, J.L.G. Nilsson, K. Svensson, S. Hjorth, T. Magnusson, A. Carlsson, B. Andersson, and H. Wikström, (+)-cis-8-Hydroxy-1-methyl-2-(di-n-propylamino)tetralin: a potent and highly stereoselective 5-hydroxytryptamine receptor agonist. J. Med. Chem. 30:2105-2109 (1987).
25. L. Björk, C. Mellin, U. Hacksell, and N.-E. Andén, Effects of the C3-methylated derivatives of 8-hydroxy-2-(di-n-propylamino)tetralin (8-OH-DPAT) on central 5-hydroxy-tryptaminergic receptors. Eur. J. Pharmacol. 143: 55-63 (1987).
26. C. Mellin, L. Björk, A. Karlén, A.M. Johansson, S. Sundell, L. Kenne, D.L. Nelson, N.-E. Andén, and U. Hacksell: Central dopaminergic and 5-hydroxy-tryptaminergic effects of C3-methylated derivatives of 8-hydroxy-2-(di-n-propylamino)tetralin. J. Med. Chem. 31:1130-1140 (1988).
27. Y. Liu, C. Mellin, L. Björk, B. Svensson, I. Csöregh, A. Helander, L. Kenne, N.-E. Andén, and U. Hacksell, (R)-And (S)-5,6,7,8-tetrahydro-1-hydroxy-N,N-dipropyl-9H-benzocyclohepten-8-ylamine. Stereoselective interactions with 5-HT_{1A}-receptors in the brain. J. Med. Chem. 32:2311-2318 (1989).
28. L. Björk, B. Backlund Höök, D.L. Nelson, N.-E. Andén, and U. Hacksell, Resolved N,N-dialkylated 2-amino-8-hydroxy-tetralins: Stereoselective interactions with 5-HT_{1A} receptors in the brain. J. Med. Chem. 32:779-783 (1989).
29. C. Mellin, J. Vallgårda, D.L. Nelson, L. Björk, H. Yu, N.-E. Andén, I. Csöregh, L.-E. Arvidsson, and U. Hacksell, A 3-D model for 5-HT_{1A}-receptor agonists based on stereoselective methyl-substituted and conformatio-nally restricted analogues of 8-hydroxy-2-(dipropyl-amino)tetralin. J. Med. Chem. in press.
30. L.-E. Arvidsson, A. Karlén, U. Norinder, L. Kenne, S. Sundell, and U. Hacksell, Structural factors of importance for 5-hydroxytryptaminergic activity. Conformational preferences and electrostatic potentials of 8-hydroxy-2-(di-n-propylamino)tetralin (8-OH DPAT) and some related agents. J. Med. Chem. 31:212-221 (1988).
31. S.-E. Hillver, L. Björk, Y.-L. Li, B. Svensson, S. Ross, N.-E. Andén, and U. Hacksell, (S)-5-Fluoro-8-Hydroxy-2-dipropylamino)tetralin: A putative 5-HT_{1A}-receptor antagonist. J. Med. Chem. 33:1741-1544 (1990).
32. H. Wikström, B. Andersson, T. Elebring, J. Jacyno, N.L. Allinger, K. Svensson, A. Carlsson, and S. Sundell, Resolved *cis*-10-hydroxy-4-*n*-propyl-1,2,3,4,5,6,10b-octahydrobenzo[f]quinoline: Central serotonin stimulating properties. J. Med. Chem. 30:1567-1573 (1987).

PRODUCTION OF BETA-LACTAMS:

A MIXTURE OF ORGANIC SYNTHESIS PROCESSES

AND ENZYMATIC PROCESSES

Svend G. Kaasgaard and Poul B. Poulsen

NOVO NORDISK A/S
DK-2880 Bagsvaerd
Denmark

MARKET OVERVIEW

Penicillin is known to everybody. It is a group of beta-lactams, which during the last 45 years, have been the most important antibiotics both regarding quantity and economy. These beta-lactams are unsurpassed because they are cheap (i.e. available for poor as well as rich), because they are efficient, and because they give very, very few side effects.

As a consequence it is a commercially important class of compounds and there is a wide interest in improving production of the beta-lactams already available on the market as well as in finding new and even more efficient beta-lactams.

This lecture will only deal with trends in process improvements in production of the well known beta-lactams.

The penicillins consist of a common beta-lactam nucleus (6-APA) upon which different side chains (R) are attached (fig. 1).

When the penicillin producing fungi grow in nature they produce mainly isopenicillin N and penicillin F but only in very low yield and they have only a limited antibiotic effect. It was soon realized, however, that a limiting factor in the penicillin production was the availability of the side chain and that addition of some carboxylic acids to the fermentation media could increase yields and give new penicillins.

Bioorganic Chemistry in Healthcare and Technology, Edited by U.K. Pandit and F.C. Alderweireldt, Plenum Press, New York, 1991

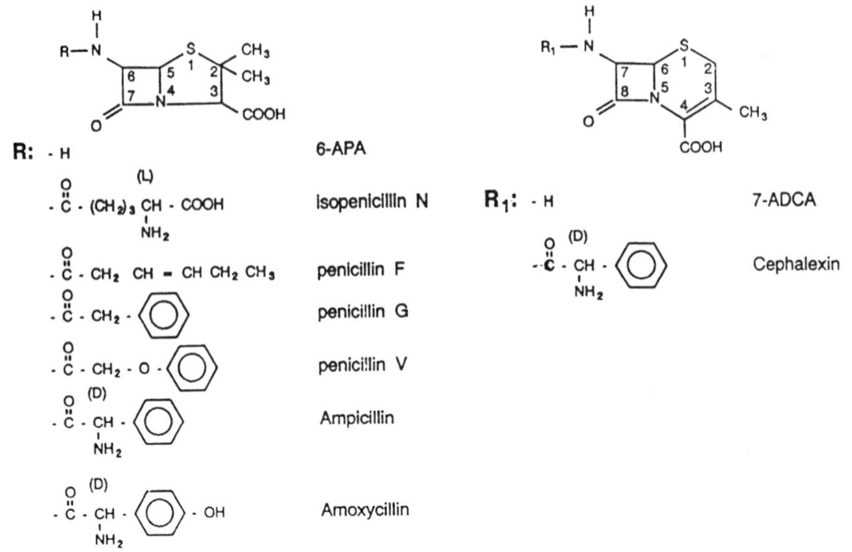

Fig. 1. Structure of the most important natural and semisynthetic beta-lactam antibiotics.

Table 1. Turnover and bulk price indications of the quantitatively most important beta-lactam antibiotics and intermediates.

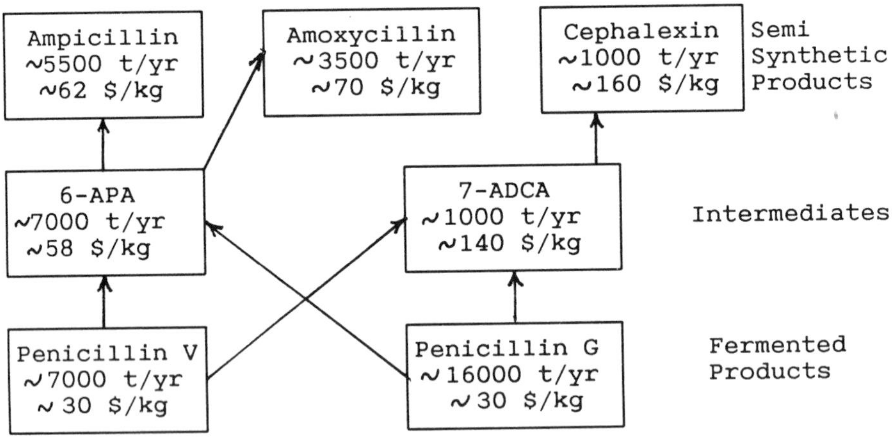

Comment: Only a few - however the 3 most important - of the semisynthetic products are shown. All amounts and prices are based upon the authors knowledge and are not to be used as company warranties. Prices e.g. fluctuate all the time.

Today the only directly fermented penicillins are penicillin G and penicillin V, produced by adding phenyl-acetic acid or phenoxyacetic acid, respectively, to the fermentation tank. They now form the basis for production of a vast range of semisynthetic penicillins and cephalosporins, among which the far most important are Ampicillin, Amoxycillin and Cephalexin. The latter being a cephalosporin. These second generation antibiotics are more efficient against a wider range of bacteria and, furthermore, they are more stable towards degradation than the fermented penicillins.

In Table 1 is shown an overview of the level of production as well as bulk price indications of the most important beta-lactam antibiotics, which indicates the industrial importance of the different compounds.

Fermentation
|
Broth Filtration
|
Solvent Extraction
|
Crystallization
|
Redissolution
|
Enzymatic hydrolysis
|
Decoloration
|
Crystallization of 6-APA
|
Drying of 6-APA

Fig. 2. Downstream processing steps involved in the production of penicillin V and its hydrolysis to 6-APA.

DESCRIPTION OF PRESENT PROCESSES

Production of penicillin V or G by fermentation

All companies use a mutant of the fungus *Penicillium chrysogenum*, for production of both penicillin V and G.

The fungus is grown in large fermentors with a size typically of more than 100 m^3 over a week. During that week a penicillin concentration of more than 30 g/l is obtained in the culture broth. It is really remarkable that strain mutation has increased the yield of this secondary metabolite with a factor of more than 500 compared to the wild type strains found in nature, and apparently yield improvement is still possible.

When the penicillin production begins to level off, the culture broth is pumped from the fermentor to the recovery section. A typical recovery (fig. 2) begins with a filtration of the broth to remove the mycelium. Thereafter, the penicillin in the filtrate is extracted into e.g. butyl acetate and back into water in order to both concentrate the penicillin solution and to remove impurities. The resulting solution is either further purified for the production of pharmaceutical grade penicillin V or G or used directly for production of 6-APA.

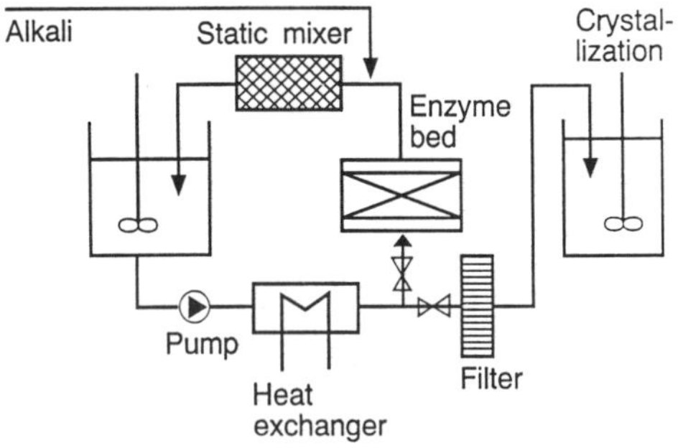

Fig. 3. Enzymatic hydrolysis of penicillin V.

Production of 6-APA from penicillin V or G

The penicillin solution is typically, as in our plant, adjusted to a concentration of approx. 10% w/w, and to a pH of 7.5 - as we use penicillin V as substrate. The pH is a little higher (around 8) if penicillin G is used. It is then completely hydrolysed enzymatically into 6-APA and the carboxylic acid side chain during typically 2 hours.

Even though the same fungus can produce both penicillin V and G, two different enzymes must be used for the hydrolysis of the two compounds. As we have penicillin V as a substrate, we use an immobilized penicillin V acylase in our 6-APA production.

In our 6-APA plant, shown schematically in fig. 3, we circulate the penicillin V solution from a holding tank through a fixed bed column, containing the immobilized enzyme, back into the holding tank through an in-line pH adjustment system. As the liberated phenoxyacetic acid is a strong acid, an efficient pH adjustment system is necessary to keep a constant pH. The contact time with the enzyme must be as short as possible in order to eliminate need for a buffer in the solution. We have been able to reduce this contact time to less than one minute although our enzyme reactor is a 1000 litre one bed reactor. When the reaction is finished, the solution may be purified by decolorizing systems before the 6-APA is crystallized at its isoelectric point, pH 4, before being dried. The phenoxyacetic acid is isolated as well and recirculated to the main fermentor.

Production of Semisynthetic beta-lactams from 6-APA

Most of the 6-APA produced today is used in the production of Ampicillin and Amoxycillin. They are produced from the 6-APA and a D-phenylglycine derivative by one of two chemical routes: The acid chloride route or the Dane-salt route, described in the following with Ampicillin as an example. In the production of Amoxycillin D-p-hydroxy phenylglycine replaces the D-phenylglycine.

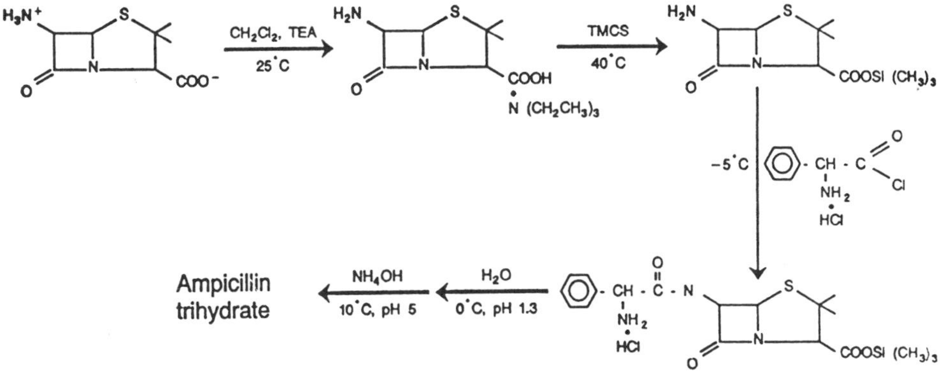

Fig. 4. Synthesis of Ampicillin by the acid chloride route.

In the acid chloride process, the amphoteric 6-APA is dissolved in methylene chloride as a triethylamine complex, the C3-carboxylic acid group is then protected by silylation after which the D-phenylglycine side chain is added as an acid chloride at low temperature, around -5° C. The formed Ampicillin is extracted into water and crystallized by adjusting the pH to the isoelectric point, around pH 5 (fig. 4).

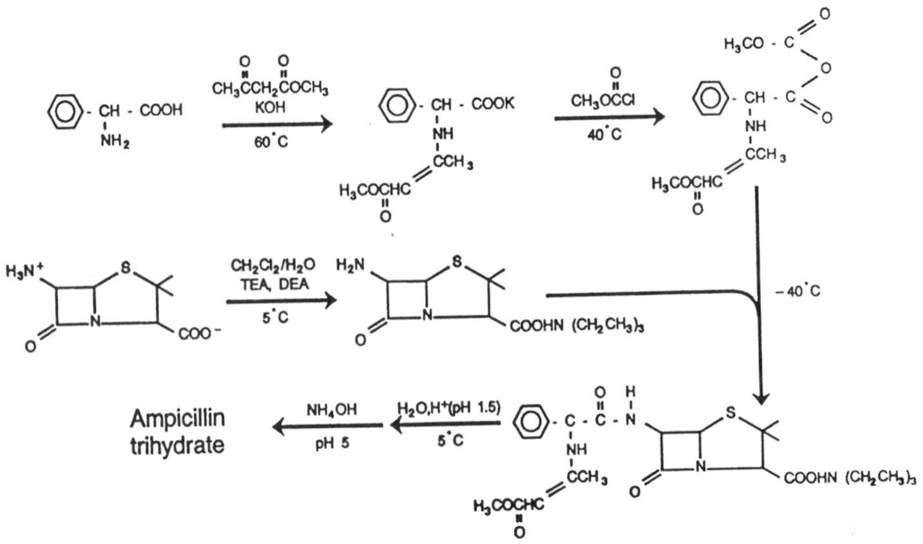

Fig. 5. Synthesis of Ampicillin by the Dane-salt route.

In the Dane-salt method the amino group of the side chain, D-phenylglycine, is protected as an enamine after which the thus formed Dane-salt is activated by adding methyl chloroformate to form a mixed anhydride. The 6-APA is dissolved in a separate tank with triethylamine in a mixture of methylene chloride and water and the condensation proceeds at -40 °C. The protection groups on the condensation product are removed and the Ampicillin extracted into the water phase by lowering the pH to 1.5. The Ampicillin is crystallized at the isoelectric point, pH 5 (fig. 5).

As can be seen the processes are quite demanding with regard to organic solvents, temperatures, and silylation, but these two processes have been used for more than 20 years and have been optimized to give an 87-91% yield and the ratio between the starting concentrations of 6-APA and the side chain is close to 1.

DISCUSSION OF PROCESS DEVELOPMENTS

Discussion of developments within penicillins to 6-APA

The beta-lactam nucleus itself, 6-APA, was first isolated from a *Penicillium chrysogenum* fermentation without added V or G side chain in the late fifties by Beecham. It has only a very weak antibiotic effect itself, but it proved to be a convenient starting point for the preparation of new penicillins by acylation of the amino group. Unfortunately, the yields in the direct fermentation of 6-APA were very low so this type of production could not be used industrially.

A chemical route to the hydrolysis of penicillin V/G was, however, soon found and 6-APA became industrially available in the early sixties.

At least one of the chemical processes consist of protecting the carboxylic group as a silyl ester and then form an iminoether as an intermediate step in a butanol containing solvent at -40 °C (fig. 6). The yields are high but the process costs for the low temperature and the regeneration of the protecting groups and the solvents are high too.

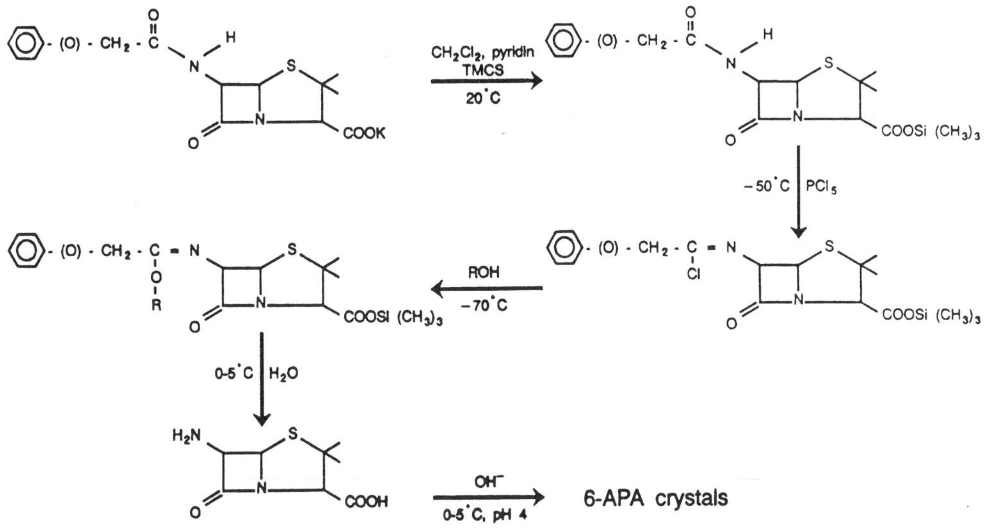

Fig. 6. Chemical hydrolysis of penicillin V/G to 6-APA.

One year after the first publication of the isolation of 6-APA, four groups simultaneously published the use of enzymes capable of hydrolysing the fermented penicillins, but it was not until 1966 that the first enzymatic production of 6-APA was introduced at Beecham. Within the next fifteen years nearly all chemical processes were taken over by enzymatic routes, so more than 95% of the 6-APA produced today is made by enzymatic processes.

A reason for this rather long exchange phase was that 6-APA production was one of the very first successful commercial applications of immobilized enzyme technology. This technology development, of course, needed some time to be optimized.

What was the driving force behind the technology change? Giacobbe (1) presented at an Enzyme Engineering Conference in 1977 a table describing 6-APA manufacturing costs which indicate some answers to the question.

Table 2. Comparison of the manufacturing costs in the production of 6-APA by a chemical or an enzymatic process.

$/kg 6-APA Components	Chemical Process	Enzymatic Process	Difference	Enzyme P / Chem P
Penicillin G	43.72	39.65	– 4.07	0.91
Chemicals	7.49	1.35	– 6.14	0.54
Immob. enzyme	-	2.67	+ 2.67	
Utilities	4.64	0.91	– 3.73	0.20
Labour	3.12	3.95	+ 0.83	1.27
Operating supplies	0.29	0.36	+ 0.07	1.24
Depreciation	6.16	3.85	– 2.31	0.63
Other overheads	2.99	2.54	– 0.45	0.85
Total	68.41	55.28	– 13.13	0.81

Basis: Plant size: 40 tons/yr, pen G : 14.35 $/bou

Ref: Glocobbe et al. (1978) (modified)

As seen from the total costs, the enzymatic process is approx. 20% cheaper than the costs in the chemical process. The enzymatic process is to be preferred as:

- Utilization of penicillin G is better (i.e. higher yield).
- Chemical costs are reduced to half.
- Utilities are much less.
- Depreciation is less due to less investments.

The only more costly items in the enzymatic process are the labour costs and operating supplies.

The process developments during the eighties within the enzymatic process have been very little commented on in the literature. We think that most commercial producers of 6-APA have been busy optimizing the process economy by using less purified substrates and by introducing new immobilized enzymes with higher activities - thereby reducing process time and thus degradation. Perhaps temperature profiles have become more commonly used in the hydrolysis process.

Another significant change during the latest years is that more countries in the third world have started production of 6-APA either based on own penicillin V/G or on imported penicillin.

Discussion of developments within 6-APA to Semisynthetic Penicillins like Ampicillin

In 1969 Cole (2) from Beecham first described an enzymatic synthesis of Ampicillin from 6-APA and a D-phenylglycine derivative. Since then there has been a wide interest in developing this enzymatic route into an industrial process, in order to avoid the use of, among others, the chlorinated solvents, but so far no one has been successful.

(1) D-PGM + 6-APA $\longrightarrow$ Ampicillin + CH_3OH

(2) D-PGM + H_2O $\longrightarrow$ D-PG + CH_3OH

(3) Ampicillin + H_2O $\longrightarrow$ 6-APA + D-PG

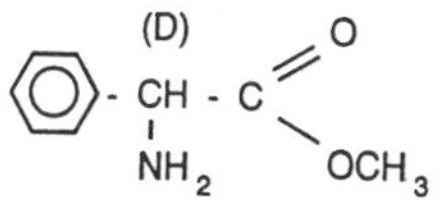

D-Phenylglycine methylester
(D-PGM)

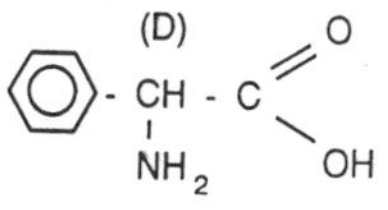

D-Phenylglycine
(D-PG)

Fig. 7. Enzymatic synthesis of Ampicillin from D-phenylglycine methylester and 6-APA.

The principle, however, is quite simple:

6-APA is dissolved in water together with an activated form of the side chain (e.g. D-phenylglycine methylester), the enzyme is added and the reaction is allowed to proceed at constant pH and at approx. 35 °C, after which the Ampicillin is isolated by precipitation.

The reaction is catalysed by the same enzyme which is used for the hydrolysis of penicillin G, e.g. a penicillin G acylase from *E. coli*.

The problem is that a large surplus of the activated side chain is needed to drive the reaction shown in fig. 7. Furthermore, the optimal concentrations quoted in the literature are much to low for any practical purpose and the yields described are not satisfactory (i.e. approx. 75%, based on the starting concentration of 6-APA).

The reason for these problems is that the enzymes, besides catalysing the Ampicillin formation (1), also catalyse the hydrolysis of the activated side chain (2) and hydrolyse the formed Ampicillin (3).

Furthermore, it is difficult to purify the Ampicillin from the reaction mixture as the Ampicillin, itself an amphoteric ion, has to be separated from two other amphoteric ions, D-PG and 6-APA, which have pK_a's very near each other.

One way to circumvent this problem could be to reduce the water activity by running the synthesis in an organic solvent and thus reducing the hydrolysis of both the methyl ester and the Ampicillin formed. As more and more information is being published now regarding the behaviour of enzymes and amphoteric ions in such environments in general, I think an industrial process for an enzymatic synthesis of the semisynthetic penicillins will stand good chances to become a reality during the next decade. Which brings me to the last point of my talk, the future of the beta-lactam industry.

GUESS ON FUTURE PROCESS ROUTES

A guess on process routes year 2000 to Ampicillin

The first question one may ask is: Are there still incentives to change the existing processes from penicillin G/V to Ampicillin?

We think, the answer is: Yes.

Especially as chemical processes involving the use chlorinated solvents are becoming more and more restricted. It would be nice if this process could be run in a pure aqueous process.

Another sub-optimal point is the side chain problem. When penicillin V and G is being produced by fermentation, the microorganism synthesizes L-alpha-aminoadipic acid and L-cysteine and L-valine. It combines the three aminoacids to isopenicillin N (fig. 8).

Phenylacetic acid or phenoxyacetic acid is taken up from the broth and exchanged with L-alpha-aminoadipic acid, which is partly reused, and the formed penicillin V or G is transported into the broth.

Penicillin V/G is isolated as described earlier and hydrolysed to 6-APA plus liberated side chain which is again partly reused. Finally, a third side chain (D-phenylglycine) is coupled to 6-APA and Ampicillin is formed.

Is the most efficient way to produce Ampicillin really to work with two 'wrong' side chains before the desired one is coupled to 6-APA?

Obviously it would be nice to answer: No; however, what can we do?

To change the existing processes into one more efficient, would necessarily involve modification of the biosynthetic pathway in the microorganism. With the success in mind that genetic engineers have obtained in other segments, i.e. construction of a pathway to human insulin in yeast, it is natural that there has been interest in changing the penicillin pathway too.

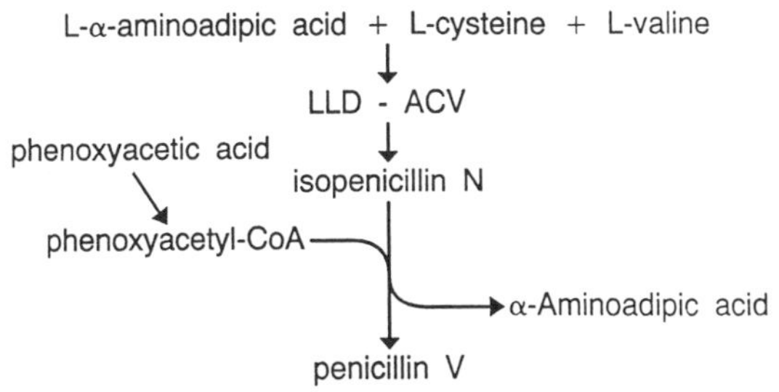

Fig. 8. Biosynthetic pathway of penicillin V in Penicillium chrysogenum

Several companies have, according to their patent activities, allocated rather large resources to solve this puzzle.

It is too early to say what the outcome will be of these attempts. We would, however, think that Ampicillin stands a good chance in being fermented directly in year 2000. If this happens and if the process economy can be optimized to beat the existing process economy this will result in a revolution within the beta-lactam industry.

The need for fermented penicillin V and G would diminish dramatically! - the need for 6-APA will, in the worst consequence, disappear completely! and the companies involved in the new technology - direct fermentation of Ampicillin - will take over the market shares from the companies utilizing existing technologies.

Today it is difficult to estimate whether, when, or how fast this revolution will occur, however, we think there is a good chance that the process engineers within the beta-lactam area will have years ahead of them with existing challenges and with nice fruits when the challenges are met.

REFERENCES

1. Giacobbe, F., Cecere, F., and Iasonna, A., 1978. in "Enzyme Engineering" G.B. Broun, G. Manecke, and L. Wingard, eds., Vol. 4, p245, Plenum, N.Y.

2. Cole, M., 1969, Penicillins and other Acylamino Compounds Synthesized by the Cell-Bound Penicillin Acylase of Escherichia coli. Biochem. J., 115:747.

MECHANISM OF ACTION OF β-LACTAMASES AND DD-PEPTIDASES

Jean-Marie Frère, Bernard Joris, Françoise Jacob, André Matagne, Didier Monnaie, Marc Jamin, Médar Hadonou, Catherine Bourguignon-Bellefroid, Louis Varetto, Jean-Marc Wilkin, Alain Dubus, Christian Damblon, Maggy Adam, Philippe Ledent, Fabien De Meester and Moreno Galleni

Center for Protein Engineering and Laboratoire d'Enzymologie, Université de Liège, Institut de Chimie, B6, Sart Tilman, B-4000 Liège, Belgium

INTRODUCTION : ß-LACTAM ANTIBIOTICS AND RESISTANCE PHENOMENA

The introduction of penicillins as antibacterial agents probably represents one of the major breakthroughs of chemotherapy during the present century. However, the bacterial world did not remain without reaction and resistance to strains started to appear almost as soon as penicillin utilisation became popular. This resulted in an endless race between the chemists and the microbiologists on one side and the bacteria on the other, the formers discovering and synthesizing new compounds while the latters found new tricks to escape their lethal action so that the ß-lactam family now comprises molecules of widely different structures of which the ß-lactam ring remains the only common feature (Figure 1A).

The resistance mechanisms which have presently been characterized rest on three unrelated but sometimes synergistic factors. 1) The appearance of modified target enzymes, the peptidoglycan synthetising DD-transpeptidases capable of fulfilling their physiological task, but less sensitive to the antibiotics[1]; 2) the secretion of ß-lactamases, soluble enzymes which efficiently hydrolyse the endocyclic amide bond of the ß-lactam (Figure 1B); 3) a decrease of the permeability of the outer membrane of Gram-negative bacteria, which presents an increased barrier to the diffusion of ß-lactams towards their targets[2]. It seems that the third factor only becomes important when it amplifies the effects of the second[3] since diffusion generally occurs much faster than the synthesis of new targets and the following analysis will center on the two types of enzymes involved in the story : DD-peptidases and ß-lactamases.

1. The Target Enzymes : the DD-Peptidases

DD-peptidases catalyse the last step of peptidoglycan biosynthesis[4], the closure of the peptide bridges which confer mechanical strength to that cell-wall polymer. The reaction can be schematised as

$$\text{R-D-Ala-D-Ala} + \text{R'-NH}_2 \longrightarrow \text{R-D-Ala-NH-R'} + \text{D-Ala}$$

where R is linked to one polysaccharide strand and R' to another. If that reaction is impaired, the peptidoglycan molecule cannot fulfil its

Bioorganic Chemistry in Healthcare and Technology, Edited by U.K. Pandit and F.C. Alderweireldt, Plenum Press, New York, 1991

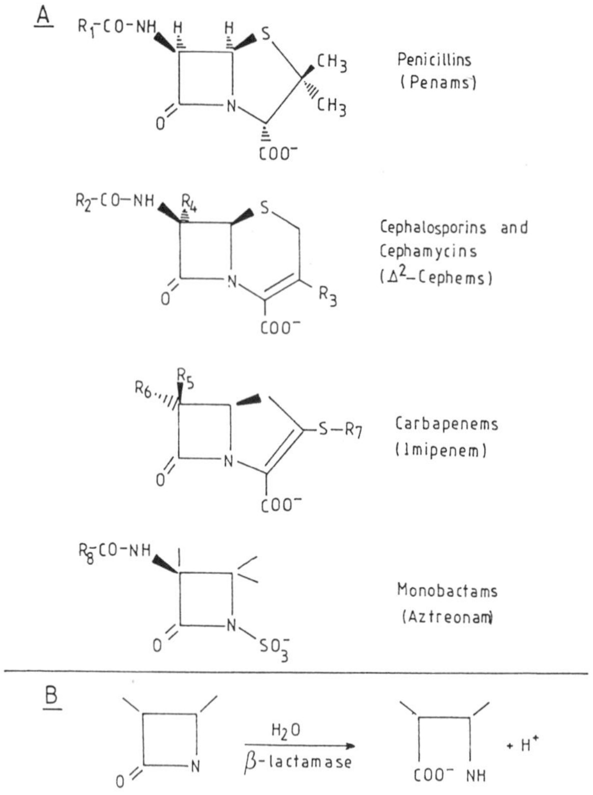

Fig. 1. Some members of the ß-lactam family (A) and the action of ß-lactamases (B).

functional role and cell lysis occurs. Most DD-peptidases are membrane-bound and that does not facilitate the study of their structures and properties. In many cases, their enzymatic activity has only been indirectly demonstrated by showing that their inactivation resulted in the formation of non-functional peptidoglycan. In the most favourable cases, an enzymatic activity could be demonstrated in vitro with the help of the natural peptidoglycan precursors or of small synthetic peptides. Some bacteria of the Actinomycetales family even excrete soluble DD-peptidases[5] which have served as very useful models for penicillin-sensitive enzymes. One of them, the Streptomyces R61 DD-peptidase is the only penicillin target whose 3-D structure is presently being solved at high resolution[6]. Although the physiological function of those soluble enzymes remains mysterious, they can closely mimick in vitro the transpeptidase activity of the physiological transpeptidases. For instance, they catalyse the formation of peptide polymers, very similar to the peptide cross-bridges in peptidoglycan[7]. The detailed study of those enzymes has allowed the collection of a large wealth of data and the general principles which could be deduced were shown to qualitatively apply to the physiological, membrane-bound enzymes. All those enzymes are inactivated by ß-lactam antibiotics following a three-step model[8]

$$E + C \underset{}{\overset{K}{\rightleftharpoons}} EC \overset{k2}{\longrightarrow} EC^* \overset{k3}{\longrightarrow} E + P(s) \qquad \text{(model 1)}$$

where E is the Enzyme, C the ß-lactam, EC a non-covalent complex, EC* a covalent adduct and P(s) an irreversibly transformed antibiotic molecule which can, in some cases, be degraded into several pieces[9]. K is the dissociation constant of EC, k2 and k3 are first-order rate constants.

Both EC and EC* are unable to carry out the transpeptidation reaction and the efficiency of the antibiotics does not seem to rely on a small value of K, but on a large k2 and a small k3 values which results in the immobilisation of a very large proportion of enzyme as the EC* adduct at the steady-state. The side-chain of a serine residue was found to be involved in that adduct[10] later characterized as an acyl-enzyme resulting from the nucleophilic attack of the serine hydroxyl on the carbon of the ß-lactam carbonyl

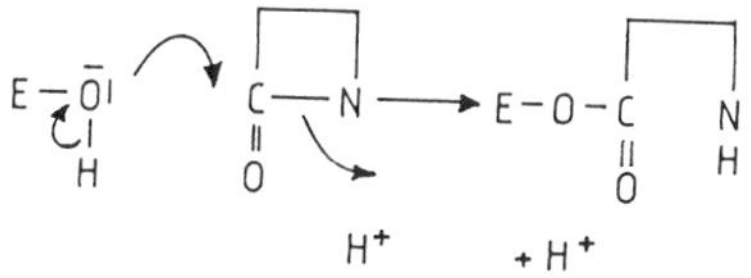

The same serine residue was also identified as the "active serine" of the enzyme[11] involved in the catalysis of the transpeptidation reaction

$$\text{E-OH} + \text{R-D-Ala-D-Ala} \longrightarrow \text{D-Ala} + \text{E-O-D-Ala-R}$$

$$\text{E-O-D-Ala-R} \xrightarrow{R'NH_2} \text{E-OH} + \text{R-D-Ala-NH-R'}$$

The validity of model 1 and of the acylation of an essential serine residue by ß-lactams could also be extented to all DD-transpeptidases and membrane-bound penicillin-binding proteins.

2. ß-Lactamases

The fact that both types of enzymes recognized the same ß-lactam-containing molecules led to the suggestion that ß-lactamases were evolutionary related to DD-peptidases[12]. However, the ß-lactamases failed to interact with the peptide substrates of the latter enzymes and the first attempts to align sequences of members of the two families did not yield very encouraging results. But in the late seventies, a serine residue was shown to be acylated by specific inactivators in various ß-lactamases[13]. The subsequent determination of the primary structure of a large number of ß-lactamases, either directly or *via* the d-nucleotide sequence of the gene, resulted in the recognition[14] of three distinct classes of active-serine enzymes, labelled A, C and D, class B consisting of a small number of Zn^{++}-containing proteins which will not be further discussed here.

The inactivators used for labelling the active-site serine residues all contained a ß-lactam ring and behaved as very poor substrates (model 1 with very small k3) or as mechanism-based inactivators[15]. The interactions with the latters can be depicted by model 2 where EC** is a novel acyl-enzyme where the inactivator molecule has rearranged

$$E + C \overset{K}{\rightleftharpoons} EC \xrightarrow{k2} EC^* \xrightarrow{k3} E + P$$

$$EC^* \xrightarrow{k4} EC^{**} \xrightarrow{k5} E + P' \qquad \text{(model 2)}$$

A large number of distinct interactions have been described, characterized by different relative values of k3, k4 and k5. In the simplest cases, k3 and k5 are negligible and all the enzyme irreversibly accumulates as the

EC** adduct but, quite often, k3 is larger than k4 so that a substantial number of turn-overs occur before complete inactivation. A possible recovery of the activity then depends upon the k5 value.

Extrapolation of those data to good substrates indicated a mechanism involving an acyl-enzyme intermediate and the validity of model 1 was recognized for describing the catalytic pathway of ß-lactamases. Further demonstration of that validity was also performed by showing an increase of the reaction rate in the presence of alternative nucleophiles[16] and, finally, by isolating the acyl-enzyme formed with good substrates[17].

DD-peptidases and ß-lactamases thus react with ß-lactam-containing compounds according to the kinetic pathway represented by model 1, the main difference being, strictly speaking, only quantitative : with DD-peptidases, very low values of k3 (10^{-3} to $10^{-6}s^{-1}$) result in an extremely slow hydrolysis of the acyl-enzyme while deacylation of ß-lactamases is generally very fast (10 to 4000 s^{-1}). Consequently, that quantitative difference results in qualitatively distinct behaviours, since DD-peptidases are inactivated by ß-lactams while ß-lactamases efficiently hydrolyse the same molecules. Strikingly, however, and as noted above, ß-lactamases mainly those belonging to class C, can form very stable acyl enzymes (k3 ≈ $10^{-3}s^{-1}$) with some substrates, making them even more DD-peptidase-like[18].

3. Common Substrates : Esters and Thioesters

Another element in the comparison was contributed by the finding that simple esters[19] and thioesters[20] could be efficiently hydrolysed by both ß-lactamases and DD-peptidases. Various compounds of general structure

$$R1\text{-}CO\text{-}NH\text{-}\underset{R2}{CH}\text{-}CO\text{-}X\text{-}\underset{R3}{CH}\text{-}COO^-$$

were synthesized where X was O or S and R1 C_6H_5-, C_6H_5-CH_2 or

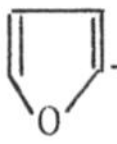-CH=CH-. When R2 and R3 were not H, the chiral center thus created had to exhibit the D configuration. Many of those depsipeptides were substrates of ß-lactamases of classes A, C and D and of the DD-peptidase of Streptomyces R61 and Actinomadura R39. Surprisingly, the two latter enzymes acted quite efficiently on non-chiral compounds where R2 = R3 = H. For instance, with the Streptomyces R61 enzyme, a kcat/Km value of 100,000 $M^{-1}s^{-1}$ was found[20] for

$$C_6H_5\text{-}CO\text{-}NH\text{-}CH_2\text{-}CO\text{-}S\text{-}CH_2\text{-}COO^-$$

making it a better substrate than the classical peptide Ac_2-L-Lys-D-Ala-D-Ala which had been considered as the best substrate for that enzyme (kcat/Km = 5000 $M^{-1}s^{-1}$). An interesting difference between the two molecules was that, at substrate saturation, acylation was rate-limiting with the peptide and deacylation with the depsipeptide, with which accumulation of the acyl-enzyme intermediate could be visualized.

With the depsipeptides, ß-lactamases were even shown to catalyse simple transpeptidation[19] and transesterification[21] reactions in the presence of suitable nucleophilic acceptors such as D-alanine or D-phenylalanine (for the esters and the thioesters) or glycolate or D-lactate (for the thioester; M. Adam, M. Jamin and J.M. Frère, unpublished results).

4. Three-Dimensional Structures

The major progress in the field was however contributed by the results of structure determinations by X-ray diffraction. When the structures of class A ß-lactamases were compared to that of the Streptomyces R61 DD-peptidase, very similar organisations of the secondary structure elements were observed[22,23], despite rather large differences in the Mr values (about 30,000 for the formers and 37,500 for the latter) and sequence similarities barely above the random noise level. More recently, a class C ß-lactamase (Mr = 39,000) was also found to be built according to the same general pattern[24]. All those enzymes were composed of an α/ß and an all-α domains, with the active serine residue situated in a pocket at the hinge between the two domains. Supported by the 3-D structure data, more meaningful sequence alignments became possible, which could be extended to proteins whose tertiary structures remained unknown[14], class D ß-lactamases and membrane-bound DD-peptidases or penicillin-binding proteins. This resulted in the identification of a small, but significant number of "conserved elements" around the active sites of the various enzymes.

1) The active serine residue is situated at the N-terminus of the first α-helix of the all-α domain. After two residues for which no conservation is observed between the various enzyme families, a lysine residue is always found, just one helix turn after the serine, so that the side-chains of both residues protrude into the active site.

2) A SDN sequence in class A ß-lactamases or a YAN sequence in class C ß-lactamases and YSN in the S. R61 DD-peptidase, located on a loop between helices α4 and α5 in the numbering of the helices of class A ß-lactamases. If the S and K residues of the α2 helix are superimposed, the OH group of the tyrosine residue of the YAN sequence in class C superimposes on the OH group of the serine residue of the SDN sequence in class A.

3) An acidic residue, glutamate in class A and class C ß-lactamases, aspartate in the R61 peptidase situated at the N-terminus of a very short helix.

4) A KTG or KSG sequence in all enzymes, but the R61 peptidase, where HTG is found, situated on the outermost strand of the ß-pleated sheet[25] and limiting the active-site cavity on the side of the α/ß domain.

Those conserved elements are schematically presented on Fig. 2.

5. Mechanisms

If the catalytic role of the active serine in the various enzymes is undisputed, that of the residues composing the other conserved elements presently remains less clear. In particular, the identification of the residue(s) which enhance the nucleophilic character of the serine hydroxyl is still an open question. In class A ß-lactamases, site-directed mutagenesis experiments[26] have demonstrated the importance of E166. Many results indicate that residue as the possible proton abstractor responsible for serine activation

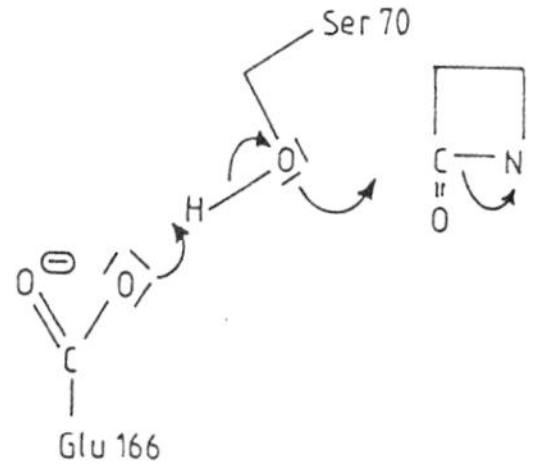

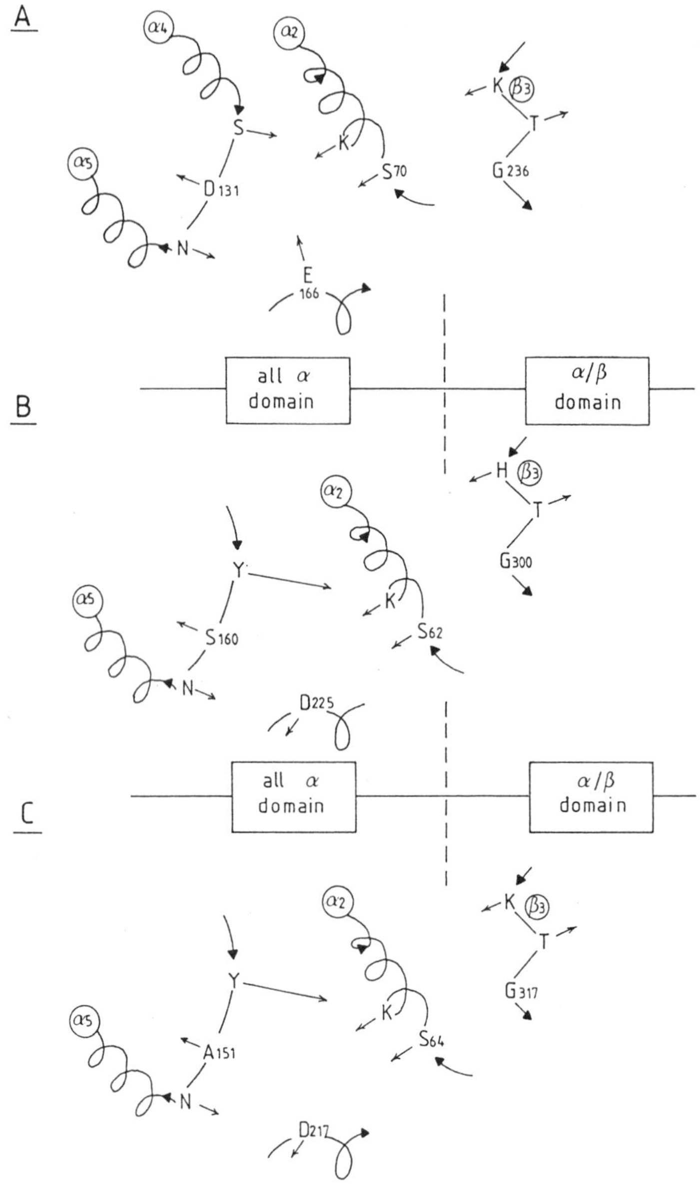

Fig. 2. Cartoon representation of the active sites of a class A ß-lactamase (A) of the S. R61 DD-peptidase (B) and of a class C ß-lactamase (C).

The heavy arrows indicate the direction of the polypeptide backbone and the light arrows that of the various side-chains. The grossly simplified representation certainly ignores other important residues, such as N170 in class A ß-lactamases, which also points into the active site and whose replacement by L results in a drastically impaired enzyme (J. Brannigan & J.M. Frère, unpublished).

The corresponding acidic residues in class C ß-lactamases and the R61 DD-peptidase do not however point into the active site[24] (Figure 2) and their replacement by a non-acidic residue does not strongly decrease the enzyme activity[27]. In the case of class C enzymes, proton abstraction might be performed by the Tyr residue of the YAN loop[24].

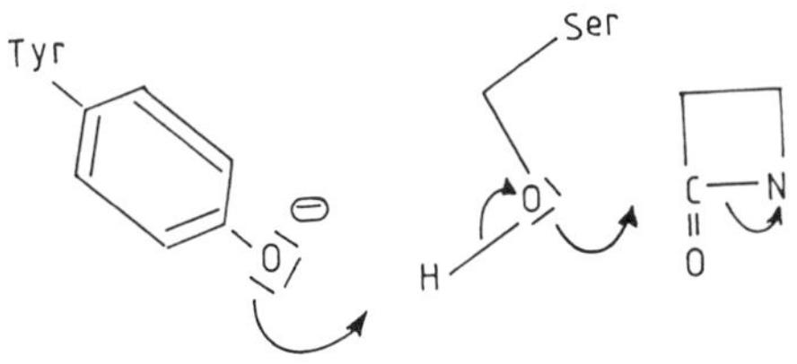

This hypothesis has three important implications.

1) The Tyr residue must be deprotonated at neutral pH, which implies a drastically decreased pK for that residue. This could be the results of interactions with the positively-charged amino groups of the two lysine residues K67 and K315.

2) Despite the underlined similarities of the 3-D structures, classes A and C ß-lactamases have evolved quite different mechanisms, where one of the major actors in catalysis would come from rather different parts of the proteins. Indeed, residue S130 in class A ß-lactamase cannot possibly play the role which is proposed for Y in class C. First, lowering of the pK of a serine hydroxyl to a value of 4-5, as would be required by the observed pH-dependence of the kcat value, appears quite unrealistic. Moreover, that residue has been replaced by a glycine residue in one class A ß-lactamase[28] and the mutated protein retained up to 40 % of activity against some substrates.

3) Assuming that the mechanism of the S. R61 DD-peptidase is not widely different from that of the class C ß-lactamases, one must also conclude that the same divergence exists in the DD-peptidase family, since several enzymes exhibit SDN or SXN sequence in a position corresponding to the YSN of the R61 DD-peptidase.

CONCLUSIONS

The experimental evidence presently available allows us to group penicillin-sensitive DD-peptidases, penicillin-binding proteins (the physiological function of some of those is still mysterious) and the ß-lactamases of class A, C and D in one super-family of penicillin-recognizing proteins. The similarities in the known secondary and tertiary structures point to a common evolutionary origin and if the similarities in the primary structures are sometimes difficult to detect, conserved elements can be located with the help of the 3-D architectures of some proteins and extrapolated to most unknown structures.

In this respect, the striking difference in the identity and position of the proton abstractor, responsible for the activation of the essential serine residue remains quite puzzling and one can hope that the refinement of the structures determined by X-ray diffraction and the new information obtained by site-directed mutagenesis techniques will shed more light on that surprising phenomenon.

Many other problems remain to be solved, the most important being probably to find a plausible explanation for the enormous differences in the k3 values between ß-lactamases and DD-peptidases. Similarly, in the class C ß-lactamases, the k3 values exhibit spectacular variations with the structure of the antibiotic, from more than 1000 s^{-1} for cefazolin to values around $10^{-3}s^{-1}$ for cloxacillin or cefotaxime, an observation which remains completely unexplained, and which is made even more surprising by the fact that the kcat/Km values for those extremely poor substrates compare much more favourably with those of the best ones[18].

In fact, the whole question of the specificity of ß-lactamases remains completely open, as illustrated by the recent observation that the mutation of the N132 residue into a S in a class A ß-lactamase virtually destroyed its cephalosporinase activity while the modified protein retained a substantial hydrolytic capacity against some penicillins[29]. Other research opportunities have also been recently underlined by the discovery of specific inactivators of DD-peptidases which do not contain a ß-lactam ring, such as lactivicin[30] and some γ-lactams[31]. It might be appropriate to conclude the present contribution by citing our old friend Nathan Citri : "Many bacteria owe their survival to ß-lactamases. So do some of us".

ACKNOWLEDGEMENTS

This work was supported in part by the Fonds de la Recherche Scientifique Médicale (contract n°3.4537.88), an Action concertée with the Belgian Government (convention 86/91-90), the Fonds de Recherche de la Faculté de Médecine ULg and a Convention tripartite between the Région wallonne, SmithKline Beecham, U.K., and the University of Liège.

REFERENCES

1. B.G. Spratt, Hybrid penicillin-binding proteins in penicillin-resistant strains of Neisseria gonorrhoeae, Nature 332:173-176 (1988).
2. H. Nikaido, Bacterial resistance to antibiotics as a function of outer membrane permeability, J. Antimicrob. Ag. Chemother. ??:17-22 (1988).
3. J.M. Frère, B. Joris, M. Crine, and H.H. Martin, Quantitative relationship between sensitivity to ß-lactam antibiotics and ß-lactamase production in Gram-negative bacteria. II. Non-steady-state treatment and progress curves, Biochem. Pharmacol. 38:1427-1433 (1989).
4. J.M. Frère and B. Joris, Penicillin-sensitive enzymes in peptidoglycan biosynthesis, CRC Crit. Rev. Microbiol. 11:299-396 (1985).
5. J.M. Frère, J.M. Ghuysen, J. Degelaen, A. Loffet, and H.R. Perkins, Fragmentation of benzylpenicillin after interaction with the exocellular DD-carboxypeptidases-transpeptidases of Streptomyces R61 and R39, Nature (London) 258:168-170 (1973).
6. J.A. Kelly, J.R. Knox, P.C. Moews, J.M. Frère, and J.M. Ghuysen, Using X-ray diffraction results and computer graphics to design ß-lactams, Kansenshogaku Zasshi 62:182-191 (1988).
7. A.R. Zeiger, J.M. Frère, and J.M. Ghuysen, A donor-acceptor substrate of the exocellular DD-carboxypeptidase-transpeptidase from Streptomyces R61, FEBS Lett. 52:221-225 (1975).
8. J.M. Frère, J.M. Ghuysen, and M. Iwatsubo, Kinetics of interaction between the exocellular DD-carboxypeptidase-transpeptidase from Streptomyces R61 and ß-lactam antibiotics. A choice of models, Eur. J. Biochem. 57:343-351 (1975).
9. J.M. Frère, M. Leyh-Bouille, J.M. Ghuysen, M. Nieto, and H.R. Perkins, Exocellular DD-carboxypeptidases-transpeptidases from Streptomyces, Methods Enzymol. XLV:610-636 (1976).

10. J.M. Frère, C. Duez, and J.M. Ghuysen, Occurrence of a serine residue in the penicillin-binding site of the exocellular DD-carboxypeptidase-transpeptidase from S. R61, FEBS Lett. 70:257-260 (1976).
11. D.J. Tipper and J.L. Strominger, Mechanism of action on penicillins : a proposal based on their structural similarity to acyl-D-alanyl-D-alanine, Proc. Natl. Acad. Sci. USA 54:1133-1140 (1965).
12. R.R. Yocum, H. Amanuma, T. O'Brien, D.J. Waxman, and J.L. Strominger, Penicillin is an active-site inhibitor for ??? genera of bacteria, J. Bact. 149:1150-1153 (1982).
13. S.J. Cartwright and S.G. Waley, ß-Lactamase inhibitors, Medicin. Res. Reviews 3:341-382 (1983).
14. B. Joris, J.M. Ghuysen, G. Dive, A. Renard, O. Dideberg, P. Charlier, J.M. Frère, J.A. Kelly, J.C. Boyington, P.C. Moews, and J.R. Knox, The active-site serine penicillin-recognizing enzymes as members of the Streptomyces R61 DD-peptidase family, Biochem. J. 250:313-324 (1988).
15. R.F. Pratt, ß-Lactamase inhibitors, in: "Design of Enzyme Inhibitors as Drugs," M. Sandler and H.J. Smith, eds, Oxford University Press (1989).
16. V. Knott-Hunziker, S. Petursson, S.G. Waley, B. Jaurin, and T. Grundström, The acyl-enzyme mechanism of ß-lactamase action, Biochem. J. 207:315-322 (1982).
17. H. Christensen, M.T. Martin, and S.G. Waley, ß-Lactamases as fully efficient enzymes. Determination of all the rate constants in the acyl-enzyme mechanism, Biochem. J. 266:853-861 (1990).
18. M. Galleni, G. Amicosante, and J.M. Frère, A survey of the kinetic parameters of class C ß-lactamases. II. Cephalosporins and other ß-lactam compounds, Biochem. J. 255:123-129 (1988).
19. R.F. Pratt and C.P. Govardhan, ß-Lactamase-catalyzed hydrolysis of acyclic depsipeptides and acyl transfer to specific amino acid acceptors, Proc. Natl. Acad. Sci. USA 81:1302-1306 (1984).
20. M. Adam, C. Damblon, B. Plaintin, L. Christiaens, and J.M. Frère, Chromogenic depsipeptide substrates for ß-lactamases and DD-peptidases, Biochem. J. 270:525-529 (1990).
21. C.P. Govardhan and R.F. Pratt, Kinetics and mechanism of the serine ß-lactamase catalyzed hydrolysis of depsipeptides, Biochemistry 26:3385-3395 (1987).
22. J.A. Kelly, O. Dideberg, P. Charlier, J.P. Wéry, M. Libert, P.C. Moews, J.R. Knox, C. Duez, C. Fraipont, B. Joris, J. Dusart, J.M. Frère, and J.M. Ghuysen, On the origin of bacterial resistance to penicillin : comparison of a ß-lactamase and a penicillin target, Science 231:1429-1431 (1986).
23. B. Samraoui, B. Sutton, R. Todd, P. Artimyuk, S.G. Waley, and D. Phillips, Tertiary structure similarity between a class A ß-lactamase and a penicillin-sensitive D-alanyl carboxypeptidase-transpeptidase, Nature 320:378-380 (1986).
24. C. Oefner, A. Darcy, J.J. Daly, K. Gubernator, R.L. Charnas, I. Heinze, C. Hubschwerlen, and F.K. Winkler, Refined crystal structure of ß-lactamase from Citrobacter freundii indicates a mechanism for ß-lactam hydrolysis, Nature 343:284-288 (1990).
25. J.A. Kelly, J.C. Boyington, P.C. Moews, J.R. Knox, O. Dideberg, P. Charlier, M. Libert, J.P. Wéry, C. Duez, B. Joris, J. Dusart, J.M. Frère, and J.M. Ghuysen, Crystallographic comparisons of penicillin-binding enzymes and studies of antibiotic binding, in: "Frontiers of Antibiotic Research," H. Umezawa, ed., Academic Press, New York (1987).
26. R. Gibson, H. Christensen, and S.G. Waley, Site-directed mutagenesis of ß-lactamase I single and double mutants of Glu166 and Lys73, Biochem. J., in press.

27. K. Tsukamoto, R. Kikura, R. Ohno, and T. Sawai, Substitution of aspartic acid-217 of Citrobacter freundii cephalosporinase and porperties of the mutant enzymes, FEBS Lett. 264:211-214 (1990).
28. F. Jacob, B. Joris, S. Lepage, J. Dusart, and J.M. Frère, Role of the conserved amino acids of the SDN loop in a class A ß-lactamase, Biochem. J., in press.
29. F. Jacob, B. Joris, O. Dideberg, J. Dusart, J.M. Ghuysen, and J.M. Frère, Engineering of a novel ß-lactamase by a single point mutation, Protein Engineering 4, in press.
30. Y. Nozaki, N. Katayama, H. Ono, S. Tsubotani, S. Harada, H. Okazaki, and Y. Nakao, Binding of a non-ß-lactam antibiotic to penicillin-binding proteins, Nature 325:179-??? (1987).
31. J.E. Baldwyn, C. Lowe, and C.J. Schofield, A γ-lactam analogue of penems possessing antibacterial activity, Tetrahedron Lett. 27:3461-3464 (1986).

NMR STUDIES OF THE ENZYME MECHANISMS OF VITAMIN B_{12} BIOSYNTHESIS

A. Ian Scott

Center for Biological NMR, Department of Chemistry
Texas A&M University, College Station, Texas 77843-3255 USA

Introduction

For the past 20 years our laboratory has been engaged in the elucidation of the vitamin B_{12} biosynthetic pathway.[1]. By 1977 it had been established that 5-aminolevulinic acid (ALA), porphobilinogen (PBG) uro'gen III and three intermediates, Factors I-III, now known to be transformed as the precorrins 1-3, were sequentially formed on the way to the corrin nucleus as summarized in Scheme 1. The ensuing decade has witnessed an exponential increase in the rate of the acquisition of pure enzymes using recombinant DNA techniques and these methods together with the exploration of new NMR pulse sequences have not only cast light on the mechanisms of the B_{12} synthetic enzymes but have given us considerable optimism with regard to the discovery of the remaining intermediates and of the enzymes which produce them.

The Enzymes of Tetrapyrrole Synthesis: PBG Deaminase and Uro'gen III Synthase

Previous work with deaminase [2-11] had established that a covalent bond is formed between substrate and enzyme, thus allowing isolation of covalent complexes containing up to 3 PBG units (ES_1-ES_3). Application of ^{3}H-NMR spectroscopy to the mono PBG adduct (ES-1) revealed a rather broad ^{3}H chemical shift indicative of covalent bond formation with a cysteine thiol group at the active site.[2] However, with adequate supplies of pure enzyme available in 1987 from the cloning of *hemC*[6] we were able to show that a novel cofactor, derived from PBG during the biosynthesis of deaminase, is covalently attached to one of the four cysteine residues of the enzyme in the form of a dipyrromethane which, in turn, becomes the site of attachment of the succeeding four moles of substrate during the catalytic cycle. Thus, at pH < 4, deaminase (**5**) rapidly develops a chromophore (λmax 485 nm)

Bioorganic Chemistry in Healthcare and Technology, Edited by U.K. Pandit and F.C. Alderweireldt, Plenum Press, New York, 1991

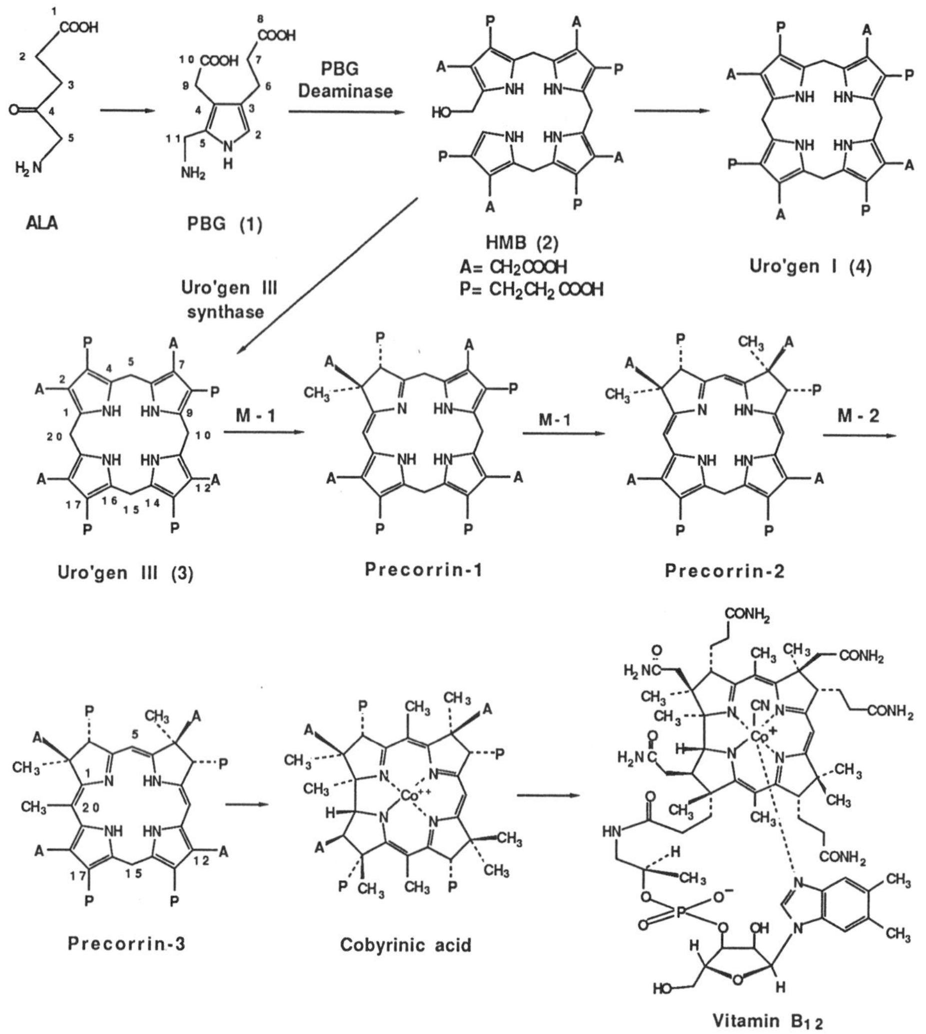

Scheme 1

diagnostic of a pyrromethene (**6**), whilst reaction with Ehrlich's reagent generates a chromophore typical of a dipyrromethane (λmax 560 nm) changing to 490 nm after 5-10 min. The latter chromophoric interchange was identical with that of the Ehrlich reaction of the synthetic model pyrromethane (**7**) and can be ascribed to the isomerization shown (Scheme 2) for the model system (**7**). Incubation of *E. coli* strain SASX41B (transformed with plasmid pBG 101: *hemA*$^-$ requiring ALA for growth) with 5-^{13}C-ALA afforded highly enriched enzyme for NMR studies. At pH8, the enriched carbons of the dipyrromethane (py-CH_2-py) are clearly recognized at 24.0 ppm (py-CH_2py), 26.7 ppm (py-CH_2X), 118.3 ppm (α-free pyrrole) and 129.7 ppm (α-substituted pyrrole). The signals are sharpened at pH12 and the CH_2-resonance is shifted to δ29.7 in the unfolded enzyme. Comparison with synthetic models reveals that a shift of 26.7 ppm is in the range expected for an α-thiomethyl

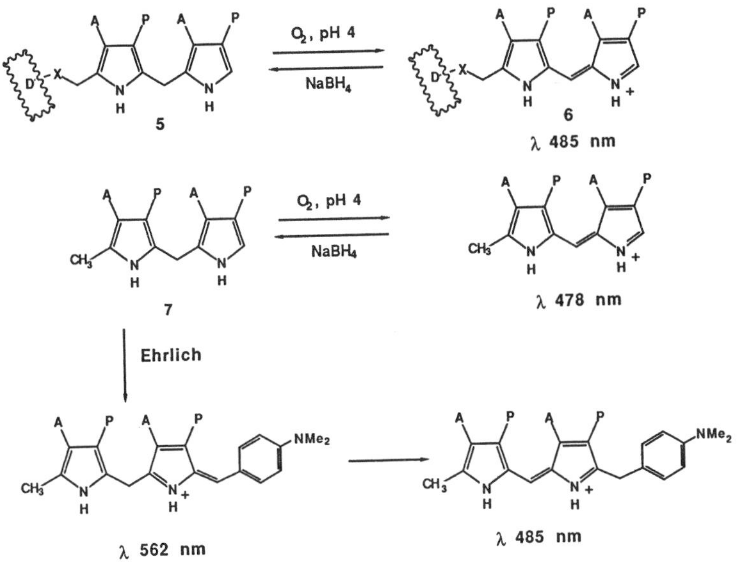

Scheme 2

pyrrole (py-CH_2 SR). Confirmation of the dipyrromethane (rather than oligo pyrromethane) came from the ^{13}C INADEQUATE spectrum taken at pH12 which reveals the expected coupling only between py-CH_2-py (δ 24.7) and the adjacent substituted pyrrole carbon (δ 128.5 ppm). When the enriched deaminase was studied by INVERSE INEPT spectroscopy, each of the 5 protons attached to ^{13}C-nuclei were observed. A specimen of deaminase was covalently inhibited with the suicide inhibitor [2,11-$^{13}C_2$]-2-bromo PBG (**8**) to give a CMR spectrum (pH12) consistent only with structure **9**. The site of covalent attachment of substrate (and inhibitor) is therefore the free α-pyrrole carbon at the terminus of the dipyrromethane in the native enzyme, leading to the structural and mechanistic proposal for deaminase shown in Scheme 3.

It was also possible to show that 2 moles of PBG are incorporated autocatalytically into the apoenzyme (obtained by cloning into an overexpression vector in *E. coli* which does not make PBG) before folding[13] and that the first (kinetic) encounter of PBG deaminase with substrate involves attachment of PBG (with loss of NH_3) to the α-free pyrrole position of the dipyrromethane to form the ES_1 complex[14,15] (Scheme 3). The process is repeated until the "tetra PBG" (ES_4) adduct (**10**) is formed. At this juncture site-specific cleavage of the hexapyrrole chain (at →) releases the azafulvene bilane (**11**) which either becomes the substrate of uro'gen III synthase, or in the absence of the latter enzyme, is stereospecifically hydrated[3] to HMB (**2**) at pH12, or is cyclized chemically to uro'gen I (**4**) at pH ≤ 8. It is quite remarkable that within the short period from mid-1987[14,15,16a,17a] to January 1988 three groups cloned and expressed deaminase and by spectroscopic methods defined the dipyrromethane active site cofactor in a series of independent [14,16a,17a] and collaborative[15] publications. The ^{13}C-labeling defines the number of PBG units (two) attached in a head-to-tail motif to the native enzyme at pH8 and reveals the identity of the nucleophilic group (Cys-SH) which anchors the dipyrromethane (and hence the growing oligopyrrolic chain) to the

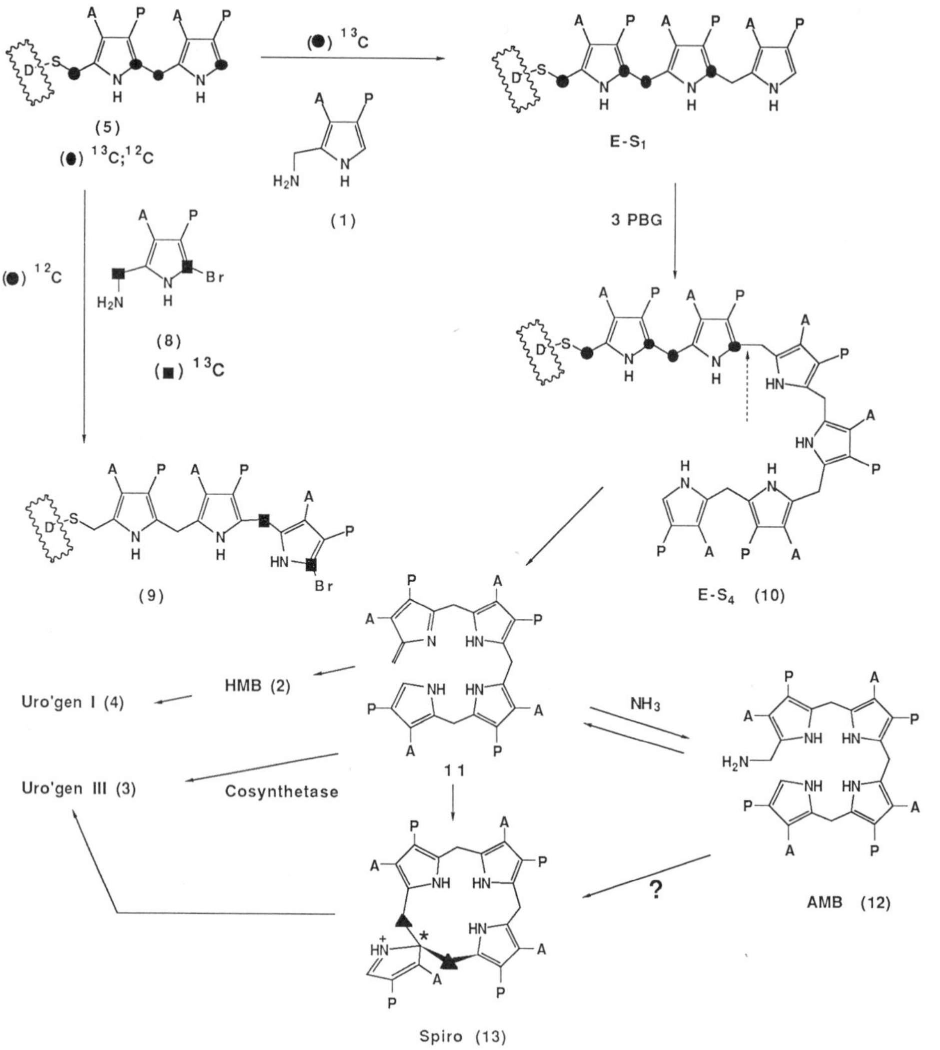

Scheme 3

enzyme. Site specific mutagenesis[7] and chemical cleavage[16,17] were employed to determine that Cys-242 is the point of attachment of the cofactor. Thus, replacement[7] of cysteine with serine at residues 99 and 242 (respectively) gave fully active and inactive specimens of the enzyme respectively.

The use of the α-carbon of a dipyrromethane unit as the nucleophilic group responsible for oligomerization of 4 moles of PBG (with loss of NH_3 at each successive encounter with an α-free pyrrole) is not only remarkable for the exquisite specificity and control involved, but is, as far as we know, a process unique in the annals of enzymology in that a substrate is used not only once, but twice in the genesis of the active site cofactor! Even more remarkable is the fact that the apoenzyme is automatically transformed to the active holoenzyme by addition of two substrate PBG units without the intervention of a second enzyme. Crystallization and X-ray diffraction studies of both the native enzyme and several of its genetically altered versions are now in progress.

Uro'gen III Synthase - The Ring D Switch

We now turn briefly to the rearranging enzyme Uro'gen III synthase. The early ideas of Bogorad[18] involving the aminomethyl bilane (AMB;**12**) as substrate were helpful in finally tracking down the elusive species which, after synthesis by PBG deaminase, becomes the substrate for Uro'gen III synthase. This was shown to be HMB (**2**) and at this stage (1978) the lack of activity of AMB (**12**) as substrate relegated the latter bilane to an interesting artefact produced quantitatively by addition of ammonia to deaminase incubations (Scheme 3). However with the acquisition of substantial quantities of pure Uro'gen III synthase obtained by cloning the genes *hemC* and *D* together[19] and overexpression in *E. coli*, the substrate specificity of the synthase (often called cosynthetase) has been reinvestigated. Ever since its conception[20] by Mathewson and Corwin in 1961, the spiro compound (**13**) (Scheme 3) has been a favorite construct with organic chemists, since both its genesis through α-pyrrolic reactivity and its fragmentation - recombination rationalize the intramolecular formation of Uro'gen III from the linear bilane, preuro'gen (HMB;**2**). A careful search for the spiro-compound (**13**) was conducted at subzero temperatures in cryosolvent (-24° C; ethylene glycol/buffer) using various ^{13}C-isotopomers of HMB as substrate.[19] Although the synthase reaction could be slowed down to 20 hours (rather than 20 sec.) no signals corresponding to the quaternary carbon (*δ ~80) ppm or to the α-pyrrolic methylene groups (▲; δ 35-40 ppm) could be observed. During these studies, however, it was found that AMB (**12**) served as a slow but productive substrate for Uro'gen III synthase at high concentrations (1 mmole in enzyme and substrate concentrations) and that if care is not taken to remove the ammonia liberated from PBG by the action of deaminase, not only is the enzymatic formation of AMB observed[21,22] but in presence of Uro'gen III synthase, the product is again Uro'gen III. This raises the interesting question of whether, under certain physiological conditions, the true substrate for Uro'gen III synthase is in fact AMB formed at the locus of deaminase by the ammonia released from PBG. Regardless of which version of the bilane system (HMB, AMB) is used in the experiment, the lack of observation of any intervening species free of enzyme strongly suggests that the intermediate is enzyme-bound and therefore difficult to detect. Present research is directed towards the solution of this problem by more sensitive, low temperature spectroscopy.

We recently proposed[19] a novel alternative to the "spiro intermediate" (**13**) hypothesis for the mechanism of ring D inversion mediated by uro'gen III synthase which uses a self assembly concept involving lactone formation as portrayed in Scheme 4. We suggested that generation of the azafulvene (**12**) as before is followed not by carbon-carbon bond formation (→ **13**) but by closure of the macrocyle *via* the lactone (Scheme 4). The regio-specificity for this step is reflected by the failure[5] of synthetic bilanes lacking the acetic acid side chain at position 17 in ring D to undergo enzyme-catalyzed rearrangement to the type-III system together with the observation that a "switched" bilane carrying a propionate at position 17 (and an acetate at 18) does indeed give a certain amount of (rearranged) uro'gen I (but also Uro'gen III) enzymatically,[5,22] indicating that lactone formation could also be

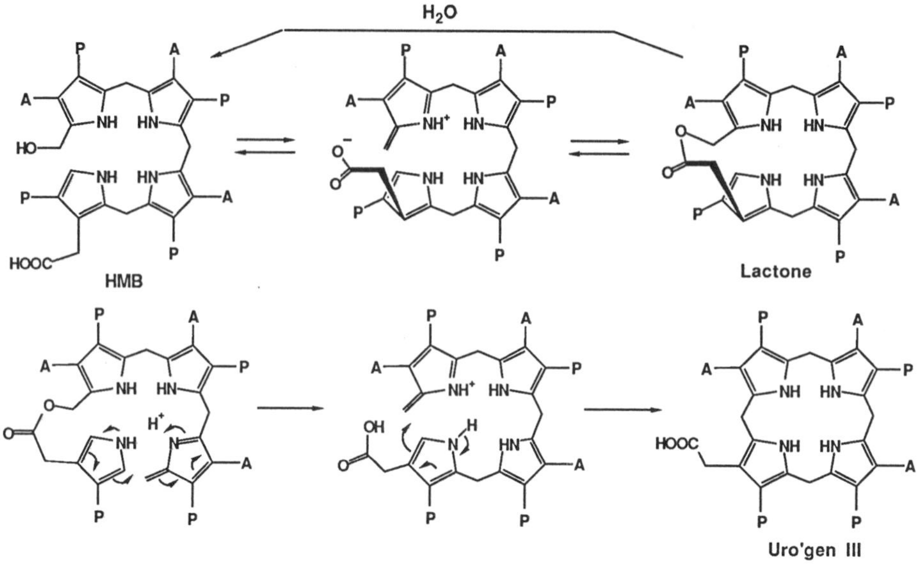

Scheme 4

achieved (but less efficiently) by a propionate group at the 17-position. The subsequent chemistry is quite similar to the fragmentation-recombination postulated for the spiro system except that a "twisted lactone" becomes the pivotal intermediate species and the sequence proceeds as shown in Scheme 4. Although this novel hypothesis is rather difficult to test experimentally, specific chemical traps for the lactones combined with the use of ^{18}O labeling are being used to confirm or refute the possible role of such macrocyclic lactones as the basis for the mechanism of Uro'gen III synthase.

Temporal Resolution of the Methylation Sequence

In spite of intensive search, no new intermediates containing four or more methyl groups (up to a possible total of eight) have been isolated, but the biochemical conversion of Factor III to cobyrinic acid must involve the following steps[23,24] (see Scheme 5), not necessarily in the order indicated: (1) The successive addition of five methyls derived from S-adenosyl methionine (SAM) to *reduced* Factor III (precorrin-**3**). (2) The contraction of the permethylated macrocycle to corrin. (3) The extrusion of C-20 and its attached methyl group leading to the isolation of acetic acid.[25-28] (4) Decarboxylation of the acetic acid side chain in ring C (C-12). (5) Insertion of Co^{+++} after adjustment of oxidation level from Co^{++}. In order to justify the continuation of the search for such intermediates whose inherent lability to oxygen is predictable, ^{13}C pulse labeling methods were applied to the cell-free system which converts Uro'gen III (**3**) to precorrin-**2** and thence to cobyrinic acid, a technique used previously in biochemistry to detect the flux of radio-labels through the intermediates of biosynthetic pathways.

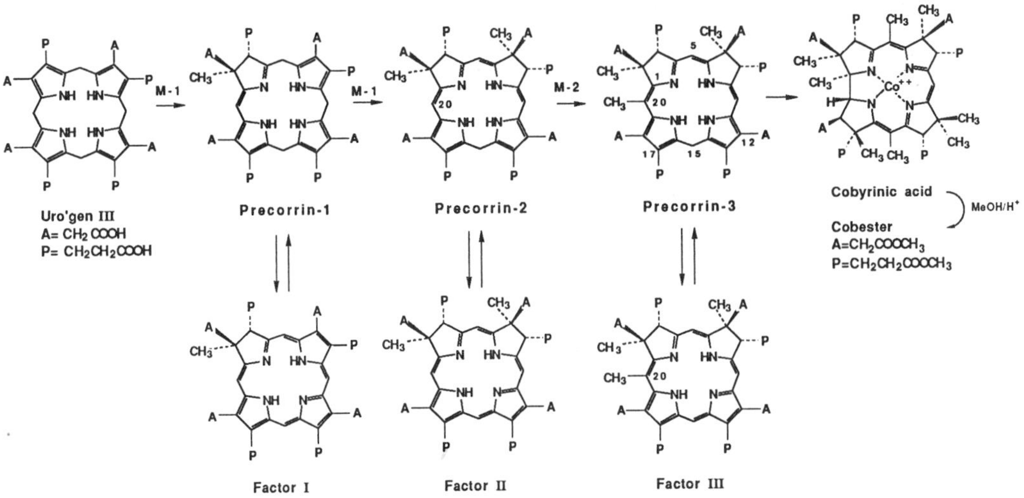

Scheme 5

The Pulse Experiments

The methylation sequence can be resolved temporally provided that enzyme-free intermediates accumulate in sufficient pool sizes to affect the intensities of the resultant methyl signals in the CMR spectrum of the target molecule cobyrinic acid when the cell-free system is challenged with a pulse of $^{12}CH_3$-SAM followed by a second pulse of $^{13}CH_3$-SAM (or vice versa) at carefully chosen intervals in the total incubation time (6-11 hours). By this approach it is possible to "read" the biochemical history of the methylation sequence as reflected in the dilution (or enhancement in the reverse experiment) of $^{13}CH_3$ label in the methionine-derived methyl groups of cobyrinic acid after conversion to cobester, whose ^{13}C-NMR spectra has been assigned.[29,30] Firstly, precorrin-**2** is accumulated in whole cells containing an excess of SAM in the absence of Co^{++}. The cells are then disrupted and Co^{++} added immediately, followed by a pulse of $^{13}CH_3$-SAM (90 atom%) after 4 hours. After a further 1.5 hr. cobyrinic acid is isolated as cobester. The ^{13}C NMR spectrum of this specimen defines the complete methylation sequence, beginning from precorrin-**2**, as C-20 > C-17 > C-12α > C-1 > C-5 > C-15, with a differentiation of 25% (± 5%) in the relative signal intensities for the C_5 and C_{15} methyl groups[31,32] further confirmed by hetero-filtered ^1H spectroscopy of the ^{13}C-enriched sample. Methylation at C-20 of precorrin-**2** to give precorrin-**3** is not recorded in the spectrum of cobester since C-20 is lost on the way to cobyrinic acid, together with the attached methyl group, in the form of acetic acid.[25-28] The sequence C-17 > C-12α > C-1 had previously been found in *Clostridium tetanomorphum*[24] and further differentiation between C-5 and C-15 insertion suggests the order C-15 > C-5 in this organism for the last two methylations, i.e. opposite from the *P. shermanii* sequence.

It is now apparent that several discrete methyl transferases are involved in the biosynthesis of cobyrinic acid from uro'gen III, since enzyme-free intermediates must accumulate in order to dilute the ^{13}C label. A rationale for these events is given in Scheme 6, which takes the following facts into account: (a) the methionine derived methyl group at C-

20 of precorrin-**3** does not migrate to C-1 and is expelled together with C-20 from a late intermediate (as yet unknown) in the form of acetic acid (b) neither 5,15-norcorrinoids[33] nor descobalto-cobyrinic acid[34] are biochemical precursors of cobyrinic acid (c) regiospecific loss of ^{18}O from [1-^{13}C, 1-$^{18}O_2$]-5-aminolevulinic acid-derived cyanocobalamin from the ring A acetate occurs,[35] in accord with the concept of lactone formation, as portrayed in Scheme 6, where precorrin-**5** is methylated at C-20 followed by C-20 → C-1 migration and lactonization. If the latter mechanism is operative, the C-20 → C-1 migration must be stereospecific, since precorrin-**3** labeled at C-20 with $^{13}CH_3$ is transformed to cobyrinic acid with complete loss of label, a less attractive alternative being direct methylation at C-1.

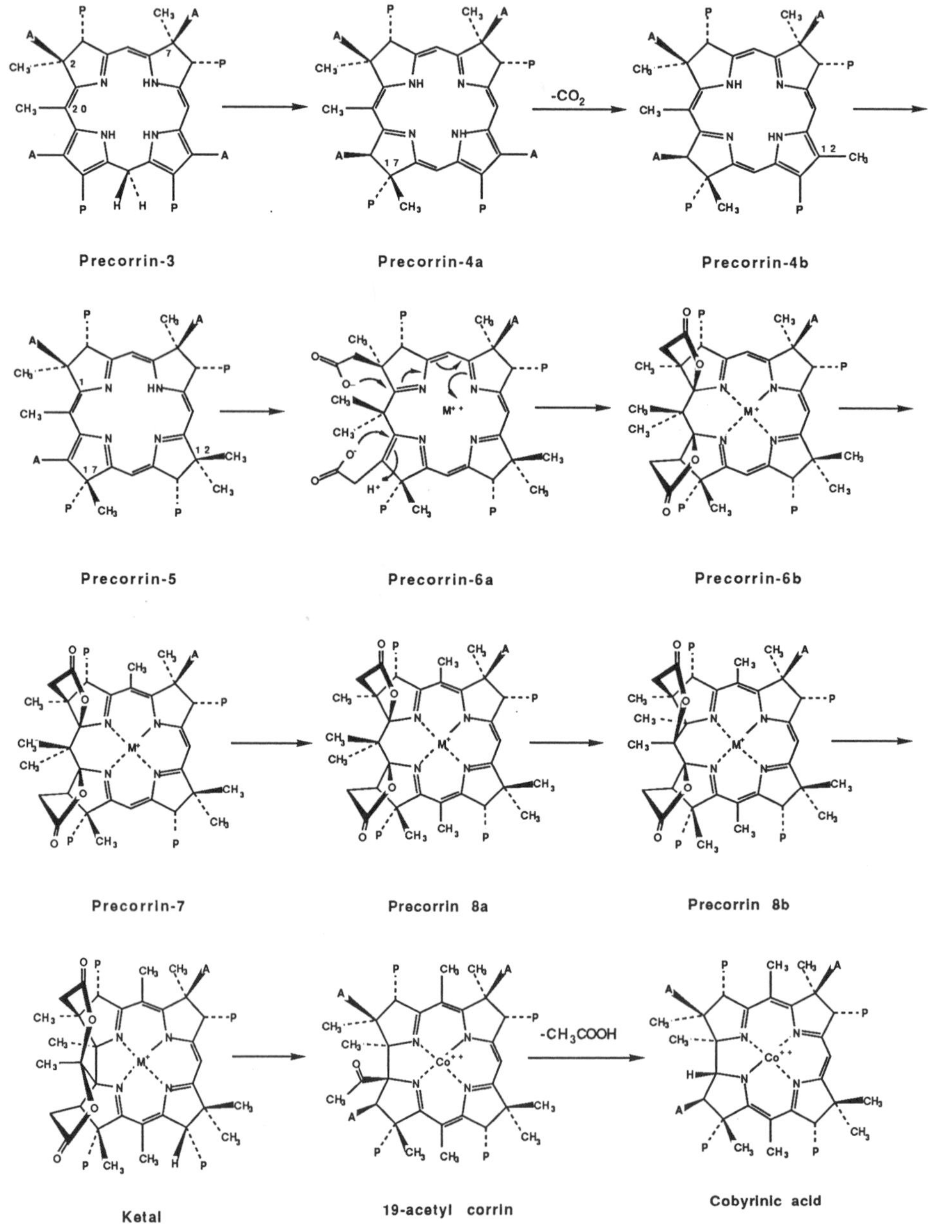

Scheme 6

It has also proved possible to define the point in the biosynthetic sequence where ring C-decarboxylation occurs, using synthetic [5,15-$^{14}C_2$]-12-decarboxylated uro'gen III (**14**) as a substrate for the non-specific methylases of *P. shermanii* to prepare the 12-methyl analogs (**15**) and (**16**) of factors II and III respectively (Scheme 7). It was shown by these studies[32] and independently by ^{13}C-labeling[36] that these ring C-decarboxylated analogs[37] are not substrates for the enzymes of corrin biosynthesis, leading to the conclusion that decarboxylation occurs at some stage after the fourth methylation (at C-17) and by mechanistic analogy, before the fifth methylation at C-12.

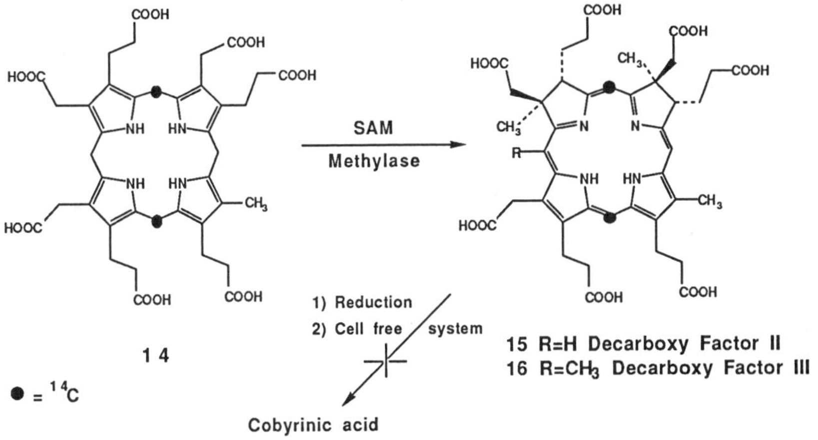

Scheme 7

We now return to the construction of a working hypothesis for corrin biosynthesis. The formulations precorrins **6b, 7, 8a, 8b** take into account the idea of lactone formation using rings A and D acetate side chains. The migration C-20 → C-1 could be acid or metal ion catalyzed and the resultant C-20 carbonium ion quenched with external hydroxide or by the internal equivalent from the carboxylate anion of the C-2 acetate in ring A (precorrin **8a** → **8b**) as suggested by the results of labeling with ^{18}O discussed above.[35] In any event, the resultant dihydrocorphinol-bislactone precorrin-**8b**, is poised to undergo the biochemical counterpart of Eschenmoser ring contraction[38] to the 19-acetyl corrin. Before this happens we suggest that the final methyl groups are added at C-5 (precorrin-7), then C-15 (precorrin-**8a**) to take account of the non-incorporation of the 5, 15-norcorrinoids.[33,39] The resultant precorrin-**8b** (most probably with cobalt in place) then contracts to 19-acetyl corrin[38,40-42] which, by loss of acetic acid, leads to cobyrinic acid. Cobalt insertion must precede methylation at C_5 and C_{15} since hydrogenocobyrinic acid does not insert cobalt enzymatically[34]. The valency change $Co^{++} \rightarrow Co^{+++}$ during or after cobalt insertion has so far received no explanation.

The Methyl Transferases

The first of the methylase enzymes catalyzes the sequential formation of Factors I and II and has been named S-adenosyl methionine Uro'gen III methyl transferase (SUMT). SUMT was first partially purified from *P. shermanii* by G. Müller[43] and recently has been overexpressed in *Pseudomonas denitrificans*.[44] In *E. coli* it was found that the *CysG* gene encodes Uro'gen III methylase (M-1) as part of the synthetic pathway to siroheme, the cofactor for sulfite reductase, and overproduction was achieved by the appropriate genetic engineering.[45] Although, SUMT and M-1 appear to perform the same task, it has beeen found that their substrate specificities differ. Thus, it has been possible to study in detail the reaction catalyzed by M-1 directly using NMR spectroscopy and to provide rigorous proof that the structure of precorrin-**2** is that of the dipyrrocorphin tautomer of dihydro-Factor II (dihydrosirohydrochlorin). Uro'gen III (enriched from [^{13}C-5-ALA] at the positions shown in Scheme 8) was incubated with M-1 and [$^{13}CH_3$]-SAM. The resultant spectrum of the

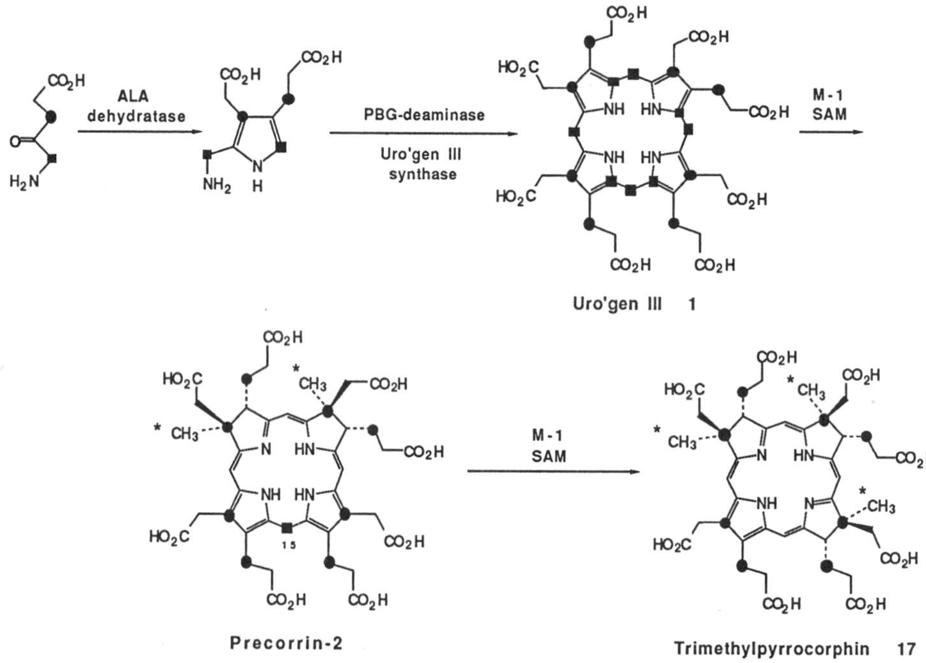

Scheme 8

precorrin-**2** revealed an sp^3 enriched carbon at C-15, thereby locating the reduced center. By using a different set of ^{13}C-labels (• from ^{13}C-3 ALA) and [$^{13}CH_3$]-SAM) the sp^2 carbons at C_{12} and C_{18} were located as well as the sp^3 centers coupled to the pendant ^{13}C-methyl groups at C_2 and C_7. This result confirms an earlier NMR analysis[23] of precorrin-**2** isolated by anaerobic purification of the methyl ester, and shows that no further tautomerism takes place during the latter procedure. The two sets of experiments mutually reinforce the postulate that precorrins-**1**, -**2**, and -**3** all exist as hexahydroporphinoids and recent labeling experiments[46] have provided good evidence that precorrin-**1** is discharged from the methylating enzyme (SUMT) as the species with the structure shown (or an isomer).

However, prolonged incubation (2 hr.) of Uro'gen III with M-1 provided a surprising result for the UV and NMR changed dramatically from that of precorrin-**2** (a dipyrrocorphin) to the chromophore of a pyrrocorphin, hitherto only known only as a synthetic tautomer of hexahydroporphyrin. At first sight, this event seemed to signal a further tautomerism of a dipyrrocorphin to a pyrrocorphin catalyzed by the enzyme but when $^{13}CH_3$-SAM was added to the incubation, it was found that a third methyl group signal appeared in the 19-21 ppm region of the NMR spectrum. When Uro'gen III was provided with the ^{13}C labels (•) (as shown in Scheme 8) 3 pairs of doublets appeared in the sp^3 region (δ 50-55 ppm) of the pyrrocorphin product. The necessary pulse labeling experiments together with appropriate FAB-MS data finally led to the structural proposal (**17**)[47,48] for the novel trimethyl pyrrocorphin produced by "overmethylation" of the normal substrate, uro'gen III, in presence of high concentration of enzyme. Thus M-1 has been recruited to insert a ring C methyl and synthesizes the long sought "natural" chromophore corresponding to that of the postulated precorrin-**4** although in this case the regiospecificity is altered from ring D to ring C. This lack of specificity on the part of M-1 was further exploited to synthesize a range of "unatural" isobacteriochlorins and pyrrocorphins based on isomers of Uro'gen III. Thus, Uro'gen I produces 3 methylated products corresponding to precorrin I, precorrin-**2** (**18**), and the type-I pyrrocorphin (**19**) (Scheme 9). Uro'gens II and IV can also

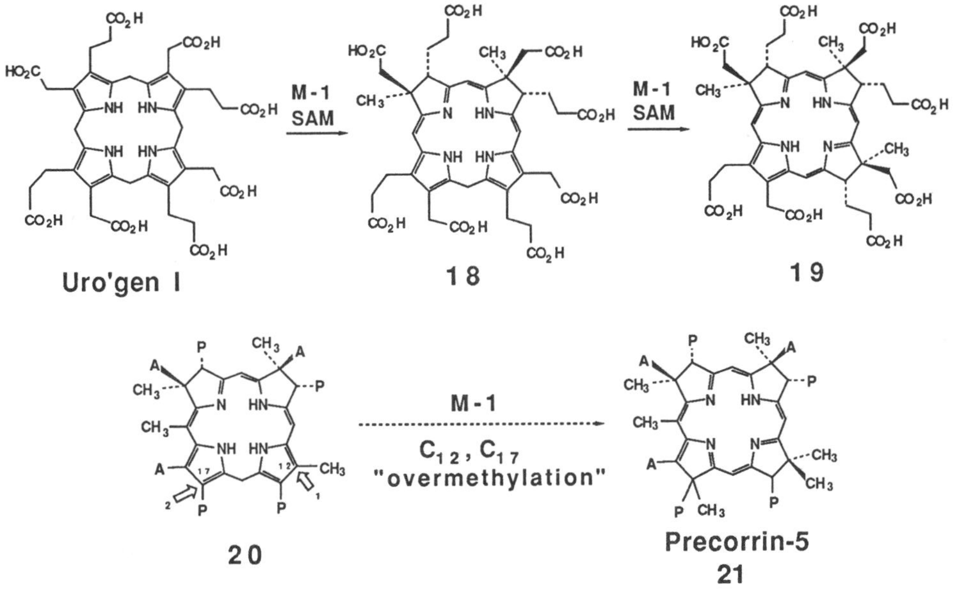

Scheme 9

serve as substrates for M-1 but not for SUMT! These compounds are reminiscent of a series of tetramethyl type I corphinoids, Factors S_1-S_4 isolated from *P. shermanii*[49,50] which occur as their zinc complexes. When uro'gen I was incubated with SUMT,[51] isolation of Factor II of the type I (Sirohydrochlorin I) family revealed a lack of specificity for this methyltransferase also, although in the latter studies no pyrrocorphins were observed. This may reflect a control mechanism in the *P. denitrificans* enzyme (SUMT) which does not "overmethylate" precorrin-2 as is found for the *E. coli* M-1 from whose physiological function is to manufacture sirohydrochlorin. The fact that *E. coli* does not synthesize B_{12} could reflect an evolutionary process in which the C-methylation machinery has been retained, but is only required to insert the C-2 and C-7 methyl groups.

The sites of C-methylation in both the type I and III series are also reminiscent of the biomimetic C-methylation of the hexahydroporphyrins discovered by Eschenmoser[52] and the regiospecificity is in accord with the principles adumbrated[53] for the stabilizing effect of a vinylogous ketimine system. In principle the methylases of the B_{12} pathway, which synthesize both natural and unnatural pyrrocorphins and corphins can be harnessed to prepare several of the missing intermediates of the biosynthetic pathway, e.g. **(20)** → **(21)**. It is of note that the instability of pyrrocorphins such as precorrin 4 towards oxygen rationalizes our inability to isolate any new intermediates under aerobic conditions (> 5 ppm O_2). In fact the strictly anaerobic transformation of precorrin 3 to a mixture of precorrin 4 and the ring C methylated pyrrocorphin using SAM and M-1 has recently been achieved in our laboratory.

Evolutionary Aspects and Further Outlook

Genetic mapping of the loci of the B_{12}-synthesizing enzymes has been reported for *Pseudomonas denitrificans*.[44] This complements a most interesting study on the genetics of *Salmonella typhimurium* which cannot make B_{12} when grown aerobically.[54] A mutant requiring methionine, cobinamide or cyanocobalamin when grown anaerobically produces B_{12} *de novo* thus leading to the isolation of other mutants blocked in B_{12} synthesis including one which cannot make Factor II required for siroheme production. All of the cobalamin mutations lie close together on the chromosome and a cluster of several methyl transferases maps at 42 min. Thus rapid progress can be expected in the isolation of the remaining biosynthetic enzymes from *Salmonella.*

Until quite recently it had been assumed that the Shemin pathway (glycine-succinate) to ALA was ubiquitous in bacterial production of porphyrins and corrins. However it is now clear that in many archaebacteria (e.g. *Methanobacterium thermoautotrophicum*,[55] *Clostridium thermoaceticum*[56-57]) the C_5 pathway from glutamate is followed. Phylogenetically the C_5 route is conserved in higher plants, and it appears from recent work[58] that *hemA* of *E. coli* (and perhaps of *S. typhimurium*) encodes the enzyme for the glutamate → ALA conversion, i.e. the C_5 pathway is much more common than had been realized. In *C. thermoaceticum* it has been shown[57] that the B_{12} produced by this thermophilic archaebacterium is synthesized from ALA produced in turn from glutamate. Although *E. coli* does not seem to be able to synthesize B_{12}, the enzyme M-1 (M_r 54,000), a close relative of SUMT (M_r 36,000), is expressed as part of the genetic machinery (*CysG*) for making siroheme, as discussed above.

Eschenmoser has speculated that corrinoids resembling B_{12} could have arisen by prebiotic polymerization of amino nitriles and has developed an impressive array of chemical models[53] to support this hypothesis, including ring contraction of porphyrinoids to acetyl corrins, deacetylation and the C-methylation chemistry discussed earlier, which provide working hypotheses for the corresponding biochemical sequences. A primitive form of corrin stabilized by hydrogen,[53] rather than by methyl substitution may indeed have existed $>4 \times 10^9$ years ago, before the origin of life[59] or the genetic code[60] and could have formed the original "imprint" necessary for the evolution of enzymes which later mediated the insertion of methyl groups to provide a more robust form of B_{12}. Since B_{12} is found in primitive anaerobes and requires no oxidative process in its biogenesis (unlike the routes to heme and chlorophyll which are oxidative) an approximate dating of B_{12} synthesis would be $2.7 - 3.5 \times 10^9$ years, i.e. after DNA but before oxygen-requiring metabolism.[61]

If B_{12} indeed were the first natural substance requiring Uro'gen III as a precursor, the question arises "why type III?" Since the chemical synthesis of the Uro'gen mixture from PBG under acidic conditions leads to the statistical ratio of Uro'gens [I 12.5%; II 25%; III 50%; IV 12.5%] containing a preponderance of Uro'gen III, natural selection of the most abundant isomer could be the simple answer. It has been suggested[53] that the unique juxtaposition of two adjoining acetate and side chains in the type III isomer (which does not obtain in the symmetrical Uro'gen I) may be responsible for a self assembly mechanism requiring these functions to hold the molecular scaffolding in place via lactone and ketal formation as portrayed in Scheme 6. These ideas are being tested by ^{18}O labeling and by studying the possible biotransformation of types I, II, and IV uroporphyrinogens to "unnatural" corrinoids.

It is anticipated that the powerful combination of molecular biology and NMR spectroscopy which has been essential in solving the problems in B_{12} synthesis posed by the assembly and intermediacy of Uro'gen III and the subsequent C-methylations leading to precorrin-**3**, will again be vital to the solution of those enigmas still to be unraveled in the fascinating saga of B_{12} biosynthesis.

ACKNOWLEDGEMENTS

The work described in this lecture has been carried out by an enthusiastic group of young colleagues whose names are mentioned in the references. Financial support over the last 20 years has been generously provided by the National Institutes of Health, National Science Foundation, and the Robert A. Welch Foundation. It is a special pleasure to thank Professors Gerhard Müller (Stuttgart) and Peter Jordan (London) for their continued stimulating collaboration.

REFERENCES

1. A. I.Scott, *Acc. Chem. Res.* **11**, 29 (1978).
2. G. Burton, P. E. Fagerness, S. Hosazawa, P. M. Jordan, and A. I. Scott, *J. Chem. Soc. Chem. Comm.*, 204 (1979).
3. A. R. Battersby, C. J. R. Fookes, K. E. Gustafson-Potter, G. W. J. Matcham, E. McDonald, *J. Chem. Soc. Chem. Comm.*, 1115 (1979).
4. P. M. Jordan, G. Burton, H. Nordlow, M. M. Schneider, L. M. Pryde, A. I. Scott, *J. Chem. Soc. Chem. Comm.*, 204 (1979).
5. Review, F. J. Leeper, *Nat. Prod. Reports*, **2**, 19 (1985).
6. A. I. Scott, *J. Heterocyc. Chem.*, **14**, S-75 (1987).
7. A. I. Scott, C. A. Roessner, N. J. Stolowich, P. Karuso, H. J. Williams, S. K. Grant, M. D. Gonzalez, T. Hoshino, *Biochemistry*, **27**, 7984 (1988).
8. P. M. Anderson, R. J. Desnick, *J. Biol. Chem.*, **255**, 1993 (1980).
9. A. Berry, P. M. Jordan, J. S. Seehra, *FEBS Lett.* **129**, 220 (1981).
10. A. R. Battersby, C. J. R. Fookes, G. Hart, G. W. J. Matcham, P. S. Pandey, *J. Chem. Soc. Perkin Trans. I*, 3041 (1983).
11. P. M. Jordan, A. Berry, *Biochem. J.*, **195**, 177 (1981).
12. J. N. S. Evans, G. Burton, P. E. Fagerness, N. E. Mackenzie, A. I. Scott, *Biochemistry*, **25**, 905 (1986).
13. A. I. Scott, C. A. Roessner, K. D. Clemens, *FEBS Lett.*, **242**, 319 (1988).
14. A. I. Scott, N. J. Stolowich, H. J. Williams, M. D. Gonzalez, C. A. Roessner, S. K. Grant, C. Pichon, *J. Am. Chem. Soc.*, **110**, 5898 (1988).
15. P. M. Jordan, M. J. Warren, H. J. Williams, N. J. Stolowich, C. A. Roessner, S. K. Grant, A. I. Scott, *FEBS Lett.*, **235**, 189 (1988).
16. (a) M. J. Warren and P. M. Jordan, *FEBS Lett.*, **225**, 87 (1987).
    (b) M. J. Warren and P. M. Jordan *Biochemistry*, **27**, 9020 (1989).
17. (a) G. J. Hart, A. D. Miller, F. J. Leeper, and A. R. Battersby, *J. Chem. Soc. Chem. Comm.*, 1762 (1987).
    (b) A. D. Miller, G. J. Hart, L. C. Packman, and A. R. Battersby, *Biochem. J.*, **254**, 915 (1988).
    (c) G. J. Hart, A. D. Miller, and A. R. Battersby, *Biochem. J.*, **252**, 909 (1988).
    (d) U. Beifus, G. J. Hart, A. D. Miller, and A. R. Battersby, *Tetrahedron Letters*, **29**, 2591 (1988).
18. R. Radmer and L. Bogorad, *Biochemistry*, **11**, 904 (1972).
19. A. I. Scott, C. A. Roessner, P. Karuso, N. J. Stolowich, and B. Atshaves, manuscript in preparation; A. I. Scott, *Acc. Chem. Res.*, 1990, September issue.
20. J. H. Mathewson, A. H. Corwin, *J. Am. Chem. Soc.*, **83**, 135 (1961).
21. S. Rosé, R. B. Frydman, C. de los Santos, A. Sburlati, A. Valasinas, and B. Frydman, *Biochemistry*, **27**, 4871 (1988).

22. A. R. Battersby,.J. R. Fookes, G. W. J. Matcham, E. McDonald, and R. Hollenstein, *J. Chem. Soc., Perkin Trans I*, 3031 (1983).

23. A. R. Battersby, K. Frobel, F. Hammerschmidt, and C. Jones, *J. Chem. Soc. Chem. Comm.*, 455-457 (1982).

24. H. C. Uzar, A. R. Battersby, T. A. Carpenter, F. J. Leeper, *J. Chem. Soc. Perkin Trans. I*, 1689-1696 (1987) and references cited therein.

25 C. Nussbaumer, M. Imfeld, G. Wörner, G. Müller, and D. Arigoni, *Proc. Nat. Acad. Sci. U.S.*, **78**, 9-10 (1981).

26. A. R. Battersby, M. J. Bushnell, C. Jones, N. G. Lewis, and A. Pfenninger, *Proc. Nat. Acad. Sci. U. S.*, **78**, 13-15 (1981).

27. L. Mombelli, C. Nussbaumer, H. Weber, G. Müller, and D. Arigoni, *Proc. Nat. Acad. Sci. U.S.*, **78**, 11-12 (1981).

28. G. Müller, K. D. Gneuss, H.-P. Kriemler, A. J. Irwin, and A. I. Scott, *Tetrahedron (Supp.)* **37**, 81-90 (1981).

29. L. Ernst, *Liebigs Ann. Chem.*, 376-386 (1981).

30. A. R. Battersby, C. Edington, C. J. R. Fookes, and J. M.Hook, *J. Chem. Soc., Perkin, I.*,, 2265-2277 (1982).

31. A. I. Scott, N. E. Mackenzie, P. J. Santander, P. Fagerness, G. Müller, E. Schneider, R. Sedlmeier, G. Wörner, *Bioorg. Chem.* **13**, 356-362 (1984).

32. A. I. Scott, H. J. Williams, N. J. Stolowich, P. Karuso, M. D. Gonzalez, G. Müller, K. Hlineny, E. Savvidis, E. Schneider, U. Traub-Eberhard, G. Wirth, *J. Am. Chem. Soc.*, **111**, 1897-1900 (1989).

33. C. Nussbaumer, D. Arigoni, Work described in C. Nussbaumer's Dissertation No. 7623,
E. T. H. (1984).

34. T. E. Podschun, G. Müller, *Angew. Chem. Int. Ed., Engl.*, **24**, 46-47 (1985).

35. K. Kurumaya, T. Okasaki, M. Kajiwara, *Chem. Pharm. Bull.*, **37**, 1151 (1989).

36. F. Blanche, S. Handa, D. Thibaut, C. L. Gibson, F. J. Leeper, A. R. Battersby, *J. Chem. Soc. Chem. Comm.*, 1117-1119 (1988).

37. A. R. Battersby, K. R. Deutscher, B. Martinoni, *J. Chem. Soc. Chem. Comm.* 698-700 (1983).

38. V. Rasettti, A. Pfaltz, C. Kratky, A. Eschenmoser, *Proc. Natl. Acad. Sci., U.S.A.*, **78**, 16-19 (1981).

39. B. Dresov, L. Ernst, L. Grotjahn, V. B. Koppenhagen, *Angew. Chem. Int. Ed. Engl.*, 20, 1048-1049 (1981).

40. A. I. Scott, N. E. Georgopapadakou, and A. J. Irwin, unpublished.

41. A. I. Scott, M. Kajiwara, and P. J. Santander, *Proc. Nat. Acad. Sci. U. S.*, **84**, 6616-6618 (1987).

42. A. R. Battersby, C. Edington, and C. J. R. Fookes, *J. Chem. Soc. Chem. Comm.*, 527-530 (1984).

43. G. Müller in Vitamin B_{12}, B. Zagalak and W. Friedrich, eds, de Gruyter, New York, 1979, 279-291, who described a methylase from *P. shermanii* capable of using Uro'gen I as substrate. The product was not characterized.
44. B. Cameron, K. Briggs, S. Pridmore, G. Brefort, J. Crouzet, *J. Bacteriol.* **171**, 547-557 (1989).
45. M. J. Warren, C. A. Roessner, P. J. Santander, and A. I. Scott, *Biochem. J.*, **265**, 725-729 (1990).
46. R. D. Brunt, F. J. Leeper, I. Orgurina, and A. R. Battersby, *J. Chem. Soc., Chem. Comm.*, 428 (1989).
47. M. J. Warren, N. J. Stolowich, P. J. Santander, C. A. Roessner, B. A. Sowa, and A. I. Scott, *FEBS Lett.*, **261**, 76-80 (1990).
48. A. I. Scott, M. J. Warren, C. A. Roessner, N. J. Stolowich, and P. J. Santander, *J. Chem. Soc., Chem. Comm.*, in press (1990).
49. G. Müller, J. Schmiedl, E. Schneider, R. Sedlmeier, G., Worner, A. I. Scott, H. J. Williams, P. J. Santander, N. J. Stolowich, P. E. Fagerness and N. E. Mackenzie, *J. Am. Chem. Soc.*, **108**, 7875-7877 (1986).
50. G. Müller, J. Schmiedl, L. Savidis, G. Wirth, A. I. Scott, P. J. Santander, H.J. Williams, N. J. Stolowich, and H.-P. Kriemler, *J. Am. Chem. Soc.*, **109**, 6902-6904 (1987).
51. A. I. Scott, H. J. Williams, N. J. Stolowich, P. Karuso, M. D. Gonzalez, F. Blanche, D. Thibaut, G. Müller, and G. Wörner, *J. Chem. Soc., Chem. Comm.*, 522 (1989).
52. C. Leumann, K. Hilpert, J. Schreiber, and A. Eschenmoser, *J. Chem. Soc., Chem. Comm.*, 1404-1407 (1983).
53. A. Eschenmoser, *Angew. Chem., Int. Ed. Engl.*, **27**, 5 (1988).
54. R. M. Jeter, B. M. Olivera, J. R. Roth, *J. Bacteriol.*, **159**, 206-213 (1984).
55. J. G. Zeikus, *Adv. Microb. Physiol.*, **24**, 215-299 (1983).
56. J. R. Stern and G. Bambers, *Biochemistry,* **5**, 1113-1118 (1966).
57. T. Oh-hama, N. J. Stolowich, A. I. Scott, *FEBS Lett.*, **228**, 89-93 (1988).
58. L. Jian-Ming, C. S. Russell, S. D. Coslow, *Gene,* **82**, 209-217 (1989).
59. K. Decker, K. Jungermann, R. K. Thauer, *Angew. Chem., Int. Ed. Engl.*, **9**, 153-162 (1970).
60. M. Eigen, B. F. Lindemann, M. Tietze, R. Winkler-Oswatitsch, A. Dress, and A. von Haeseler, Science, **244**, 673-679 (1989).
61. S. A. Benner, A. D. Ellington, and A. Taver *Proc. Nat. Acad. Sci.*, **86**, 7054 (1989).

SYNTHESIS OF HOMOCHIRAL PHARMACEUTICALS USING EFFECTIVE ENANTIOSELECTIVE BIOCATALYSTS

H.J. KOOREMAN[a] and J.H.G.M MUTSAERS[b]

[a]International Bio-Synthetics B.V.
P.O. Box 1820
2280 DV Rijswijk
The Netherlands

[b]Gist-brocades nv
R&D Department
P.O. Box 1
2600 MA Delft
The Netherlands

INTRODUCTION

Scientists, both in academia and industry, as well as regulatory authorities become more and more aware of the importance of enantioselectivity in drug innovation. About 25% of the medicines used to-day are racemic mixtures. The therapeutic activity nearly always resides in one of the isomers (the so-called eutomer), while the other isomer (distomer) can have an unwanted interaction with the human body; at best the distomer is a metabolic burden. Essentially the same holds true for other interactions between synthetic organic compounds and biological systems e.g. the impact of agrochemicals on ecosystems.

This undesirable situation originates from the fact that, in general, it is difficult to separate a racemic mixture into its enantiomers by using classical organic chemical methods, although in a limited number of cases this approach has led to an efficient formation of homochiral (single isomer) compounds.

Other options for enantioselective synthesis are:

using the 'chiral pool', i.e. taking simple homochiral compounds like lactic acid or tartaric acid as starting material. Unfortunately the number of available 'chiral synthons' is rather restricted.
In this paper we will introduce new cost-effective routes to two representatives of such versatile homochiral starting substances, R- and S-dioxolane-2,2-dimethyl-4-methanol -more commonly called R- and S-isopropylidene glycerol (R- and S-IPG)- which were up to now too expensive for larger scale application.

Bioorganic Chemistry in Healthcare and Technology, Edited by U.K. Pandit and F.C. Alderweireldt, Plenum Press, New York, 1991

using asymmetric chemical catalysts or bio- technological conversions. Both the chemical and the biotechnological methods have already led to very promising results.
In this presentation the latest outcome of our work on a newly developed biocatalyst, carboxyl esterase (or naproxen esterase), will illustrate the effectiveness of the biotechnological approach.

The two biotechnological projects mentioned above were worked out on behalf of International Bio-Synthetics (IBIS) by joint research teams of Gist-brocades Research (Delft, The Netherlands) and the Sittingbourne Research Centre of Shell (Sittingbourne, UK). IBIS is a joint venture of the Royal-Dutch Shell Group of Companies and Gist-brocades.

R- AND S-IPG AS VERSATILE C_3-SYNTHONS

R-IPG has been described in literature[1] as being formed from ascorbic acid. S-IPG is somewhat better known and can be prepared from D-mannitol[2]. These processes are tedious, of relatively low efficiency and result in the unstable aldehydes which require borohydride reduction to the R- respectively the S-IPG. Both enantiomers, but especially the former, are expensive chemicals. This is a serious barrier for the extensive use of these, in principle, versatile intermediates. As published earlier[3] the IBIS research team recently developed an efficient new route to R-IPG utilizing microbes that are able to discriminate between the two enantiomers. A broad range of micro-organisms have been isolated capable of oxidizing out of the racemic IPG the S-isomer to the corresponding R-isopropylidene glyceric acid (R-IPGA), leaving the R-IPG virtually untouched (Figure 1). R-IPG can easily be isolated from the reaction mixture in a high yield and with an optical purity >98% (e.e.). The chemical purity is somewhat less due to a contamination with the isomeric 1,3-dioxane.
R-IPGA is initially isolated with an optical purity of about 90% (e.e.), but this can be improved to 98% by further purification.The chemical reduction of this acid leads to S-IPG in a high yield (Figure 2).

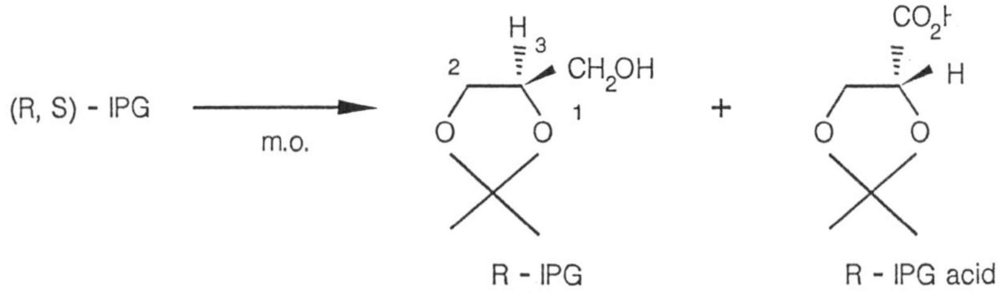

Micro - organisms : Rhodoccus erythropolis
Corynebacterium equi
Nocardia coralina
Mycobacterium rhodochrous
Nocardia canicruria

FIGURE 1

FIGURE 2

A second - microbial- route to S-IPG is very similar to the production of the R-isomer. We isolated a series of micro-organisms capable this time to convert the R-isomer leaving the S-IPG (Figure 3)[4].
Very recently another research group published a closely related approach[5].
There is abundant evidence these homochiral synthons can be used for the preparation of many other chiral building blocks[6] as well as for the synthesis of a range of medicinally important compounds[7], inter alia betablocking agents like S-metoprolol and S-atenolol.

NAPROXEN ESTERASE

Substituted propionic acids

Some of the most important non-steroidal anti-inflammatory (NSAI) drugs have a substituted propionic acid structure. Naproxen, produced in about 1000 tons/year quantities, is used in the homochiral S-configuration. Others like ibuprofen and ketoprofen are sold as the racemic mixture (Figure 4).

Very recently an increased interest in S-ibuprofen could be perceived[8].
Probably present production of S-naproxen and (on pilot scale) S-ibuprofen uses classical resolution.

Enzymatic kinetic resolution

Several research groups attempted to develop enzymatic systems for the kinetic resolution of R,S-naproxen or -ibuprofen esters. Some lipases were found to be reasonably efficient, but have some serious draw-backs:

- preference for higher alkyl esters
- low specific activity[9].

Extensive screening by our team led to the isolation and identification of a Bacillus subtilis (Thai I-8) strain that produced a more suitable enzyme. As shown by HPLC-SEC-chromatography the cell lysate contained two hydrolysing enzymes:

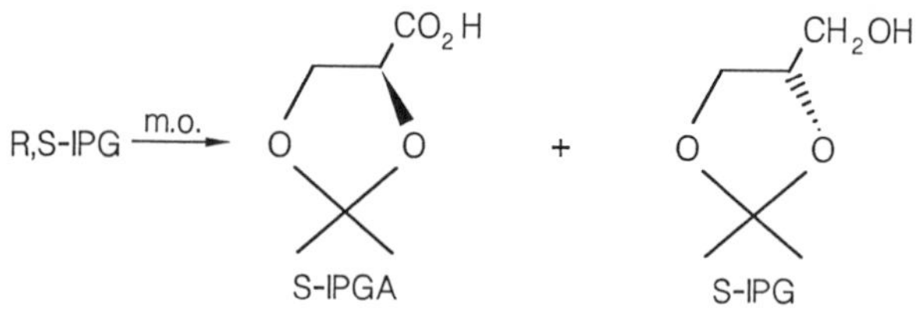

FIGURE 3

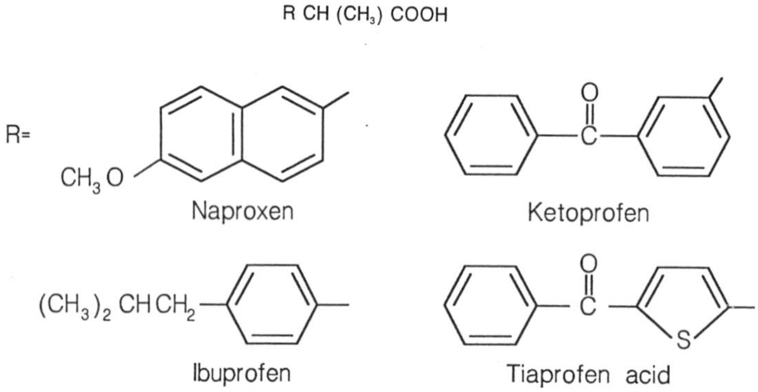

FIGURE 4

Esterase production in Bacillus Thai I-8

	S-selectivity
Cell lysate	97.1%
Fraction 31	43.1%
Fraction 38	99.0%

esterase production level: 6 units/liter

FIGURE 5

- a highly selective esterase (fraction 38), analyzed by incubation with naproxen methyl ester, and
- less selective, disturbing lipase (fraction 31), analyzed by incubation with p-nitrophenyl laurate.

The results of this experiment are summarised in Figure 5.

In the next phase the gene coding for the esterase was isolated and cloned into a productive strain, B. subtilis I 85/pNAPT-7. As is shown in Figure 6 the productivity of this strain is 800 x higher than that of the wild strain Thai I-8[10], while at the same time the enantioselectivity of the system has significantly been improved.
The difference in specific activity between our esterase and the lipase mentioned above can be demonstrated by comparing the conditions and reaction times of two related experiments:

Sih[9]: 100 mg product, 10 mg lipase, 22°C, 5 days.

IBIS/GB: 100 mg product, 4 mg esterase, 40°C, <4 hr.

Scope of application

The naproxen esterase system can be used for the resolution of a range of related compounds, like ibuprofen and tiaprofenic acid, but -more surprisingly- also for aryloxy propionic acids e.g. S-fluazifop and S-diclofop. Unexpectedly, the absolute spatial arrangement of these isomers is opposite to the one of the aryl propionic acids (Figure 7)

The herbicidal activity resides in the R-configuration.
As exemplified with a range of alkyl esters of ibuprofen (Figure 8) the relative activity and the enantioselectivity of the esterase are inversely related to the length of the alkyl substituent.

Also for naproxen it can be shown that the enzyme gives a better performance with the methyl than with the ethyl ester.

Improving the robustness

During tests at higher substrate concentrations (commercial conditions) irreversible inactivation of the enzyme has been noticed. E.g. carboxyl esterase obtained from Bacillus Thai I-8 was almost completely inactivated within one hour when 30 g/l naproxen ester was added (pH = 9, T = 40°C and Tween 80 (TM) medium). The esterase as such (without the ester being present) is stable under similar conditions for several hours. It has been shown the enzyme was inactivated by the naproxen formed during the hydrolysis. It was hypothesized the acidic product e.g. naproxen, would react with the amino groups of basic amino acids at the surface of the enzyme, thereby allowing the hydrophobic bulk of the naproxen to interfere with the folding of the protein structure. The resulting unfolding will be noticed as increased susceptibility of the enzyme to proteolytic breakdown.

Esterase production after cloning

	units/mg protein	S-select	esterase production level
Bacillus pNAPT-7	1.6	99.4%	5000 units/liter
Original strain (Thai I-8)	0.002	97.1%	6 units/liter

FIGURE 6

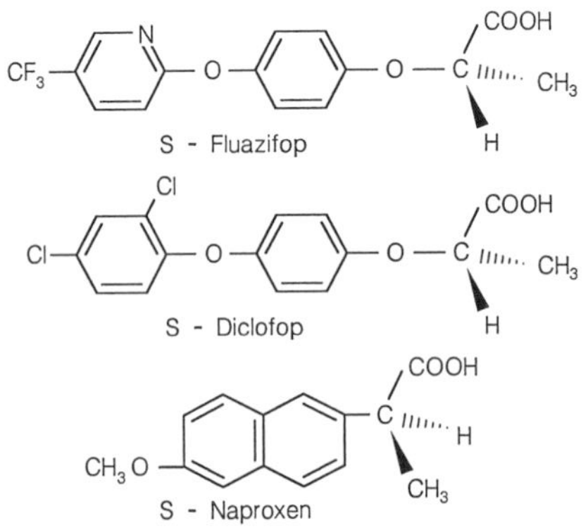

FIGURE 7

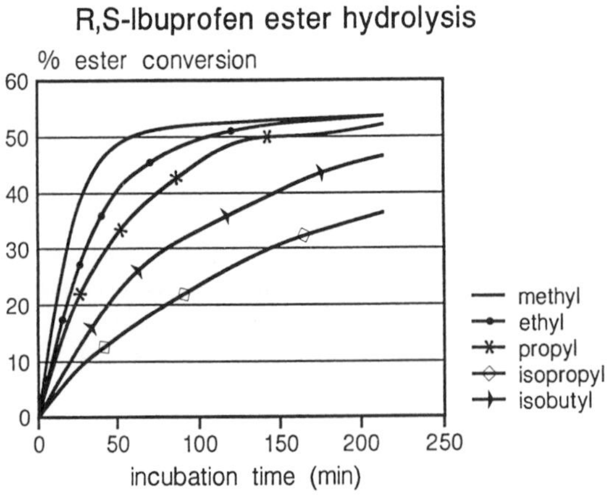

FIGURE 8

By substituting (via protein engineering) or chemically modifying the amino acids involved (e.g. lysine) new enzyme preparations can be produced which are more resistant to the 'stress' conditions.

Prior to the protein engineering experiments the amino acid sequence of the carboxylic esterase had to be determined[11].
A number of mutants of the esterase were made by substituting lysine residues by a glutamine residue. As expected, the majority of the constructed mutants showed decreased stability or activity. However, several mutant enzymes exhibited increased stability while retaining the activity of the enzyme: see Figure 9.

Modification of the carboxyl esterase by simple treatment with formaldehyde or a related reagent led to an even more impressive result as exemplified in Figure 10.

The untreated enzyme is completely inactivated on incubation with naproxen for 1.5 hours at 40°C. Formaldehyde treated enzyme preparations proved to be completely stable on incubation with naproxen (15 mg/ml), while showing only partial loss of enzyme activity.
The influence of the treatment of the esterase with different stabilizing agents on the conversion of R,S-naproxen methyl ester is shown in Figure 11.

Summarizing the modifications described above are imperative for a successful application of the carboxyl esterase under commercial conditions (high substrate concentrations). Especially the simple treatment with formaldehyde gives a very satisfactory result.

The process

In this presentation the main emphasis has been on the enzymatic kinetic resolution of the R,S-esters of substituted propionic acids. The total process of transforming e.g. racemic naproxen or ibuprofen into the S-enantiomer obviously involves other, chemical steps that are, however, less complicated. Following a scheme as given in Figure 12 the racemic product can very efficiently be converted in high yield into the therapeutically active S-enantiomer.

CONCLUSIONS

Pharmaceutical products are increasing in structural complexity. Due to this the present,important share of chiral structures (40-50%) in the group of synthetic drugs will only expand. It is to be expected that new product development will more and more focus on the synthesis of the eutomer, the enantiomer bearing the wanted therapeutic activity. Thus, there is a strong need for an improved stereoselective synthetic arsenal. Organic chemists are presently trying to extend their skill pool of enantioselective methods. Although chemical and biotechnological methods will be competitive in a number of cases, we envision both approaches will most often show to be complementary.
In this paper we have described two examples of successful application of biotechnology in organic synthesis; in both instances using a microbial or enzymatic system to discriminate between two enantiomers.

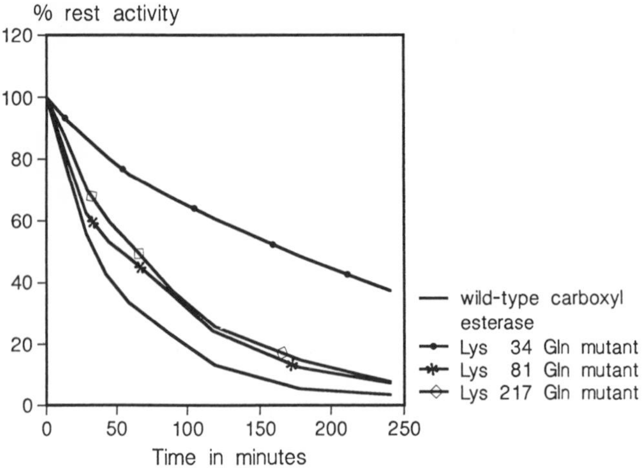

FIGURE 9

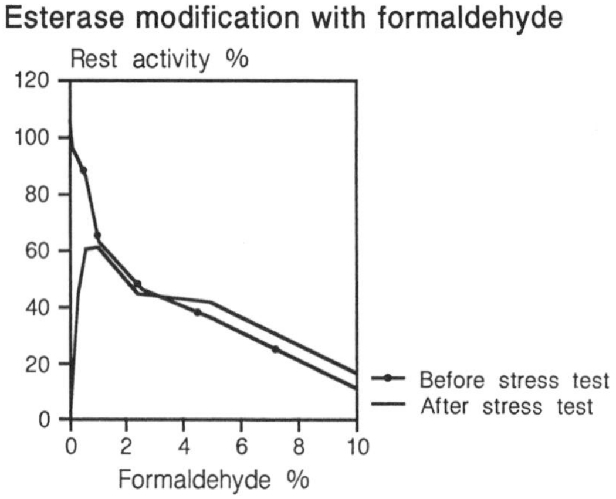

FIGURE 10

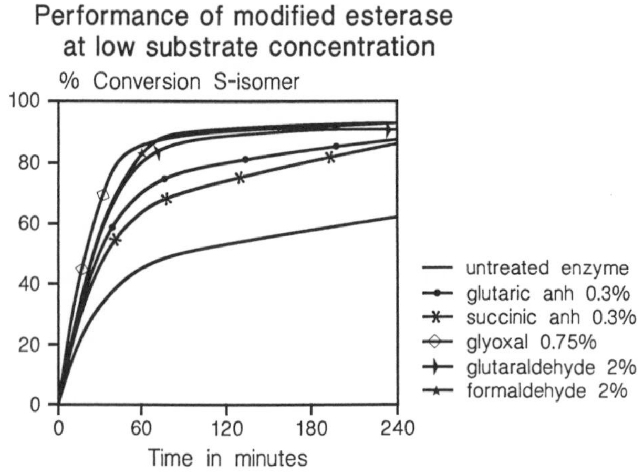

FIGURE 11

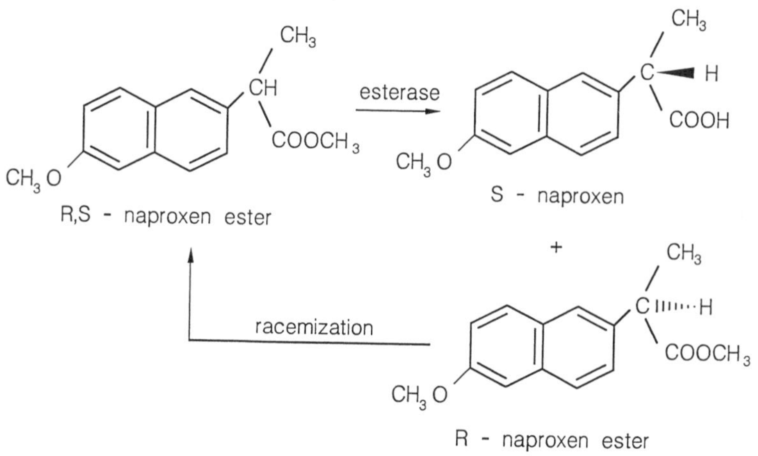

FIGURE 12

Obviously, the opportunities for biocatalysis are far more comprehensive. Our research teams are also exploring other strategies e.g. stereoselective, biochemical reduction or oxidation of prochiral starting materials. This exciting development will contribute to the need of purer and safer drugs and pesticides to the benefit of the internal as well as the external environment.

ACKNOWLEDGEMENT

The research work described was performed by mixed R&D teams of Sittingbourne Research Centre of Shell (Sittingbourne, Kent, UK) and Gist-brocades (Delft, The Netherlands) on behalf of International Bio-Synthetics, Rijswijk, The Netherlands.

REFERENCES

1. C. Hubschwerken, Synthesis, 1986, 962 and references therein.
2. D. Y. Jackson, Synth. Commun. 18, 337 (1988) and references therein.
3. H. J. Kooreman, in: Trends in Drug Research (V. Claassen, Ed.) 401-412, Elsevier.
4. Gist-brocades nv, Patent application filed; to be published.
5. W. J. H. van den Tweel et al, 5th Eur. Congress on Biotechnology, Copenhagen, 1990, Abstract Book, p. 283.
6a. J. Jurczak, S. Pikul and T. Bauer, Tetrahedron, 42, 447 (1986), and references therein.
b. S. Takano, Y. Iwabuchi and K. Ogasawara, J. Chem. Soc. Chem. Commun., 1527, (1988).
c. Modern Synthetic Methods. vol. 2, p. 241, 1980.
d. M. Yodo, Y. Matsushita, E. Ohsugi and H. Harada Chem. Pharm. Bull, 36, 902, (1988).

7a. G. Hirth, R. Barner, Helv. Chim. Acta, 65, 1059, (1982).
b. J. H. van Boon, Synthesis, 5, 399, (1982).
c. A. Holly, Coll. Czech. Chem. Comm., 43, 3103, (1978).
d. P. Dostert, M. Langlois, P. Geurret, J. F. Ancher, B. Bucher and G. Mocquet, Eur. J. Med. Chem., 15, 199, (1980).
8. G. Geisslinger et al, Eur. J. Clin. Pharmacol. 38, 493 (1990).
9. C. J. Sih et al, A facile enzymatic resolution for the preparation of (+)-S-2-(6-methoxy-2-naphthyl)propionic acid (Naproxen), Tetrahedron Letters, vol. 27 no. 16; 1763-1766 (1986).
10. W. J. Quax et al, Molecular cloning of stereoselective esterase from Bacillus species, Proceedings 4th European Congress on Biotechnology, 1987; volume 1: 519.
11. Gist-brocades nv, patent application filed; to be published.

CHEMICAL INTERVENTION IN CATALYSIS BY ENZYMES

Colin J. Suckling

Department of Pure and Applied Chemistry
University of Strathclyde, 295 Cathedral St
Glasgow, G1 1XL, Scotland

INTRODUCTION

One of the major motivations of bioorganic chemistry has been the establishment of the molecular mechanisms of biological processes. The most extensive application of such knowledge has been in the field of medicinal chemistry, one of the most successful fields of scientific health care. The approach of chemists to the study of biological reaction mechanisms is an extension of the principles of physical organic chemistry to macromolecular systems. As is well known, the clearest expression of this has come in the study of the mechanisms of action of enzymes. Similarly, enzyme inhibition has become the major area for mechanism based drug design. In such studies, the emphasis has been upon the achievement of selectivity and potency; the former property can be related to the catalytic action of a target enzyme and both selectivity and potency to the binding interactions of the inhibitor with the enzyme. Most work on mechanism based enzyme inhibition, therefore, has as its goal the discovery of lead compounds for the development of new drugs. However at a basic scientific level, the use of mechanism based enzyme inhibitors can contribute both to knowledge concerning enzyme mechanism, and to an understanding of the biological processes that occur within a cell. Indeed the view of many medicinal chemists is that the better understanding of biological mechanisms at the molecular level is the key to advances in medicinal chemistry. The chemist's intervention in catalysis by enzymes can be both investigative in new knowledge and productive of leads to new drugs. In this paper, I shall describe an example of each.

THE INHIBITION OF DIHYDROFOLATE REDUCTASE

A long standing research interest at Strathclyde has been the inhibition of enzymes involved in the biosynthesis and metabolism of folic acid and its reduced derivatives [1]. Most recent work has centred on the enzyme dihydrofolate reductase (DHFR) and its inhibition by 7,7-spirocyclopropylpteridines [2]. This enzyme is one of the most well characterised enzymes as a result of extensive crystallographic and nmr studies [3,4]. Consequently it has become one of the prime candidates for investigative studies of molecular recognition. Many important drugs act as inhibitors of DHFR and their interactions with the enzyme have been examined in detail. Bacterial, protozoal, and neoplastic diseases can all be treated with appropriate inhibitors of DHFR most of which contain a 2,4-diaminopyrimidine or pteridine and are hence substrate analogues.

Bioorganic Chemistry in Healthcare and Technology, Edited by U.K. Pandit and F.C. Alderweireldt, Plenum Press, New York, 1991

The principal components of recognition of diaminopyrimidine drugs to the active site have been identified as protonation of the diaminopyrimidine ring by an aspartic acid at the active site together with a hydrophobic interaction in antibacterial drugs such as trimethoprim or a specific binding of the glutamate side chain in anticancer drugs such as methotrexate. Drugs containing the natural amino-oxo substitution in the pyrimidine ring, however, have not been developed.

Although the diaminopyrimidine and diaminopteridine based DHFR inhibitors have been outstandingly successful drugs, there are problems with toxicity [5]. Pyrimethamine, for example, when used for the treatment of toxoplasmosis in particular can cause a more general interference with folate metabolism and lead to haematological toxicity. Methotrexate is intrinsically highly toxic and its clinical success rests upon combination with rescue therapy using 6-formyltetrahydrofolate (leucovorin) to maintain one carbon metabolism. This toxicity together with the potential of achieving high selectivity using an enzyme inhibitor that became active only through binding and activation at the active site (a latent or suicide inhibitor [5]) encouraged us to apply the chemistry of cyclopropanes to this enzyme [6]. The cyclopropane ring is small and its introduction to generate a substrate analogue should not make binding of the substrate analogue improbable.

The original mechanism proposed for the mechanism of action of this enzyme involved protonation of N-5 imine thereby activating it to reduction by hydride from NADPH [7]. On the basis of this concept, we designed, prepared, evaluated the spirocyclopropylpteridine 1 (figure 1). This molecule was shown to be a time-dependent irreversible inhibitor of DHFR and was the first inhibitor that was a close substrate analogue with the natural amino-oxo substitution pattern in the pyrimidine ring to be discovered [2].

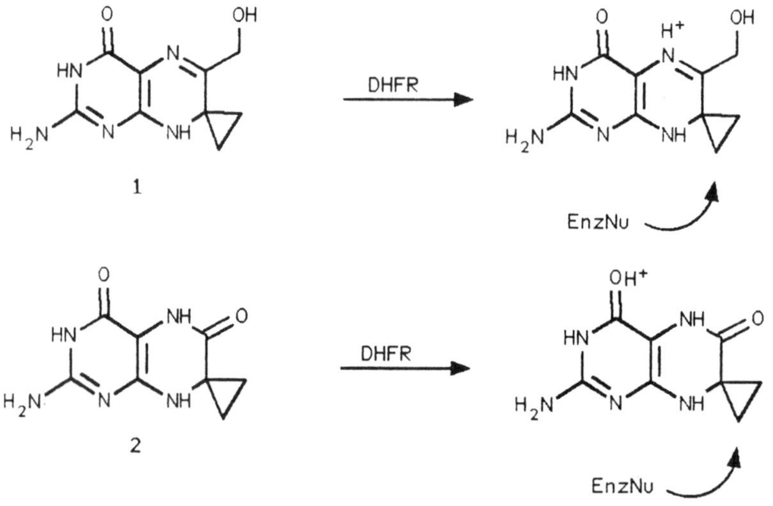

Figure 1. Structures and enzyme-mediated activation of dihydrofolate reductase inhibitors.

Whilst this work was in progress, however, results from crystallographic studies and stereochemical data forced a review of the mechanistic interpretation upon which the original design of or inhibitors had been based [7]. In the revised mechanism, protonation was viewed as taking place on the 4-oxo substituent of the pteridine ring; by rotation of a hydrogen bond, this proton could then be viewed as being transferred to the required destination on N-5. No independent reaction-based support for this view was, however, available. One possible avenue to explore was the possibility that a latent inhibitor might also be activated by protonation of a 4-oxo substituent. In order to investigate this possibility, it was necessary to block the possibility of direct protonation on N-5; this was achieved in compound 2. This molecule was shown to be a time-dependent irreversible of DHFR also [8]. Clearly the enzyme recognises both types of pteridines 1 and 2 and is able to activate them presumably to nucleophilic ring opening. However with dihydropteridines, other mechanisms such as nucleophilic addition to C-6 cannot be ruled out with the evidence presently available.

This result is of interest in providing oblique support for a description of the mechanism of reduction by DHFR but in doing so, it poses significant problems of molecular recognition. Both inhibitors lack either a hydrophobic substituent or the glutamate side chain of the common drugs mentioned above. It is not therefore possible to describe directly by analogy with the drugs the way in which the pteridines 1 and 2 bind to DHFR; this question is currently under investigation. We can, however, suggest a concept that could be of significance in medicinal chemistry, a concept that has been recognised under the designation 'molecular similarity' [9]. In designing latent or suicide inhibitors, the extent of molecular similarity between the substrate and the inhibitor can be limited to a few essential features. Indeed this view has been mirrored in the discovery of some new substrates for DHFR [10]. The fullest expression of such a concept would be a molecule not belonging to the same structural series as the substrate, but nevertheless capable of being activated by the enzyme leading to irreversible inhibition. We accidentally discovered such a case in parallel studies on enzymes of pyrimidine biosynthesis [11].

INHIBITORS OF STEROID METABOLISM

The focus of the second example in this paper is chemical intervention to investigate the metabolic pathways within a cell in the context of diseases of cholesterol imbalance. The relationship between cholesterol homeostasis and disease states such as atherosclerosis and gall stones is well established. Thus the enzymes that initiate the major fluxes of metabolites of cholesterol, cholesterol 7α-hydroxylase leading to bile acids and acyl coenzyme A: cholesterol acyl transferase (ACAT) leading to mobile cholesterol in plasma, have become significant targets for chemotherapeutic intervention. Indeed several potential inhibitors of ACAT have been described and their clinical potential has been considered [12]. A critical factor in the success of any therapy involving either of these enzymes, however, is the effect of blocking one major pathway upon the flux of steroid through the other. It would clearly be inappropriate to redress an imbalance in one cholesterol metabolite by generating an imbalance in another. There is thus a need to investigate with precision the controls that exist in cholesterol metabolism in key cells such as liver and intestine. Our approach has been to design, synthesize, and evaluate mechanism based inhibitors, especially for cholesterol 7α-hydroxylase, and to study their effects upon cellular cholesterol metabolism. In the course of this work, we also discovered a surprising feature of molecular recognition by cholesterol metabolising enzymes.

At the time our work began, there were no highly selective inhibitors of cholesterol 7α-hydroxlase, the first and rate limiting enzyme in the biosynthesis of bile acids, and the effect of blocking this pathway on the overall metabolism of a cell was unknown. Cholesterol 7α-hydroxylase is a member of the cytochrome P-450 class. As has been established for antifungal agents such as ketoconazole, which inhibits a cytochrome P-450 enzyme in lanosterol biosynthesis by coordinating with the haem iron through the an imidazole substituent [13], one strategy for inhibiting cholesterol 7α-hydroxylase would be the introduction of a donor group into the ring B region of the steroid structure. This was embodied in our work in a compound known for short as azacholesterol (figure 2. 3) [14].

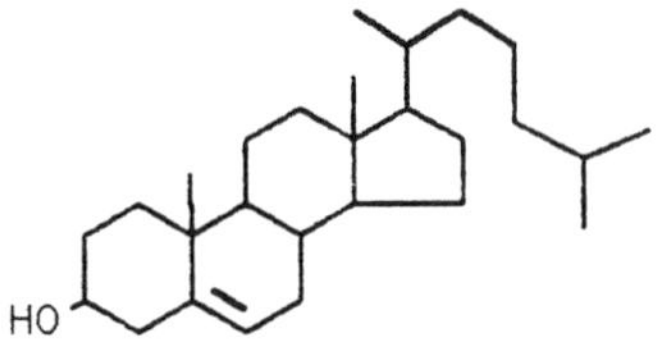

cholesterol

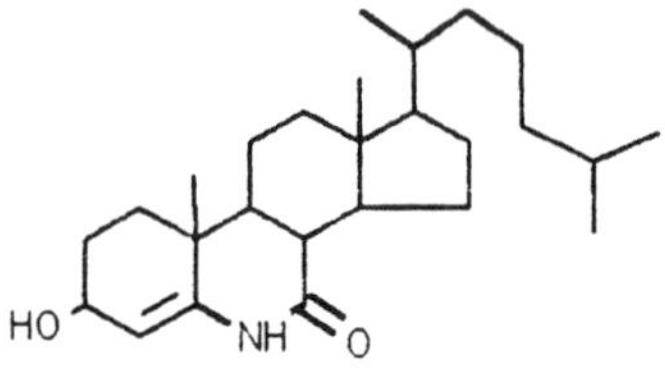

azacholesterol 3

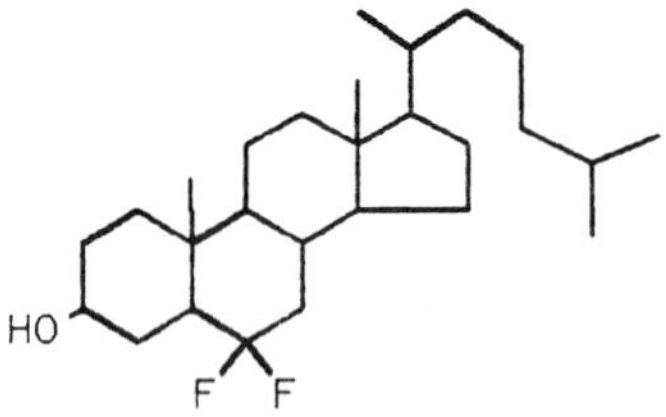

4

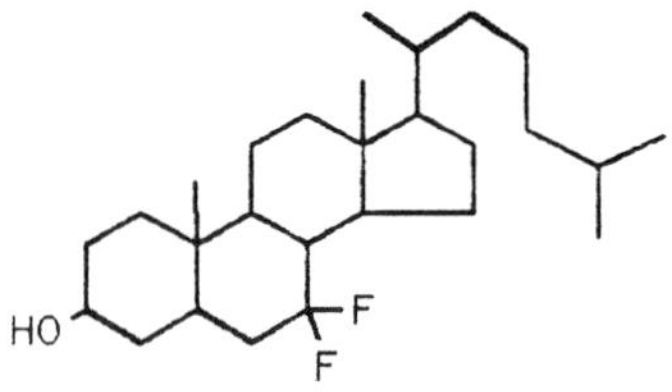

5

Figure 2. Structures of cholesterol analogues.

In addition to blocking the active site crudely by coordination, the mechanism of turnover can be considered as a basis for inhibitor design. We reasoned that if 7α-hydroxylase operated either by a radical mechanism or by insertion, the inclusion of fluoro substituents at the 6-position would generate an inhibitor of the enzyme. Accordingly, we prepared 6,6-difluorocholestanol **4**, and, for comparison, 7,7-difluorocholestanol **5** [15]. The fluorine atoms provide not only a chemical mechanistic control of the reactivity of the sterols but also introduce the possibility of probing cellular interactions further by ^{19}F nmr. The biological results to be described have been obtained in collaboration with Dr K.E. Suckling's laboratories at SmithKline Beecham.

The liver cell is believed to have a generally accessible metabolic pool of cholesterol together with substrate pools associated with particular metabolic pathways such as bile acid formation. Thus, if 7α-hydroxylase is inhibited selectively, the question to be asked is what is the fate of the excess cholesterol that would have followed the bile acid pathway? The emphasis must be placed upon selectivity. If azacholesterol inhibited any enzymes other than 7α-hydroxylase, for example ACAT or hydroxymethylglutarylcoenzyme A reductase (HMGCoA reductase), no results of significance would emerge. Azacholesterol was found to be an inhibitor of cholesterol 7α-hydroxlase with an apparent K_i of 4 μM in the preparation used. Importantly, no significant inhibition was detected with ACAT, HMGCoA reductase, or with enzymes of lanosterol conversion to cholesterol. This is surprising in view of the fact that 7-oxocholesterol, also an inhibitor of 7α-hydroxylase, inhibits ACAT and HMGCoA reductase [16]. Ketoconazole similarly lacks selectivity of action [17]. Whilst we have no molecular evidence for the nature of the interaction of azacholesterol with 7α-hydroxylase, it is clearly ideally suited for investigating metabolic fluxes because of its selectivity of action.

Previous work in Dr K.E. Suckling's laboratories had shown that the intracellular pool of cholesterol has essentially equal access to either of the major routes open to it in hepatocytes. Further, the blocking of ACAT by a selective inhibitor of that enzyme diverted cholesterol into the bile acid pathway. In contrast, when 7α-hydroxylase was inhibited in similar experiments in hepatocytes, the excess cholesterol was not taken up into the esterification pool and no significant increase in the export of cholesterol ester from the cells was observed. In fact, newly synthesised cholesterol from exogenous mevalonate was found to be secreted into the cell culture medium. These observations suggest that newly synthesised cholesterol that has entered the substrate pool for hydroxylation is no longer available for esterification and is evidence for the compartmentation of cholesterol metabolism in the hepatocyte [18]. As mentioned in the introduction, such information is important in the rational design of therapeutic strategies for diseases of cholesterol imbalance.

The fluorosteroids were designed for similar studies to those described in the previous paragraph but with the added potential of being nmr probes for interactions between the biological receptors and the sterols. Both 6,6- and 7,7-difluorocholestanol were found to be acceptable substitutes for cholesterol in its physicochemical role as a rigidifier of bilayer membranes [work in collaboration with Dr D Reid, SmithKline Beecham]. They were also investigated as inhibitors of several cholesterol-transforming enzymes including ACAT and cholesterol 7α-hydroxylase [19]. As expected, the 6,6-isomer **3** was found to be a specific inhibitor of the 7α-hydroxylase, and was also shown to be a substrate for ACAT. On the other hand, surprisingly, the 7,7-isomer **4** was not accepted as a substrate or inhibitor by either enzyme. There is thus a stark contrast in the recognition of the two isomers by the active sites of enzymes although the less discriminating bilayer membrane model did not distinguish between the two.

To obtain some insight into the basis of this molecular recognition phenomenon, we consulted the electrostatic potential surfaces of the molecules (figure 3) [work in collaboration with Dr P. Bladon, Strathclyde]. No firm conclusions can be drawn from these results, but they do indicate how the two fluorosteroids may differ significantly from each other in the position of the negative potential associated with the fluorine atoms. As shown in figure 3, the 6,6-isomer shows negative potential in the region of the π-bond of cholesterol, although to a greater extent. On the other hand, the 7,7-isomer's negative potential bulge is substantially displaced around the 'bay' of ring B of the steroid skeleton and quite diferent from that of cholesterol. It appears that this difference can be recognised by two different cholesterol-transforming enzymes which both reject the 7,7-isomer. This result poses major questions about the way in which enzymes bind non-polar molecules such as cholesterol with the high specificity required for its selective transformation. It is also interesting to enquire which of the fluorines, 7α or 7β, or perhaps both, is responsible for the rejection of the 7,7-isomer. The compounds required to investigate this point are currently being synthesised.

Nmr spectoscopy has been promoted as a non-invasive technique for investigating cellular metabolism. The sensitivity of the ^{19}F nucleus makes it attractive as a probe provided that suitable compounds can be prepared. The evaluation of the properties of 6,6- and 7,7-difluoro-cholestanol was extended to investigate their potential as surrogates for

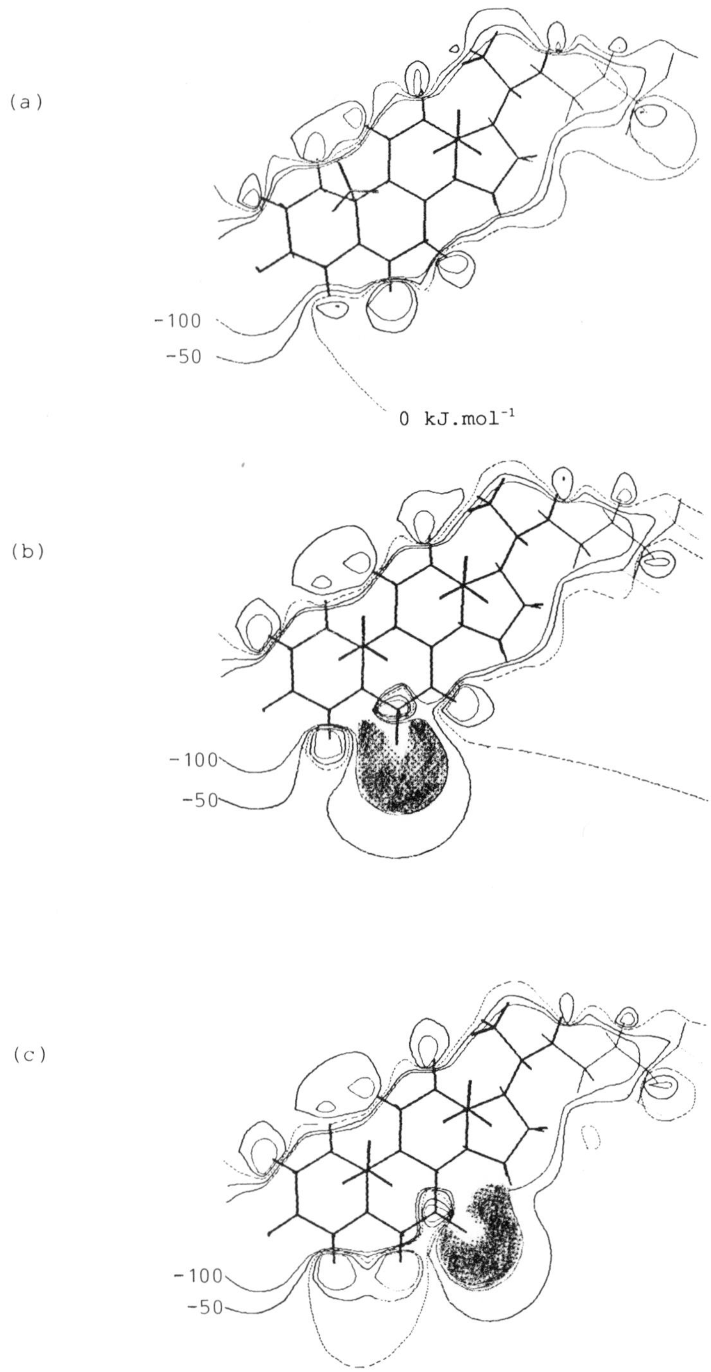

Figure 3. Potential maps of cholesterol, 6,6- and 7,7-difluorocholestanol in the mean plane of the ring system. (a) cholesterol (b) 6,6-difluorocholestanol, (c) 7,7-difluorocholestanol. The relevant regions of high negative potential are shaded. Potentials were calculated using semiempirical methods.

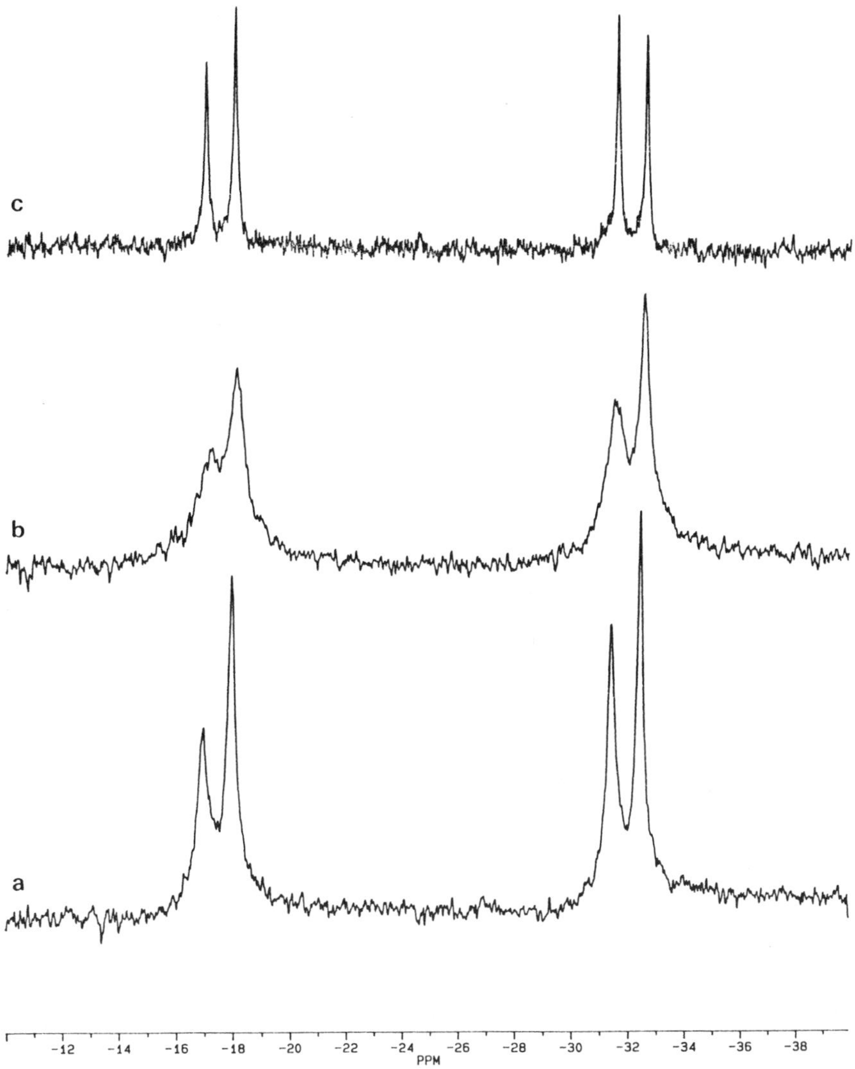

Figure 4. Broadband proton decoupled spectra of 3-oleyl-6,6-difluoro-cholestane incorporated into human hDL at (a) 40°C and (b) 25°C. Trace (c) represents the same material at 40°C after a few minutes' incubation with 1 mg 3-doxylcholestane. The resulting signal diminished only slightly over a 24 hr period.

cholesterol in studies of cellular metabolism by nmr. Our results obtained in collaboration with Dr David Reid (SmithKline Beecham) indicate the potential of this method; however we have yet to apply it to cellular metabolism. We have shown, for example, that the chemical shift of the fluorines in ring B results in considerable shifts of the order of 1 ppm to higher field upon esterification. Such shifts are observed only in aqueous dispersions and not in chloroform. Thus we can anticipate that it will be possible to follow the esterification state of mobile cholesterol surrogates in cells by ^{19}F nmr.

The most interesting experiment so far studied the incorporation of 3-oleyl-6,6-difluorocholestanol into human high density lipoprotein (hHDL) (figure 4). Unusual doublet profiles were obtained which suggest the coexistence of two slowly exchanging environments of the fluorosteroid ester in which it has slightly different chemical shift, coupling, and relaxation properties. Manganese (II) ions, an aqueous relaxation probe, did not significantly affect the observed spectrum. In contrast, the addition of a small quantity of a lipophilic spin probe, 3-doxylcholestane caused a rapid diminution in intensity and sharpening of the spectrum. These observations can be understood if the doxylcholestane partitions into one of the environments of the probe fluorosterol where it broadens the ^{19}F resonances beyond detection. The signals from the probe fluorosterol in the second environment, which is inaccessible to the doxylcholestane, remain unchanged and are revealed as sharp lines. It is possible that the two environments of relevance in the hHDL particle are the cholesterol rich phopsholipid outer shell, which is accessible to doxylcholestane, and the triolein rich core, which is not.

CONCLUSION

Chemical intervention in catalysis by enzymes is thus not limited to the conventional study of enzyme inhibition as the basis for drug design. Its scope extends to the investigation of a wide range of biological phenomena including cellular flux and compartmentation, to say nothing of the applications of enzymes and modified enzymes in organic synthesis. In the investigative mode, its full power will be demonstrated when selective inhibitors, usually mechanism based, are used in combination with selective spectroscopic techniques, such as nmr, and well chosen biological systems. Such a composite experimental system can be considered as a series of selective filters. The enzyme inhibitors provide controls on the metabolism to be investigated, substrate analogues bear the spectroscopically sensitive nuclei, and finally, the spectroscopic technique itself focusses attention upon significant events. This last point was shown by the studies of lipoproteins mentioned above using normal FT nmr pulse sequences but we have demonstrated in other work of direct relevance to clinically used drugs the potential of ^{1}H spin echo nmr in understanding metabolic events in cells [20].

REFERENCES

1. R. Cameron, S.H. Hicholson, D.H. Robinson, C.J. Suckling, and H.C.S. Wood, Syntheses of some functionalised 7,7-dialkyl-7,8-dihydropterins. J.Chem.Soc.,Perkin Trans.1, 2133 (1985).

2. J. Haddow, C.J. Suckling, and H.C.S. Wood, Latent inhibition of dihydrofolate reductase by a pteridine cyclopropane. J.Chem.Soc.,Perkin Trans.1, 1297 (1989).

3. J.T. Bolin, D.J. Filman, D.A. Mathews, R.C. Hamlin, and J.Kraut, Crystal structures of E.coli and L.casei dihydrofolate reductase refined at 1.7A resolution. J.Biol.Chem., 257:13650 (1982).

4. G.C.K. Roberts, Nmr and mutagenesis studies of dihydrofolate reductase. in Chemistry and Biology of Pteridines 1989, eds. H.Ch. Curtius, S. Ghisla, N. Blau, de Gruyter Berlin, 681 (1990).

5. J.J. McCormack, Reductases. in Comprehensive Medicinal Chemistry, vol. 2, ed. P.G. Sammes, Pergamon Press, Oxford, 290 (1990).

6. C.J. Suckling, The cyclopropyl group in studies of enzyme inhibition and metabolism. Angew.Chem.Int.Edn.Engl., 27:537 (1988).

7. T. Umichara, S. Tsuki, K. Tanabe, S.J. Benkovic, K. Furakawa and K. Taira, Computational studies on pterins and speculation on the mechanism of dihydrofolate reductase. Biochem.Biophys.Res.Commun., 161:64 (1988).

8. J. McGill, L. Rees, C.J. Suckling, and H.C.S. Wood, Mechanism based inhibitors of dihydrofolate reductase, in Chemistry and Biology of Preidines 1989, eds. H.Ch.Curtius, S. Ghisla, N. Blau, de Gruyter Berlin, 732 (1990).

9. C.J. Suckling, Use and limitations of models and modelling in medicinal chemistry, in Comprehensive Medicinal Chemistry, vol. 4, ed. C.J. Drayton, Pergamon Press Oxford, 83 (1990).

10. V. Thibault, M.J. Koen, and J.E. Gready. Enzymic properties of a mechanism based substrate for dihydrofolate reductase. Biochemistry, 28:6042 (1989).

11. I.G. Buntain, C.J. Suckling, and H.C.S. Wood, Irreversible inhibition of dihydroorotate dehydrogenase by hydantoins derived from aminoacids. J.Chem.Soc.Perkin Trans.1 3175 (1988).

12. K.E. Suckling and E.F. Stange, Role of ACAT in cellular cholesterol metabolism. J.Lipid.Res., 26:647 (1985).

13. J.L. Adams and B.W. Metcalfe, Therapeutic consequences of the inhibition of sterol metabolism, in Comprehensive Medicinal Chemistry, vol. 2, ed P.G. Sammes, Pergamon Press, Oxford, 340 (1990).

14. W.J. Sampson, J.D. Houghton, P. Bowers, R.A. Suffolk, K.M. Botham, C.J. Suckling, and K.E. Suckling, The effects of 6-azacholest-4-en-3β-ol-7-one, an inhibitor of cholesterol 7α-hydroxylase, on cholesterol metabolism and bile acid synthesis in primary cultures of rat hepatocytes, Biochem.Biophys.Acta 960:268 (1988).

15. L. Brown, W.J.S. Lyall, C.J. Suckling, and K.E. Suckling, The synthesis of some cholesterol derivatives as probes for the mechanism of cholesterol metabolism, J.Chem.Soc.,Perkin Trans.1, 595 (1987).

16. M.A. Scwartz and S. Margolis, Effects of drugs and sterols on cholesterol 7α-hydroxylase activity in rat liver microsomes. J.Lipid.Res. 24:28 (1983).

17. H.J. Kampen, K. Van Son, L.H. Cohen, M. Grifioen, H. Verboom, and L. Havekes, Effect of ketoconazole on cholesterol synthesis and HMGCoA reductase and LDL receptor activity in Hep G2 cells. Biochem.Pharm. 36:1245 (1987).

18. E.F. Stange, Receptor mediated endocytosis: its role in cholesterol homeostasis. Biochem.Soc.Trans., 15:189 (1987).

19. K.E. Suckling, B. Jackson, R.A. Suffolk, J.D. Houghton, and C.J. Suckling, Effects of 6,6-difluorocholestanol and 7,7-difluorocholestanol on hepatic enzymes of cholesterol metabolism. Biochim.Biophys.Acta 1002:401 (1989).

20. J. Reglinski, W.E. Smith, C.J. Suckling, M. Al-Kabban, M.J. Stewart, and I.D. Watson, Doxorubicin-induced altered glycolytic patterns in the leukemic cell studied by spin echo nmr. Clinica Chim.Acta, 175:285 (1988).

IMMOBILISATION OF REDUCTASES BY TWO-PHASE SYSTEMS

AND POLYMER MATRICES

G.L. Lemière, C. Gorrebeeck, M. Spanoghe, D. Lanens, R.A. Dommisse, J.A. Lepoivre, F.C. Alderweireldt

Antwerp University (RUCA), Laboratory of Organic Chemistry Groenenborgerlaan 171, B-2020 Antwerp, Belgium

Co-enzyme dependent enzymes such as horse liver alcohol dehydrogenase (HLAD) can be very useful, but they pose some specific problems. To make the reactions, catalyzed by these enzymes, economically acceptable, the co-enzyme has to be regenerated in situ, and when immobilized enzymes are used, the co-enzyme has to be co-immobilized.

For recycling nicotinamide adenine dinucleotide (NAD^+), which is the co-enzyme of HLAD, very useful methods have been elaborated[1]. We use the enzymatic recycling system with coupled substrates (Fig. 1).

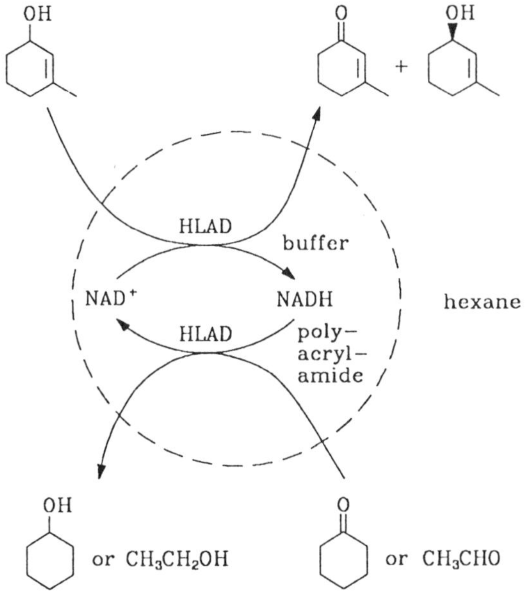

Figure 1. HLAD-Catalyzed synthesis with NAD(H) recycling in polyacrylamide gel beads placed in a water-immiscible solution of substrates.

Bioorganic Chemistry in Healthcare and Technology, Edited by U.K. Pandit and F.C. Alderweireldt, Plenum Press, New York, 1991

For the co-immobilisation of NAD^+ several methods have been elaborated: 1 covalent binding to the immobilized enzyme[2] or to the carrier[3], 2 inclusion with a semipermeable membrane, after binding of NAD^+ to a water soluble polymer[4], 3 adsorbtion of enzyme and co-enzyme to a solid material[5], 4 trapping of both in microemulsions[6] or 5 trapping in gel beads, which are placed in a water-immiscible organic solvent[7].

We are working on this last method. We have chosen for polyacrylamide gel beads containing an aqueous solution of HLAD and NAD^+ placed in a hexane solution of coupled substrates: 3-methyl-2-cyclohexenol and cyclohexanone (Fig. 1). This method was chosen since it is easy and safe. First the enzyme and co-enzyme are not chemically modified; they are simply restrained in the aqueous gel beads, since they cannot cross the water-hexane interface, whereas the substrates and products can diffuse into the gel beads and back into the organic solvent. Further polyacrylamide is a well known versatile material and the beads of it can be synthesized easily by an inverse suspension polymerization process. As desired gel beads with varying diameters or polymer composition[8] can be prepared.

The classical way of bringing HLAD and NAD^+ into the gel beads is addition of the enzyme and co-enzyme to a solution of polymerising acrylamide. Since we have found that an important amount of the enzyme activity is lost during the polymerization process[9], we looked for another technique. The gel beads are now prepared without enzyme or co-enzyme, dried and in a following step the dry gel beads are swollen in a buffered solution of HLAD and NAD^+. Since the drying and swelling processes are completely reversible, amounts of dry beads are swollen in an amount of enzyme solution which is equal to the quantity of water which is lost by drying. In this way gel beads are prepared which are saturated by the aqueous phase, and which have absorbed all the enzyme-co-enzyme solution quantitatively.

Figure 2 shows some results obtained with equal amounts of (15,5)[8]-gel beads with different diameter. These beads were placed in rotating test tubes containing a hexane solution of 3-methyl-2-cyclohexenol and cyclohexanone. As can be seen, the reaction rate increases with increasing bead diameter. This result is opposite to what is normally expected. This is caused by a strong tendency of these beads to clogg. Small beads give compact clumps, large beads give loose aggregates. By carefull observation it can be seen that the substrate solution does not penetrate the clumps of small beads, but it fills the free space in the loose aggregates of larger beads. Analogous results are found for beads with higher or lower total monomer concentrations.

This brings us to an important aspect of enzymatic reactions in heterogeneous systems, namely the transport problems of substrates and products to the enzyme in the gel beads and back into the organic solvent. Therefore the distribution of the enzyme in the gel beads is important for the activity of these heterogeneous catalysts. It can be imagined that as a result of our swelling procedure, the enzyme will not be homogeneously distributed over the total volume of the beads. It can be expected that with increasing T%-values[8] the gel beads will become less permeable so that the enzyme cannot penetrate into the core of the beads.

Figure 3 shows some results with gel beads in a solution of substrates in hexane. Gel beads with almost the same size, but with increasing total monomer concentration are tested for their reactivity. The reaction rate increases with increasing T%-values. This suggests that in the case of the gel beads with higher T%-value the enzyme is more concentrated at the surface of the beads and can be better reached by the substrates coming from the organic solution.

In a control experiment Sephadex G-15 and G-150 beads with the same diameter were swollen in a HLAD/NAD^+ solution and placed in a substrate solution. For these commercial gels it is known that HLAD can penetrate the G-150 gel, but is completely excluded from the G-15 gel. The reactivity found for the G-15 gel was almost two times higher than for the G-150 gel. This result fully supports the explanation given for the polyacrylamide gel beads.

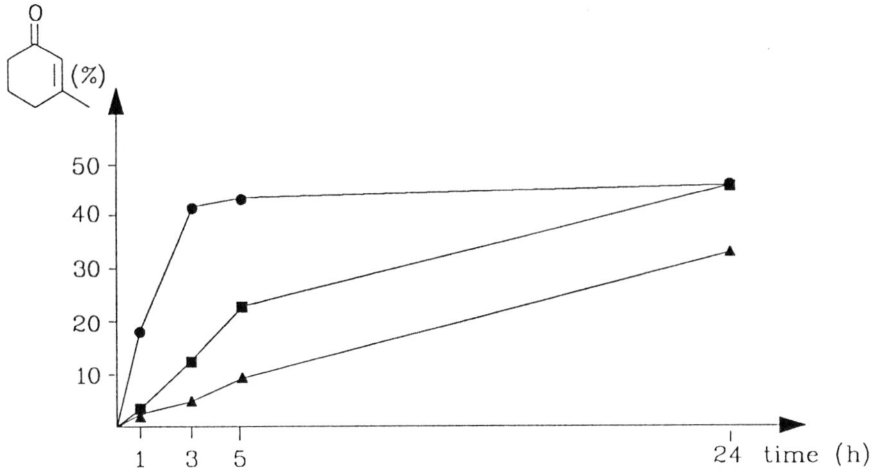

Figure 2. Influence of the bead size on the reaction rate. Diameter of swollen (15,5)-gel beads: ● 994 μm, ■ 114 μm, ▲ 73 μm.

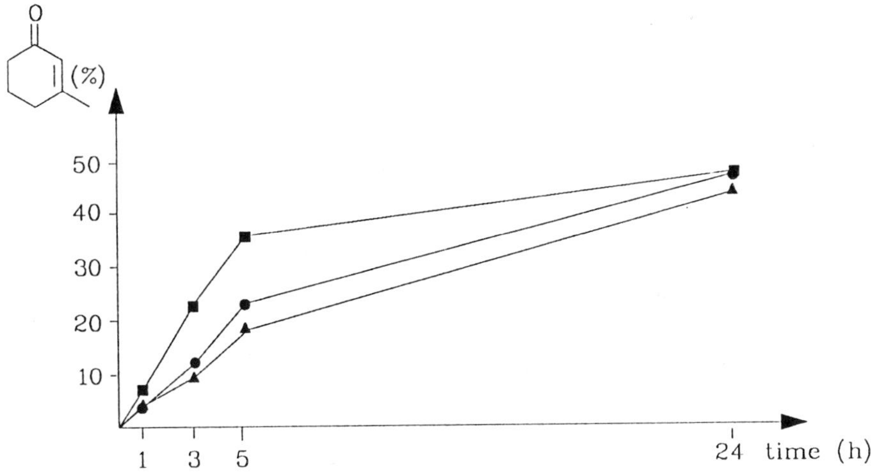

Figure 3. Influence of the total monomer concentration T% on the reaction rate. T% of the beads: ■ 20% (129 μm), ● 15% (114 μm), ▲ 10% (103 μm).

To obtain more information about the permeability of the different types of polyacrylamide gels an experiment was set up in which the gel beads were swollen in an essentially different way. Portions of dry beads were swollen in quantities of HLAD/NAD$^+$ solution which were twice the quantity necessary for saturating the beads. After swelling the beads were separated from the excess of enzyme solution. The reactivity of the beads was tested as before in a heterogeneous system. The reactivity of the residual enzyme solution was tested in a classical way. Table I gives the results. The reactivity of the gel beads are given relative to the reactivity of the (5,5)-gel; the reactivity of the residual enzyme solution is given relative to the reactivity of the original HLAD/NAD$^+$ solution.

In contrast to the preceeding experiments the activity of the gel beads decreases with total monomer concentrations increasing from 5% to 20%. For higher T%-values the reactivity is constant. Obviously, the enzyme cannot penetrate the beads with higher T%-values, but remains in the excess solution. This can clearly be seen from the activity of these solutions which increases with increasing total monomer concentrations of the corresponding beads. This last set of values further reveals that even for the softest gels the enzyme is more or less hindered to penetrate the gel beads, since in all cases the reactivity of the excess is higher than that of the original HLAD/NAD^+ solution.

Table I

Gel-	(5,5)	(10,5)	(15,5)	(20,5)	(25,5)	(30,5)
Relative reactivity[a] of the gel beads	1.00	0.75	0.42	0.26	0.26	0.26
Relative reactivity[b] of the superfluous solution	1.44	1.54	1.56	1.73	1.82	1.95

[a] Relative to the reactivity of the (5,5)-gel.
[b] Relative to the reactivity of the original HLAD/NAD^+ solution.

Finally, to confirm what is suggested by the preceeding experiments, we have tried to visualize the enzyme directly in the gel beads. Therefore beads of different total monomer concentration were swollen in a solution of HLAD which was labeled with a gadolinium diethylenetriaminepentaacetic acid complex (Gd-DTPA)[10]. Using this paramagnetic labeling it was possible to make high resolution NMR-images of these beads.

Figure 4 shows the images. As can be seen the (5,5)-gel beads are fully penetrated by the labeled enzyme. In the (10,5)-gel beads one can clearly observe a bright outer shell, which means that the enzyme has only partially penetrated the beads. In the (15,5)-gel beads only a rim of approximately 250 microns thick is seen. On the beads with higher total monomer concentration no HLAD-(Gd-DTPA) can be seen.

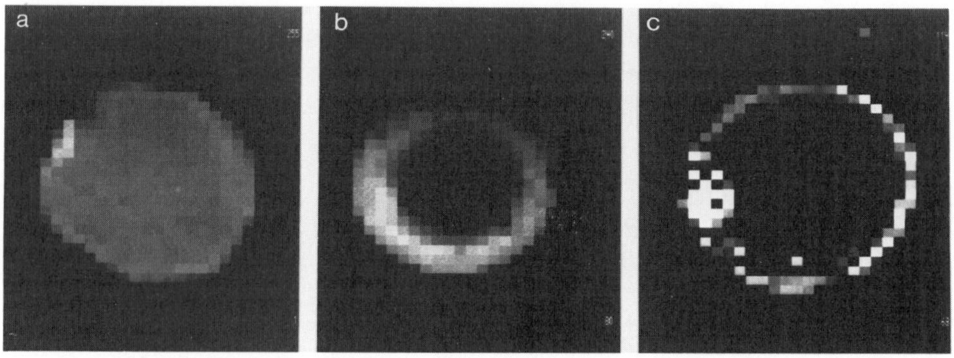

Figure 4. High resolution NMR images of polyacrylamide gel beads containing Gd-DTPA labeled HLAD : a. (5,5)-gel, b. (10,5)-gel and c. (15,5)-gel.

As a conclusion we can say that we have a very simple method to prepare polyacrylamide gel beads containing HLAD and NAD^+. It is a method which avoids the harmfull action of the polymerization process on the enzyme. It is a method which allows quantitative immobilization of enzyme and co-enzyme, both without any chemical modification. Transport problems of substrates and products can be minimized by the use of polyacrylamide gel beads with high total monomer concentration, since the enzyme is concentrated at the surface of the beads so that the enzyme can immediately be reached by the substrates from the water-immiscible solvent.

References

1. a. J.B. Jones, D.W. Sneddon, W. Higgins and A.J. Lewis, J. Chem. Soc. Chem. Commun. 856 (1972), b. S.S. Wang and C.-K. King, Adv. Biochem. Eng. **12**, 119 (1979), c. D. Mandler and I. Willner, J. Chem. Soc. Perkin 2, 805 (1986), d. H.K. Chenault and G.M. Whitesides, Appl. Biochem. Biotech. **14**, 147 (1987), e. A. Deffner and H. Simon, Ann. N.Y. Acad. Sci. **501**, 171 (1987), f. K. Otsuka, S. Aono, I. Okura and F. Hasumi, **51**, 35 (1989).
2. M.-O. Mansson, P.-O. Larrson and K. Mosbach, Eur. J. Biochem. **86**, 455 (1978).
3. a. S. Gestrelius, M.-O. Mansson and K. Mosbach, Eur. J. Biochem. **57**, 529 (1975), b. Y. Yamazaki and H. Maeda, Methods Enzymol. **136**, 21 (1987), c. H. Ukeda, K. Ono, M. Imabayashi, K. Matsumoto and Y. Osojima, Agric. Biol. Chem. **53**, 235 (1989).
4. a. Y. Morikawa, I. Karube and S. Suzuki, Biochim. Biophys. Acta, **523**, 263 (1978), b. R. Wichmann, C. Wandrey, A.F. Bückmann and M.-R. Kula, Biotechnol. Bioeng. **23**, 2789 (1981).
5. a. J. Grunwald, B. Wirz, M.P. Scollar and A.M. Klibanov, J. Am. Chem. Soc. **108**, 6732 (1986), b. A. Aoki, M. Ueda, H. Nakajima and A. Tanaka, Biocatal. **2**, 89 (1989).
6. a. K. Martinek, A.V. Levaskov, N. Klyachko, Y.L. Khmelnitski and I.V. Berezin, Eur. J. Biochem. **155**, 453 (1986), b. K.M. Lee and J.-F. Biellmann, FEBS Lett **223**, 33 (1987).
7. D.G. Drueckhammer, S.K. Sadozai, C.-H. Wong and S.M. Roberts, Enzyme Microb. Technol. 564 (1987).
8. The polymer composition can be characterized by the monomer concentration T% and the degree of cross-linking C%:

$$T\% = \frac{\textit{weight of acrylamide} + \textit{weight of BIS}}{\textit{volume of water}} \times 100$$

$$C\% = \frac{\textit{weight of BIS}}{\textit{weight of acrylamide} + \textit{weight of BIS}} \times 100$$

Therefore gel beads will be further characterized by two numbers between parentheses: (T%,C%).
9. R. Mertens, G.L. Lemière, J.A. Lepoivre and F.C. Alderweireldt, Biocatal **2**, 121 (1989).
10. W.W. Layne, D.J. Hnatowitch, P.W. Doherty, R.L. Childs, D. Lanteigne and J. Ansell, J. Nucl. Med. **23**, 627 (1982).

TWO APPROACHES IN THE DESIGN OF PLANT ENZYME INHIBITORS

Liliane GORRICHON

Laboratoire de synthèse et physicochimie organique,
URA CNRS 471, Université Paul sabatier,
31062 TOULOUSE CEDEX (FRANCE)

This work was engaged in collaboration with our colleagues in physiology interested in plant growth regulation mainly through two important biological pathways :

- the biosynthesis of aromatic aminoacids (the shikimate pathway),
- the lignification process.

The ability to synthesize aromatic aminoacids is restricted to plants and microorganisms (1)(2) ; so it is a good target for the definition of phytoactive compounds. The development of bioactive compounds was also of interest to identify and understand the differences which appear between plants and microorganisms in the shikimate pathway (3).

The problem in the lignification process was quite different. It was to reduce the lignin fluxes without damaging plant development. It was expected that a reduction of the lignin content would improve the digestibility and nutritional value of forage crops.

As organic chemists we decided to examine the chemical reaction sequences known in these two biological transformations, to select a target enzyme and to try to define some molecules which could act as specific inhibitors of these enzymes.

Without any detailed structural informations on the enzymes the first approach was the synthesis of substrate or "transition state" analogs. The second was the development of irreversible (or suicide) inactivators (4)(5).

Biosynthesis of aromatic aminoacids (shikimate pathway)

Many models of substrate or transition state analogs have been studied in the litterature (6)(7) since the discovery of phosphonoglycine

Bioorganic Chemistry in Healthcare and Technology, Edited by U.K. Pandit and F.C. Alderweireldt, Plenum Press, New York, 1991

as a very potent herbicide which inhibits an enzyme of the shikimate pathway : the phosphoenolpyruvate shikimate synthase (8)(9).

We were interested in the third step of the shikimate pathway and the target enzyme was the dehydroquinase (DHQase) which catalyses the syn dehydration of (-)dehydroquinic acid into (-)dehydroshikimic acid.

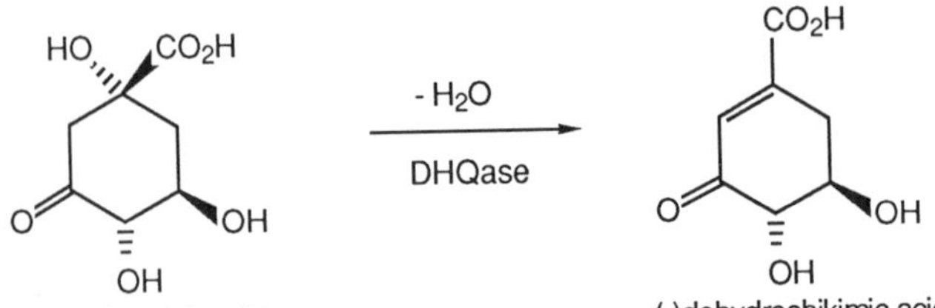

(-)dehydroquinic acid (-)dehydroshikimic acid

In this step some differences appear between plants and microorganisms. First in enzyme organisation ; all the enzymes are independent in microorganisms ; in plants the dehydroquinase is associated in a bifonctional complex to the shikimate oxidoreductase (4th step)(10).

The shikimate pathway :

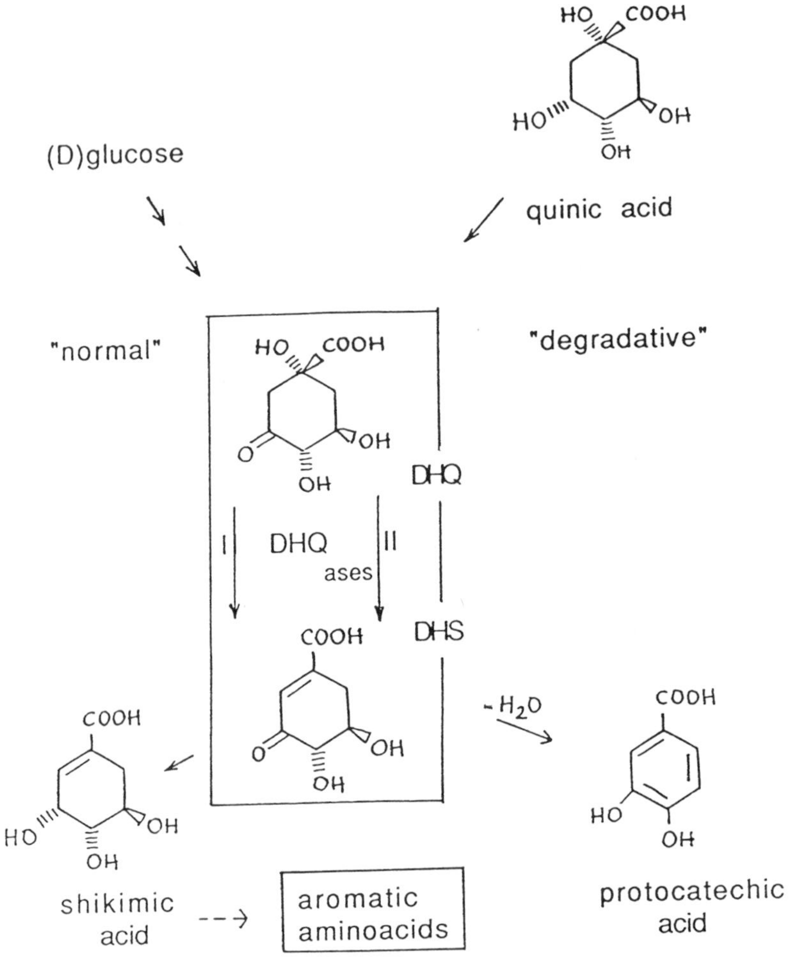

Another difference is the presence of a second dehydroquinase II ; this activity is dependent on the occurrence of quinic acid only present in plant (11).

The main steps of the mechanism (12) proposed to explain the syn dehydration catalysed by the two DHQases are :

- a covalent interaction between the C=O of the substrate and a lysine NH_2 residue in the enzyme,
- a selective deprotonation of the C-2 pro(R) hydrogen,
- the β-elimination of the C_1-OH.

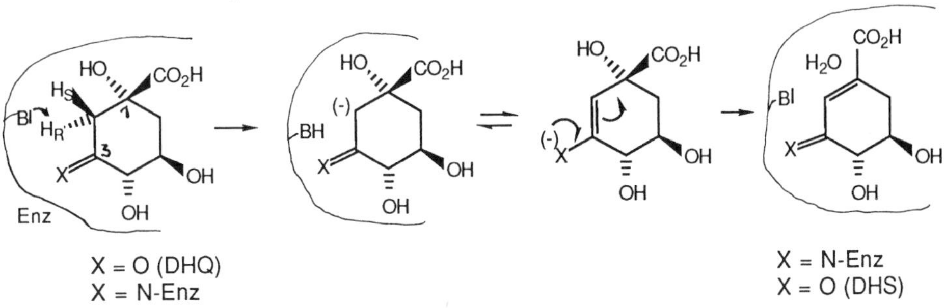

Some models of the first step in this mechanism vere consi-dered. As we were still unsuccessful in this approach, only one example among those we tried will be described.

Maintaining on the substrate the carboxylate on C_1 and the three hydroxyl groups (C_1-OH, C_5-OH and specially the C_4-OH known to interact with the enzyme) we postulated that C_3 hydroxylated quaternary compounds might mimic the first intermediate gem aminoalcohol formed in the lysine NH_2 attack on the substrate.

Enzymic mechanism (DHQase) :

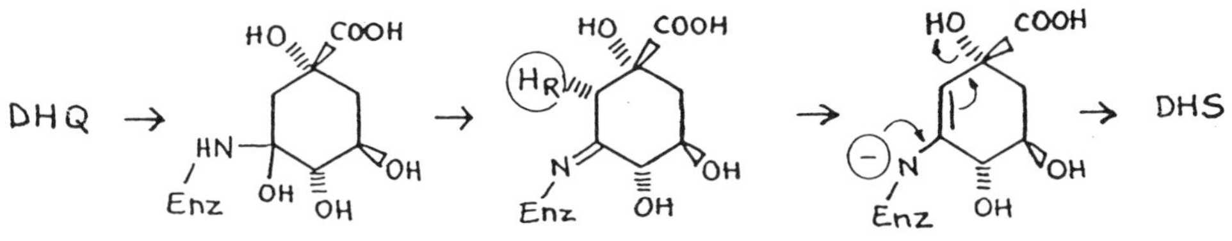

Modelisation :

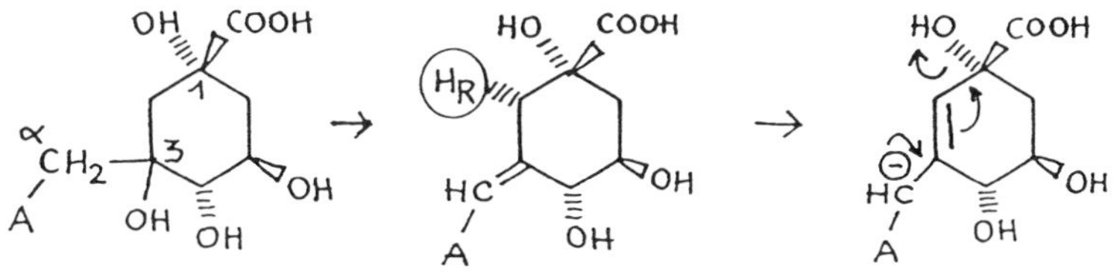

A : electron withdrawing group => H_α acidity ↗

An electron withdrawing group A was introduced to increase the acidity of the α-methylenic hydrogens and facilitate the elimination of water.

The synthesis of these compounds starts with the esterification and protection of (-)dehydroquinic acid and then nucleophilic addition on the ketonic function of organometallic reagents : ester enolates, metallated sulphone, propargylic reagent.

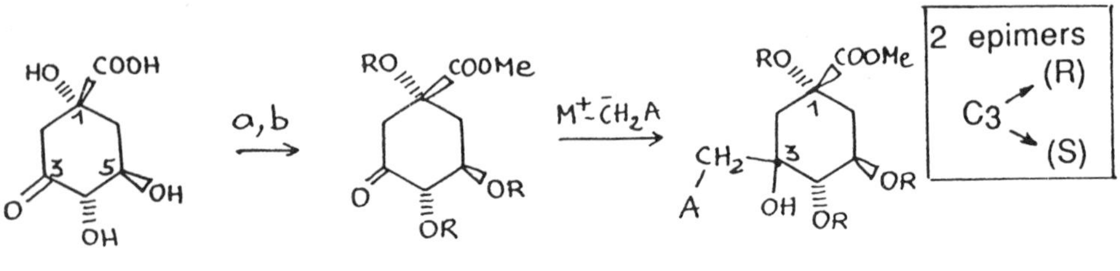

(-)dehydroquinic acid

Me-DHQ (R = $SiMe_3$)

a) MeOH, amberlite, H^+, reflux, 50%
b) Me_3SiCl, HMDS, 0°, 80%

A		yields ( )*	
$-CO_2Et$	I	57	(63)
$-SO_2Ph$	II	65	(65)
$-C{\equiv}CH$	III	30	(59)

* *from Me-DHQ effectively reacted*

Two diastereoisomers were expected (R or S on C_3). The reaction is stereospecific with Li derivatives at -78°C or when M = MgBr at 20°C. Only the isomer with the C_3-OH and COOMe in the syn position was obtained in ~ 60% yield.

The second isomer was isolated in low yields with lithioderivatives at higher temperatures (-40° or -20°).

The configuration of these compounds was determined from 2D NMR experiments which show an interaction between the C_3-OH and the methoxy of the ester function on C_1 ; intramolecular lactonisation of the major isomer was also observed.

The deprotection of the hydroxyl groups was easily obtained in quantitative yield and the ester function removed, using pig liver esterase, just before the biological assays.

However this approach was unsuccessful and no significant inhibition of the enzyme DHQase was observed. This may be due to the steric hindrance of the electron withdrawing group A or due to the removal of the C=O function which could make the recognition of the substrate inefficient (13).

Fortunately some better biological results were obtained with lignin inhibition.

Lignification process

Lignin monomer biosynthesis is considered to begin when the shikimate pathway stops. The aromatic amino acid, phenyl alanine, is transformed to p-OH cinnamic acids and to S-CoA esters ; the esters are then reduced to aldehydes and the aldehydes to alcohols. The polymerisation of the alcohols give the lignins (14).

Lignin monomer biosynthesis

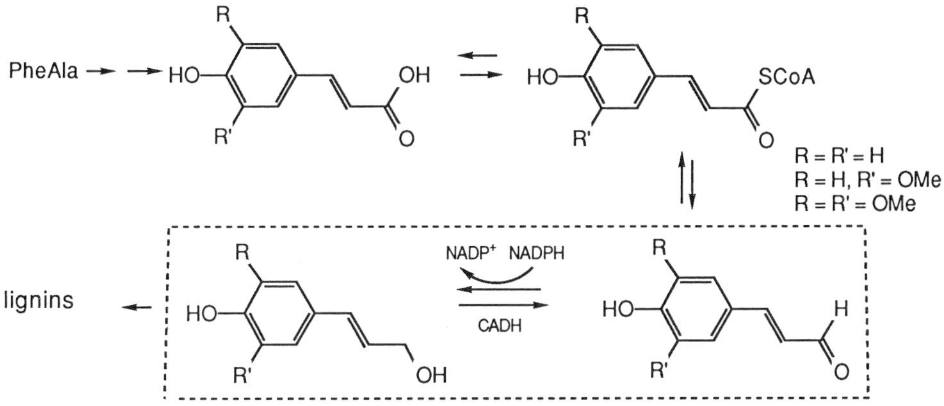

Only the enzymes involved in the last two steps are specific for lignin biosynthesis. We selected the last one : cinnamoyl alcohol dehydrogenase (CAD). CAD belongs to the well know group of NADPH dependent zinc metalloenzymes and presents some important structural similarities with other Zn dehydrogenases (15). From the information available it was reasonable to assume that one of the zinc atoms involved in the dimeric structure of cinnamoylalcohol dehydrogenase acts as a Lewis acid in the reduction.

Zn^{++}, by complexation, will activate the aldehyde C=O before the hydride attack.

Enzymic reduction of p-OH cinnamaldehydes

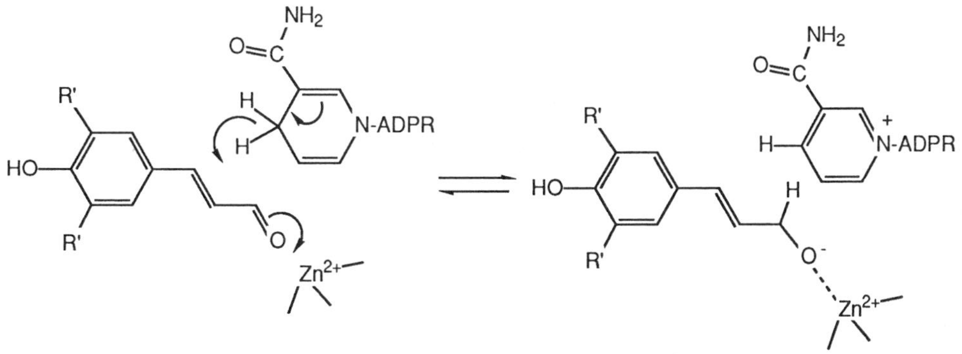

The synthesis of models was realised taking into account two requirements :

- some structural analogy with the substrate(s)
- their complexing ability to bind Zn^{++}

Reversible substrate analogs inactivators:

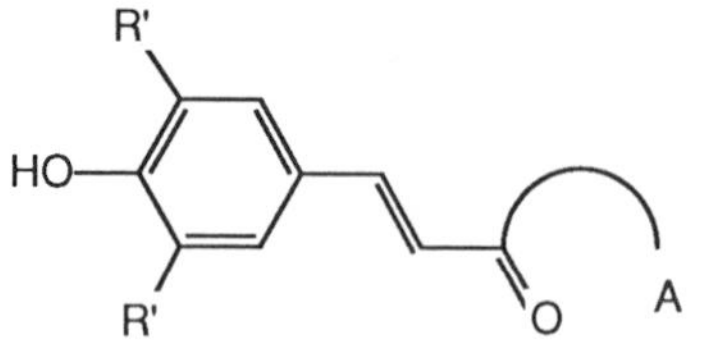

Requirement :

complexing properties of the A group / Zn^{++}

The best results were first obtained with p-OH cinnamoyl esters and amides, mainly the cinnamoyl 2-aminopyridines. These compounds have shown *in vitro* and *in vivo* inhibitory effects on CAD (16). These compounds act as competitive inhibitors of the cinnamic substrates.

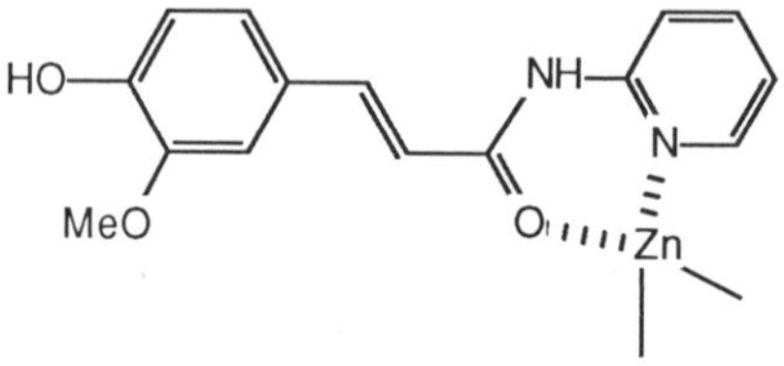

The development of bidendate ligands was further examined on simpler aromatic compounds ; mainly on β bifunctional systems ; diketones, β-diesters, keto or esteramides were examined without any success.

Ar-A⌒B A, B = CO, COOR, CON<

We reasoned that the complexing ability of these compounds to bind zinc would be greatly increased by replacing the C=O function by a S=O moiety.

The biological results were in agreement with this hypothesis and some sulphinamoyl acetates were found particularly active against the cinnamoyl alcohol dehydrogenase activity.

Sulphinamoylacetates :

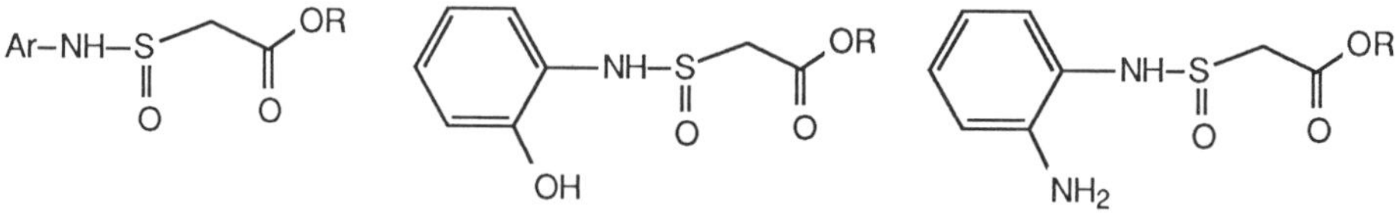

OH-PAS

NH_2-PAS

Moreover irreversible inactivation with ortho OH or NH2 substituted aryl sulphinamoylesters (OHPAS-NH_2PAS) was postulated by Boudet et al. (17)

Biological results (17)

With these two compounds inhibition was directly proportional to inhibitor concentration up to 1 m M and increased with preincubation time between the inhibitor and CAD. From dialysis experiments ; it also appears that the inhibitors and the enzyme were tightly bonded.

These two compounds have no affinity for ethanol dehydrogenase or other phenolic enzymes (except NH_2PAS on the cinnamoyl CoA reductase).

In vivo the fluxes of lignins (from poplar stems) were reduced to a larger extend than expected from *in vitro* results.

These observations agree with the hypothesis of an irreversible (pseudoirreversible type) inhibition (17) and strongly suggest that the two ortho substituted sulphinamoylesters (OHPAS, NH_2PAS) act as suicide inactivators.

Getting back to the actual chemistry we were trying to explain these results : the hypothesis is that the inactivation of the cinnamoyl alcohol dehydrogenase arise if there is formation of

- a complex between Zn^{++} and the inhibitors at the enzyme's active site
- a latent or masked electrophilic function in the ortho OH orNH_2 sulphinamoylesters. This electrophile will combine to give a strong (covalent) interaction with a basic moiety in the enzyme.

This reaction will be responsible for the irreversible (suicide) inactivation.

Complexing properties of sulphinamoylesters

They were examined with zinc salts by UV and IR spectroscopic methods in conditions previously used to modelise interactions between horse liver zinc dehydrogenase and its substrates (18)(19). In this last case the results were further confirmed by spectroscopic studies with the purified enzyme (20).

The IR results for the complexation of sulphinamoyl esters with $ZnBr_2$ in aprotic media (Et_2O) showed that complexation does not occur between the S=O and the ester function as we supposed but between the S=O and the ortho substituant (OH or NH_2, if any) (21).

Complexing properties of sulphinamoylesters with $ZnBr_2$ (Et_20):

IR

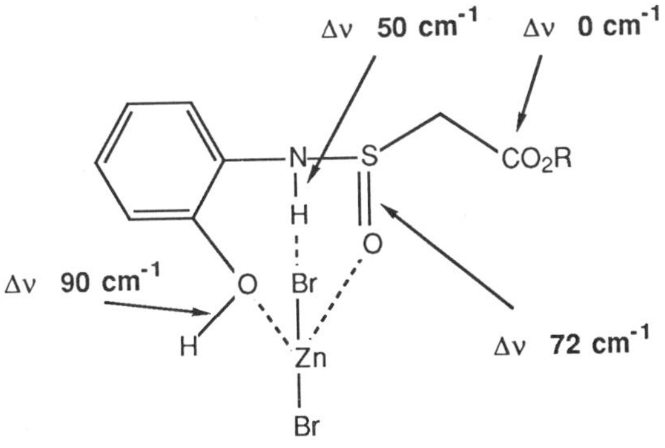

The values of the complexation constants which were determined from the UV studies also increase greatly with the ortho substituted compounds (22)

Hydrolysis and aminolysis of sulphinamoylesters

The second point is related to a latent electrophilic function in sulphinamoylesters. The hydrolysis or aminolysis reactions were examined ; they could occur at the enzyme active site to unmask the electrophilic function in the model compounds.

The mechanism of hydrolysis in aqueous basic media was studied in detail ; it agrees with an elimination reaction which finally gives the t.butoxycarbonyl methane sulphinate ion and anilines (23).

After a rapid deprotonation of the activated methylenic hydrogens the rate determining elimination of the aromatic amine takes place on the conjugate base of the substrate, leading to a sulphine intermediate :

- with a strong electron withdrawing group on the aromatic ring the elimination is first order,
- with a weaker electron withdrawing group the elimination is second order and needs the presence of a general acid catalysis.

Hydrolysis (A) and aminolysis (B) of sulphinamoylesters

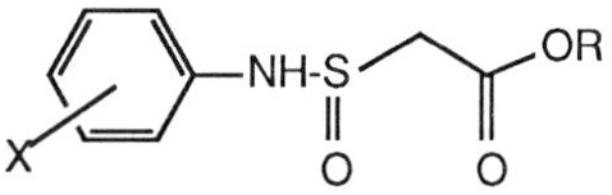

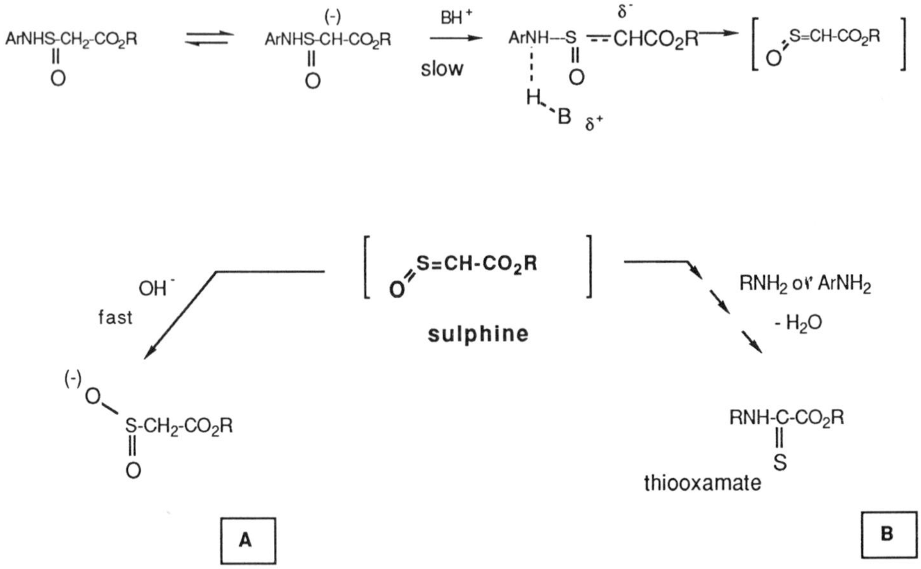

. Strong electron withdrawing groups : E_1cb mechanism (first order)
. Weaker electron withdrawing groups or donating group
=> general base catalysis : E_2 mechanism (second order)

The aminolysis of the sulphinamoylesters gave an unexpected rearrangement to thiooxamates. The mechanism postulated in the reaction proceeds through the same sulphine intermediate observed in the hydrolysis mechanism.

Are these data in agreement with the biological results ?
Compounds **1**-**4** do not inhibit the enzyme ;

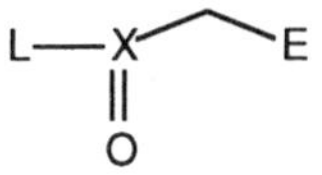

. no inhibition with :

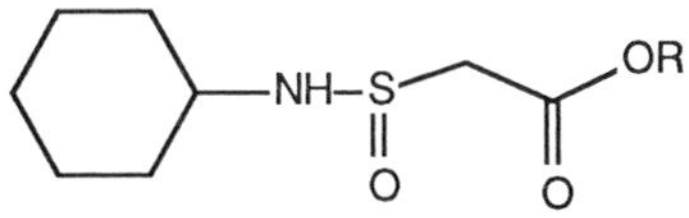

1

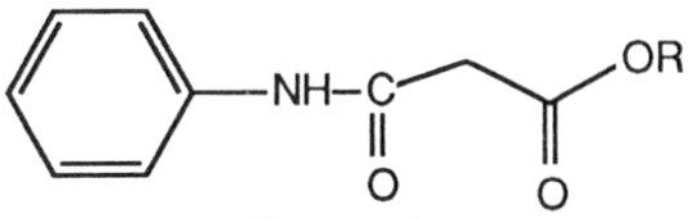

2

R = alkyle

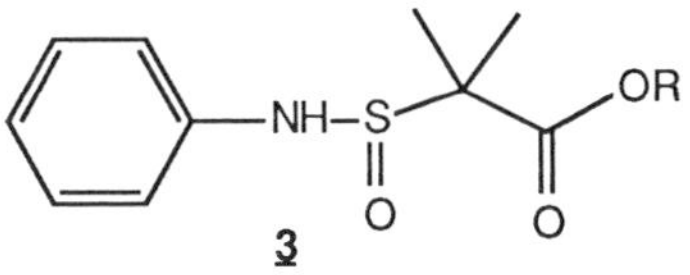

3

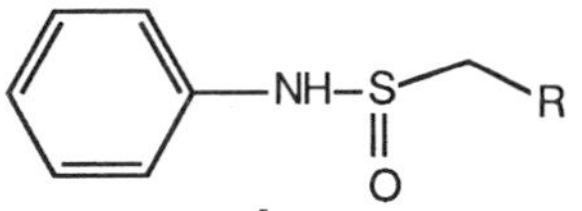

4

and the steric requirements for CAD inactivation by sulphinamides are

- the presence of an aromatic leaving group
- the S=O function
- the methylenic hydrogens α to S=O
- an electron withdrawing group α to the methylene group.

For irreversible (suicide) inhibition of CAD :

- an ortho OH or NH2 function on the aromatic ring.

These requirements can be understood from the complexation and kinetic study results :

- The ortho substituents allow correct binding of the sulphinamoylesters to zinc through the oxygen of the S=O and the OH/NH_2 function at the enzyme's active site.
- The deprotonation step observed in the hydrolysis and aminolysis mechanism could occur in the enzyme :

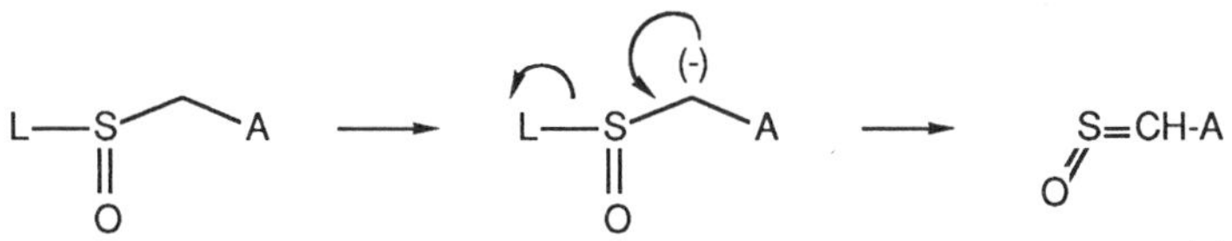

sulphine

The presence of an electron withdrawing group and the complexation between S=O and Zn^{++} would facilitate the deprotonation by lowering the pKa of the CH_2 hydrogens. With the ortho hydroxy-suphinamoylesters, the elimination reaction will finally lead to the 2-aminophenol (inactive) and the sulphine ester.

The electrophilic sulphine function would be responsible for the irreversible inactivation of the enzyme.

Possible interpretation of CAD irreversible inactivation by sulphinamoyl acetates

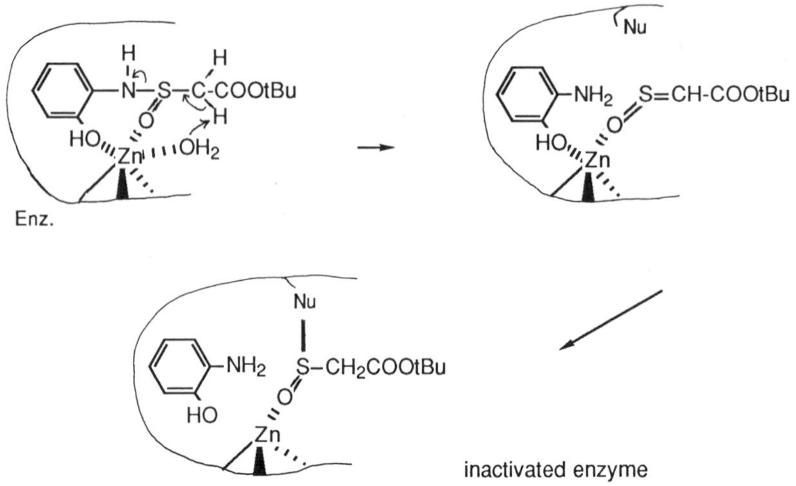

CONCLUSION

The design of plant enzyme inhibitors was examined on

- the dehydroquinase in the third step of the shikimate pathway ; the synthesis at C-3 quaternary compounds from dehydroquinic acid gives inactive compounds ;

- the cinnamoylalcohol dehydrogenase in the lignin monomer biosynthesis :
  . p-OH cinnamides and sulphinamoylesters were efficient against the enzyme activity.
  I. rreversible (suicide) inactivation was suggested from the biological results with ortho hydroxy or amino sulphinamoylesters. Complexation and kinetic results agree with an inactivation through the sulphine moiety O=S=CH which could constitute a new family of irreversible inactivators.

REFERENCES

(1) B. GANEM, *Tetrahedron* ,1978, **34**, 3353
(2) P.A. BARTLETT, *Rec. Adv. Phytochem.*, 1986, **20**, 119
(3) K. DUNCAN, S. CHAUDHURI, M.S. CAMPBELL, J.R. COGGINS, *Biochem. J.* 1986, **238**, 475
(4) H. DUGAS, *Bioorganic chemistry, A chemical approach to enzyme action*, Springer Verlag 2nd ed., 1989
(5) C. WALSH, *Tetrahedron,* 1982, **38**, 871
(6) L.M. REIMER, D.L. CONLEY, D.L. POMPLIANO, J.W. FROST, *J. Amer. Chem. Soc.*, 1986, **108**, 8010
(7) P.A. BARTLETT, C.R. JOHNSON, *J. Amer. Chem. Soc;*, 1985, **107**, 7792
(8) H. HOLLANDER, N. AMRHEIN, *Plant physiol.*, 1980, **66**, 823

(9) N. AMRHEIN, B. DENS, P. GEHRKE, H.C. STEINRUCKEN, *Plant physiol.*, 1986, **66**, 830

(10) S. CHAUDHURI, J.M. LAMBERT, L.A. McCOLL, J.R. COGGINS, *Biochem. J.*, 1986, **239**, 699

(11) A.M. BOUDET, A. BOUDET, H. BOUYSSOU, *Phytochem.*, 1977, **16**, 919

(12) M.J. TURNER, B.W. SMITH, E. HASLAM, *J. Chem. Soc. Perkin trans I*, 1975, 52

(13) T.D.H. BUGG, C. ABELL, J.R. COGGINS, *Tetrahedron Lett.*, 1988, **28**, 6783

(14) H. GRISEBACH, *Lignins in the biochemistry of plant*, P.K. STUMPF and E.E. CONN (Eds), Academic Press 1981, p. 457

(15) M.H. WALTER, J. GRIMA-PETTENATI, C. GRAND, A.M. BOUDET, C.J. LAMB, *Proc. Natl. Acad. Sci. USA*, 1988, **85**, 5546

(16) A.M. BOUDET, C. GRAND, *Lignin Synthesis Inhibitors Monogr. on Crop. Prot. Counc.* 1986, **36**, 67

(17) C. GRAND, F. SARNI, A.M. BOUDET, *Planta*, 1985, **163**, 232

(18) P.W. JAGODZINSKI, W.L. PETITCOLAS, *J. Amer. Chem. Soc.*, 1981, **103**, 234

(19) M.F. DUNN, J.F. BIELLMANN, G. BRANLANT, *Biochemistry*, 1975, **14**, 3176

(20) C.T. ANGELIS, M.F. DUNN, D.C. MUCHMORE, R.M. WING, *Biochemistry*, 1977, **16**, 2922

(21) M. BALTAS, J.D. BASTIDE, A. DE BLIC, L. CAZAUX, L. GORRICHON, P. MARONI, M. PERRY, P. TISNES, *Spectrochimica Acta*, 1985, **41A**, 789 and 793

(22) M. BALTAS, L. CAZAUX, A. DE BLIC, L. GORRICHON, P. TISNES, *J. Chem. Research* , 1988, 284 (S) and 2236 (M)

(23) M. BALTAS, L. CAZAUX, L. GORRICHON, P. MARONI, P. TISNES, *J. Chem. Soc., Perkin trans II*, 1988, 1473

THE MECHANISM OF THE UROCANASE REACTION

János Rétey

Chair of Biochemistry
Institute of Organic Chemistry
University of Karlsruhe
Willstätter Allee
7500 Karlsruhe 1 F.R.G.

INTRODUCTION

Urocanase, an ubiquitous enzyme of M_r = 120,000, catalyses the addition of water to urocanate in an unusual manner (Eqn.1).

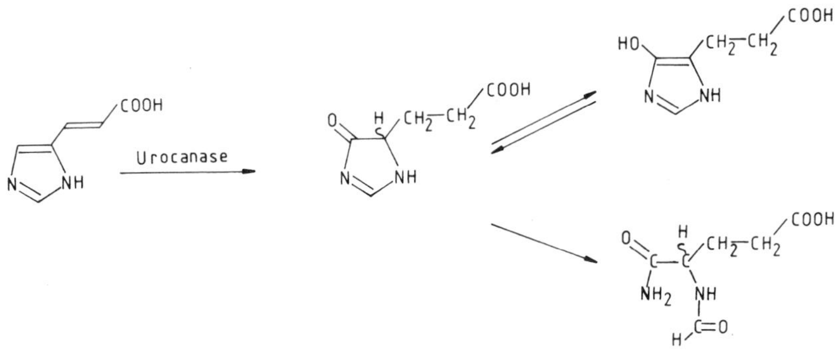

The product of the reacton is 3-(5'-hydroxyimidazol-4'-yl)-propionate which is at pH 7 in spontaneous equilibrium with its tautomeric form, 3-(5'-oxoimidazol-4'-yl)propionate. The latter is also subject to a slow spontaneous and irreversible hydrolysis to N-formyl-isoglutamine. All these transformations can be monitored directly by NMR-spectroscopy.[1]

Urocanase consists of two identical subunits each of which contains a tightly bound NAD.[2-4] Intact NAD is essential for enzymic activity. Addition of various nucleophiles or

Bioorganic Chemistry in Healthcare and Technology, Edited by U.K. Pandit and F.C. Alderweireldt, Plenum Press, New York, 1991

reduction by sodium borohydride leads to inactivation with simultaneous increase of the absorption at 330 nm.[5] Inhibition by nucleophiles is photoreversible[6] reduction is not.[4,7] Hug et al.[6] observed a similar increase in absorption upon binding the competitive inhibitor,3-(imidazol-4-yl)-propionate, to urocanase. The corresponding adduct has been isolated by Matherly et al.[8] On the basis of its ^{1}H-NMR spectrum it was proposed that the covalent bond is between the C4 atom of the nicotinamide and the π-nitrogen atom of the imidazole ring.

In this lecture experiments are described which led to a revision of proposed structure of the above adduct and suggest a chemically plausible, novel mechanism for the urocanase reaction.

Early studies on the stereospecificity and mechanism of action of urocanase

It was early recognized that both the oxygen introduced into the 5'-position and the hydrogen atoms incorporated into the side-chain originated from water.[9] When the reaction was carried out in deuterium oxide and the product was degraded by oxidation with potassium permanganate (R,R)-($^2H_2$)succinic acid was isolated.[10] This showed that both solvent protons were added to the Re face of the double bond in a syn-fashion. In contrast to this cryptic stereospecificity the urocanase product 3-(5'-oxoimidazol-4'-yl)-propionate is racemic. Both NMR and UV measurements showed that the reason for it is a spontaneous and pH-dependent equilibrium between the keto and the enol form (see Eqn.1). Monitoring the urocanase reaction in deuterium oxide by ^{1}H-NMR spectroscopy led to the discovery of the urocanase-catalysed exchange of the 5'-H atom of urocanate with solvent deuterium.[1,11-12] The exchange reaction was 2-3 times faster than the consumption of urocanate and was confirmed also by mass spectrometry.[13] Moreover, the same exchange occurs also with the competitive inhibitor and substrate analogue, 3-(imidazol-4'-yl)-propionate.[1,11,13] If the exchange is a partial reaction of the overall enzymic conversion, then the first attack must occur at the imidazole ring and not at the side chain.

Specific ^{13}C-labelling of the prosthetic NAD and revision of the structure of its adduct with 3-(imidazol-4'-yl)-propionate

In 1982 Matherly et al.[8] published the isolation of a covalent adduct between NAD and 3-(imidazol-4'-yl)-propionate that had been formed on urocanase. They proposed structure **1** for the isolated adduct in analogy to an early work of van Eys[14] on the addition of imidazoles to NAD. (In the enzymically formed unstable intermediate the dihydropyridine was actually present, it was, however, trapped and stabilized by oxidation with phenazine methosulphate.) In our opinion the published ^{1}H-NMR spectrum showed inconsistencies with structure **1**. In particular the signal of one of the imidazole protons was lacking. Matherly et al.[8] explained this discrepancy by an alleged exchange of the H2' atom of the imidazole ring by solvent deuterium in the NMR tube. A further drawback of structure **1** is that no plausible mechanism can be deduced from it for the urocanase reaction. For these reasons Gerlinger and Rétey surmised[15] that the adduct isolated by Matherly et al.[8] had the structure **2**, which would account for the lack of the ^{1}H-NMR signal in position 2' and would allow the formulation of a plausible mechanism for the overall reaction.

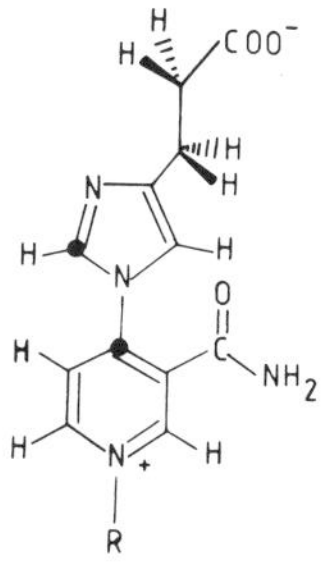

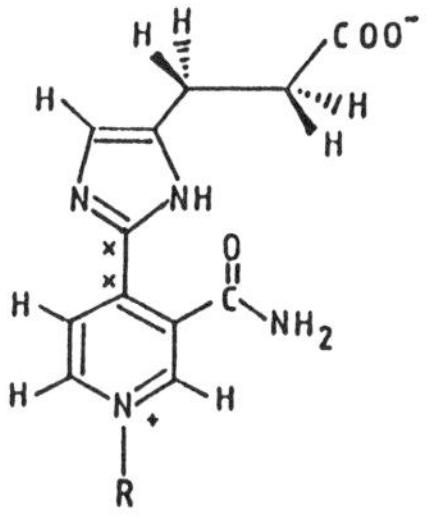

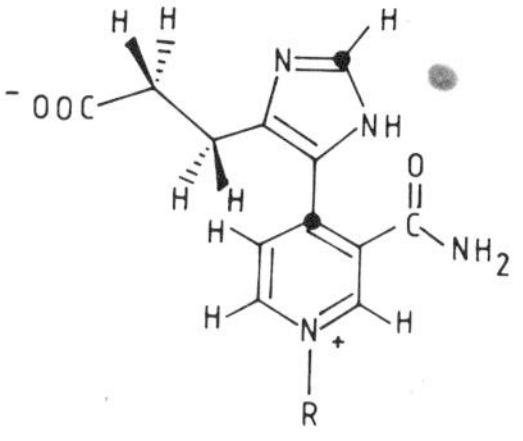

R = ADP-ribosyl

1 **2** **3**

In order to prove or disprove the proposed structures **1** and **2** Jürgen Klepp in my laboratory specifically labelled the components of the adduct with the isotope ^{13}C. Since the tightly bound NAD of urocanase does not reversibly dissociate from the protein, the need for a biosynthetic incorporation of the label emerged. Therefore, in collaboration with Dr. K. Grimm,[16] we prepared a nicotinate auxotrophic mutant of Pseudomonas putida. Simultaneously a sample of (4-^{13}C)nicotinic acid was synthesised. The mutant bacteria were then cultivated in the presence of the labelled nicotinate (1 mg/l) in a 250 l fermentor. The resulting bacteria served as a source for urocanase which was isolated in the usual manner.[3,4] Examination of the HPLC-purified urocanase by ^{13}C-NMR spectroscopy confirmed the incorporation of the label into the tightly bound NAD. The labelled urocanase was then treated with phenazine methosulphate, sodium dodecyl-sulphate and denatured by treatment with perchloric acid. The purification of the adduct was carried out following essentially the procedure of Matherly et al.[8]

A slightly modified isolation procedure was worked out with unlabelled material and the identity of our isolated adduct with that of Matherly et al.[8] was confirmed by UV and NMR spectroscopy. The ^{1}H-decoupled ^{13}C-NMR spectrum of the doubly ^{13}C-labelled adduct showed two singlets. The lack of a direct ^{13}C-^{13}C coupling excludes structure **2**. Even more telling are the ^{1}H-NMR spectra.

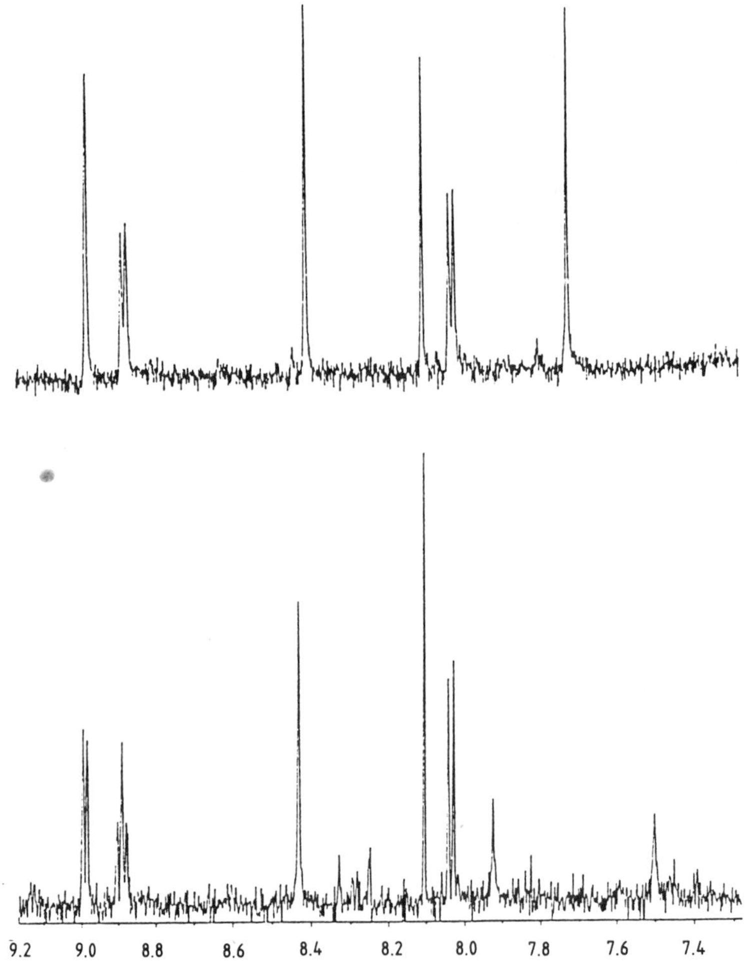

Fig.1. The aromatic region of the ^{1}H-NMR spectrum of the NAD-imidazolylpropionate adduct. (a) unlabelled, (b) ^{13}C-labelled at C4 and C'2.

In Fig.1 the aromatic region of the ^{1}H-NMR spectra both of the unlabelled (a) and doubly ^{13}C-labelled (b) adduct are shown for comparison. In spectrum (a) one finds signals for the pyridinium protons in positions 2, 6 and 5 at δ = 8.97, 8.87 and 8.02 ppm, respectively, for the two adenine protons at δ = 8.42 and 8.09 ppm as well as a singlet at δ = 7.69 ppm for the only carbon-bound imidazole proton. This spectrum is essentially identical with that published by Matherly et al.[8] The spectrum of the doubly labelled adduct (Fig.1b) provides now the crucial information for the assignment of the only imidazole proton. The signal at δ = 7.69 ppm appears as a doublet with a large coupling constant of J_{1H-13C} = 212 Hz, what assigns this signal to the H2' atom of imidazole. This is consistent with the ^{13}C-NMR data and shows that the H2' atom of the adduct does not exchange with solvent deuterium in the NMR tube. A further change in the spectrum concerns the splitting pattern of the H2 and H6 pyridinium signals and can be explained by vicinal coupling ($^3J \approx 6$ Hz) of these protons with the ^{13}C4 atom of the pyridinium ring. The lack of a signal for H5' of the imidazole ring suggests that the substitution occurred in this position and the structure of the adduct corresponds to **3**. Non-enzymic electrophilic substitutions, e.g. bromination,[17] also occur preferentially in the imidazole 4'(5')-position. Alternative structures to **3**, in which the 5'-position carries an oxygen atom and the junction to the pyridium ring is at the π-nitrogen of imidazole, have been also considered. They were rejected by comparison with the ^{1}H-NMR spectra of the unsubstituted 3-(5'-oxoimidazol-4'-yl)-propionate and 3-(5'-hydroxyimidazol-4'-yl)-propionate.[16]

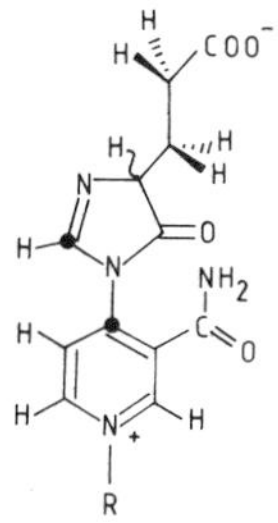

4

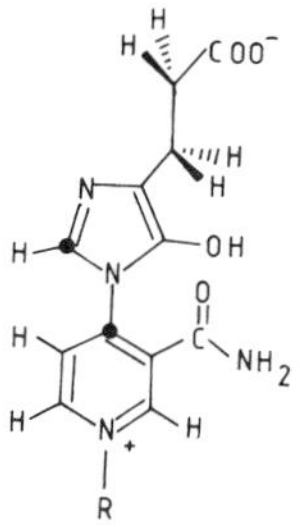

5

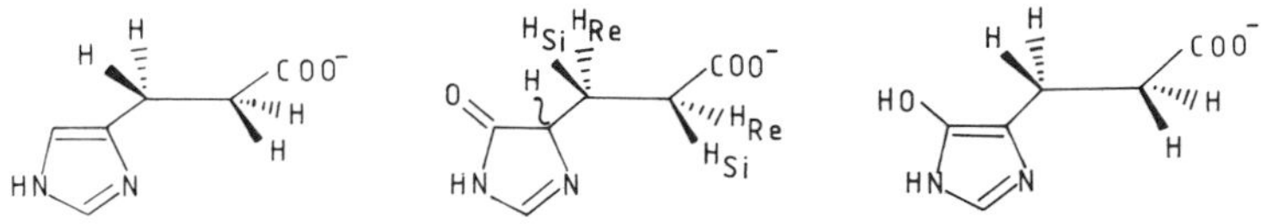

$-(CH_2)_\alpha-$	2.49 ppm	$-(CH_2)_\alpha-$	2.19 ppm 2.20 ppm	$-(CH_2)_\alpha-$	2.40 ppm
$-(CH_2)_\beta-$	2.86 ppm	$-(CH_2)_\beta-$	1.90 ppm 2.18 ppm	$-(CH_2)_\beta-$	2.70 ppm

Fig.2. Comparison of the ^{1}H-NMR-data for the side-chain methylenes of imidazolpropionate and its C5-substituted analogues.

In particular the signals of the side-chain protons are inconsistent with **4** and **5** (see Fig.2). The side-chain protons of the enzymically formed adduct give rise to two clean triplets at δ = 2.51 and 3.02 ppm. Were the substituent in position 5' of the imidazole oxygen, then one would expect an equilibrium between the keto and the enol form. In the keto form the protons of each side-chain methylene group are diastereotopic and exhibit different chemical shifts (Fig.2). Both the coupling pattern and the δ values of the side-chain proton signals of the adduct are closest to the unsubstituted 3-(imidazol-4'-yl)-propionate making an oxygen substitution in the ring unlikely.

A further evidence against the bonding of NAD to the π-nitrogen of imidazole is the lack of a coupling ($^3J_{1H-13C}$) between H2' of the imidazole and $^{13}C4$ of the nicotinamide in the ^{13}C-labelled adduct (Fig.1b). In several synthetic precursors it was observed[16] that such a geminal coupling in planar geometry exhibits $^3J_{1H-13C}$ values between 6 and 9 Hz. Thus we are left with structure **3** for the NAD-3-(imidazol-4'-yl)-propionate adduct that is in agreement with all analytical data. In contrast to structure **1** it suggests a chemically plausible mechanism for the urocanase reaction which is outlined in the following scheme:

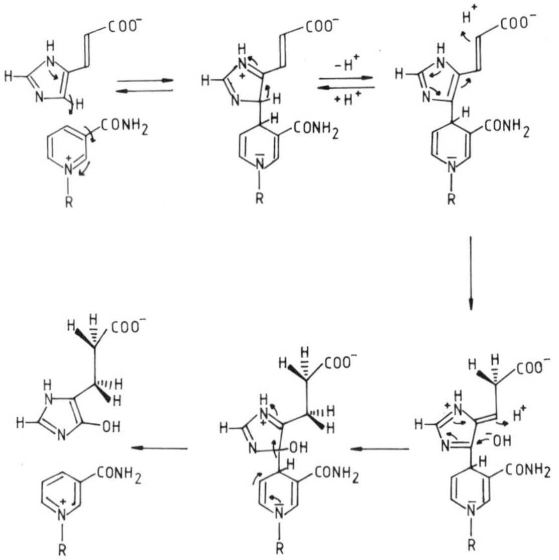

Scheme 1. Proposed mechanism of the urocanase reaction.

The first chemical step in this mechanism is the electrophilic substitution at the imidazole 5'-position by NAD. In the first intermediate the previously firmly bound H5' atom is now labilized by the neighbouring immonium group. Abstraction of the H5' atom completes the electrophilic substitution. Reversibility of the first two steps is responsible for the observed exchange of the H5' atom. If these two steps occur with the competitive inhibitor 3-(imidazol-4'-yl)-porpionate, then the second intermediate accumulates as evinced by the increase of absorption at 330 nm and by the isolation of the oxidatively stabilized adduct.[16]

If, however, the normal substrate, urocanate, is fixed by covalent binding to NAD, then stereospecific protonation of the side-chain double bond can occur, the electrons being supplied from the imidazole ring. In the next intermediate the originally nucleophilic 5'-position has become electrophilic; an "umpolung" which allows attack by an OH group followed by stereospecific addition of the second proton to the side-chain. The overall reaction is completed by an extended reto-aldol-type cleavage of the last covalent intermediate to form the product and regenerate NAD.

ACKNOWLDGEMENT

Thanks are due to my coworkers whose hard work and devotion made my report possible. I am especially grateful to Drs. Jürgen Klepp, Angelika Fallert-Müller, Erich Gerlinger und Margrit Knauer (born Oberfrank). Fruitful collaborations with Dr. W.E. Hull of the German Cancer Research Centre, Heidelberg (500 MHz NMR) and Dr. K. Grimm of the Institute of Botany, University of Karlsruhe (nicotinate auxotrophic P. putida), are acknowledged. Financial support was provided by the Deutsche Forschungsgemeinschaft and the Fonds der Chemischen Industrie.

REFERENCES

1. E. Gerlinger, W.E. Hull and J. Rétey, Eur.J.Biochem., 1981, 117, 629.
2. R.M. Egan and A.T. Phillips, J.Biol.Chem., 1977, 252, 5701.
3. V. Keul, F. Kaeppeli, C. Ghosh, T. Krebs, J.A. Robinson and J. Rétey, J.Biol.Chem., 1979, 254, 843.
4. J. Klepp and J. Rétey, Eur.J.Biochem., 1989, 185, 615.
5. D.J. George and A.T. Phillips, J.Biol.Chem., 1970, 245, 528.
6. D.H. Hug, P.S. O'Donnell and J.K. Hunter, J.Biol.Chem., 1978, 253, 7622.
7. D.H. Hug, P.S. O'Donnell and J.K. Hunter, Photobiochem. Photobiophys., 1981, 3, 175.
8. L.H. Matherly, C.W. DeBrosse and A.T. Phillips, Biochemistry, 1982, 21, 2789.
9 B. Magasanik and H.R. Bowser in: Symposium on Aminoacid Metabolism (E.D. McElroy nd B. Glass, eds.), The Johns Hopkins Press, Baltimore, 1955, p.398.
10. F. Kaeppeli and J. Rétey, Eur.J.Biochem., 1971, 23, 198.
11. E. Gerlinger and J. Rétey, FEBS Letters, 1980, 110, 126.
12. S. Sawada, K. Endo, M. Ushida, N. Esaki and K. Soda, Bull.Kyoto Univ. Education, Ser.B, 1981, 58, 11.
13. R.M. Egan, L.H. Matherly and A.T. Phillips, Biochemistry, 1981, 20, 132.
14. J. van Eys, J.Biol.Chem., 1958, 233, 1203.
15. E. Gerlinger and J. Rétey, Z.Naturforsch., Teil C, 1987, 42c, 349.
16. J. Klepp, A. Fallert-Müller, K. Grimm, W.E. Hull and J. Rétey, Eur.J.Biochem., 1990, in the press.
17. I.E. Balaban and E.L. Pyman, J.Chem.Soc., 1922, 121, 947.

THE MICROBIOLOGICAL REDUCTION/OXIDATION CONCEPT: AN APPROACH TO A CHEMOENZYMATIC PREPARATION OF OPTICALLY PURE LACTONES AND LACTONIC SYNTHONS

Robert Azerad, Didier Buisson, Sophie Maillot, and Jamal Ouazzani-Chahdi

Laboratoire de Chimie et Biochimie Pharmacologiques
et Toxicologiques, UA 400 CNRS, Université René Descartes
45 rue des Saints-Pères, 75270 - Paris Cedex 06, France

In the recent years, intensive work has been devoted to the efficient preparation of new chiral synthons by asymmetric transformation of prochiral molecules, through enzymic or microbiological catalysis. In this respect, we have developed combined chemical-microbiological reduction and oxidation (Baeyer-Villiger like) methods which allow to obtain either enantiomer of chiral substituted lactones and lactonic synthons.

It has been previously demonstrated that the (S)-ketol **2**, stereospecifically obtained by microbiological reduction (1) of the prochiral diketone **1**, is easily converted to a rearranged hydroxylated (S)-γ-butyrolactone **3** through a classical Baeyer-Villiger reaction (*Scheme 1, path a*). The dehydrated lactones **4** and **5**, specifically obtained in separate reactions, and their ozonolysis derivatives may represent useful chiral lactonic synthons.

The preparation of the corresponding enantiomeric (R)-lactones may be easily achieved through the resolution of the racemic ketol **6** (*Scheme 1, path b*) by a highly enantioselective microbiological Baeyer-Villiger-like oxidation (2), catalyzed by a ketol-induced monooxygenase (figures 1 and 2). The optically pure (R)-ketol **7** recovered from the incubation mixture, may be then oxidized in a usual chemical Baeyer-Villiger reaction to give access to the (R)-series corresponding lactones.

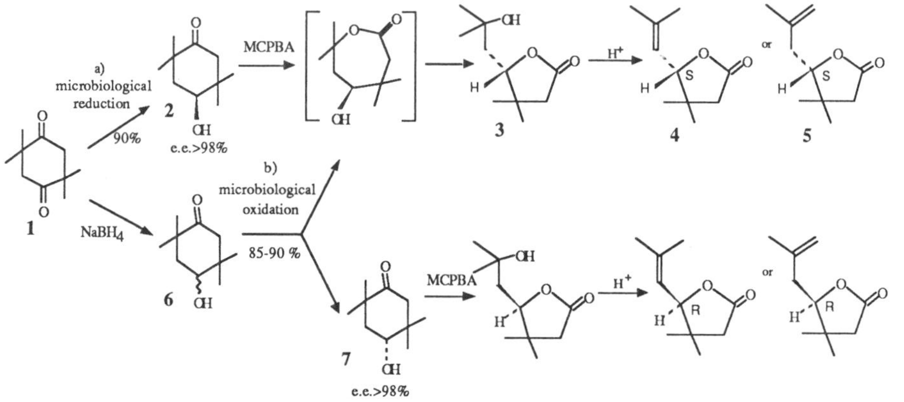

Scheme 1

Bioorganic Chemistry in Healthcare and Technology, Edited by U.K. Pandit and F.C. Alderweireldt, Plenum Press, New York, 1991

This method (M.R.O.C. = **M**icrobiological **R**eduction /**O**xidation **C**oncept) has been developed and generalized to be applied to variously substituted cyclic 1,4-diones. As an impressive illustration of the M.R.O.C potentialities, a fast, high yield and highly stereoselective synthesis of the four stereoisomers of eldanolide **8**, the wing gland pheromone of the male african sugar-cane borer, *Eldana saccharina* (Wilk) [3], is described. Only the *trans*-isomers of this pheromone have been up to now described [4]. A retrosynthetic scheme for the simple preparation of such a lactone is given in *Scheme 2*, making use of the Baeyer-Villiger oxidation of an adequately substituted 4-hydroxy-cyclohexanone followed by the rearrangement of the resulting ε-lactone.

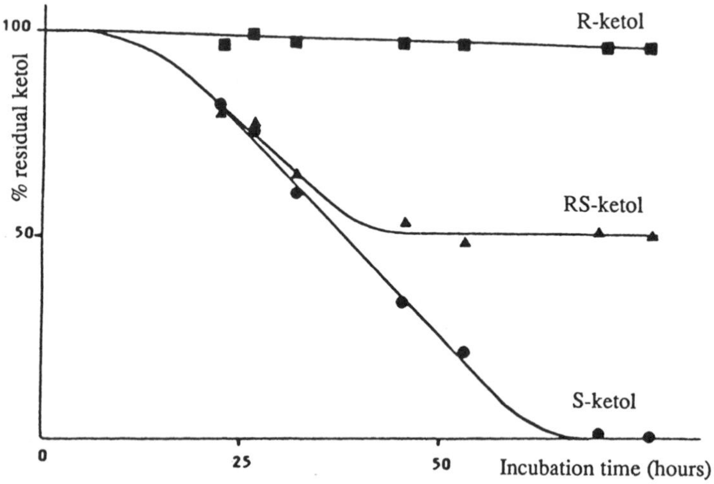

Figure 1. Kinetics of oxidation of (R)-, (S)- and (RS)-2,2',5,5'-tetramethyl 4-hydroxy-cyclohexanone by incubation with *Curvularia lunata* NRRL 2380 grown without inducer. Substrate: 1g/litre; mycelium: 137g/litre (wet weight); temperature: 27°C; orbital shaking: 370 rpm.

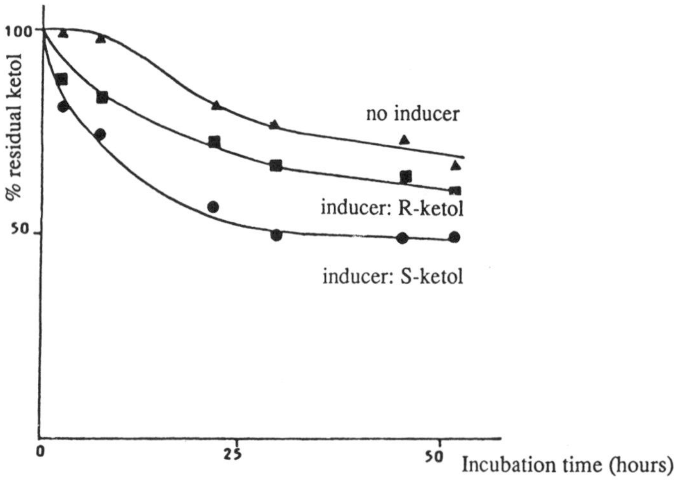

Figure 2. Kinetics of oxidation of (R,S)-2,2',5,5'-tetramethyl 4-hydroxycyclohexanone by incubation with *Curvularia lunata* NRRL 2380 grown in the absence or in the presence of (R)- or (S)-ketol as inducer.(20 mg/litre, added 24 hours before substrate)

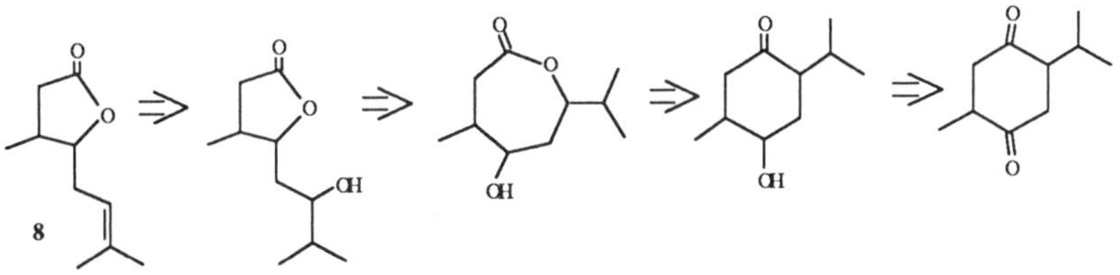

Scheme 2

From the isomeric mixture of 2-isopropyl-5-methyl-1,4-cyclohexanedione, easily prepared from thymol [5], a microbiological reduction by *C.lunata* afforded in quantitative yield the corresponding mixture (*Scheme 3*) of four optically pure 4(S)-ketols. All of them could be obtained on a small scale by silicagel chromatography; they were characterized as the products of an exclusive *regio-* and *stereoselective* reduction of the less hindered carbonyl group (adjacent to the methyl group) by GPC on a chiral column [1] and by individual *m*-chloroperbenzoic acid (MCPBA) oxidation, followed by dehydration to only two different isomeric optically pure (4S)-*cis* and *trans*-γ-lactones, the later beeing identified as enantiomeric to natural (4R)-*trans* eldanolide.

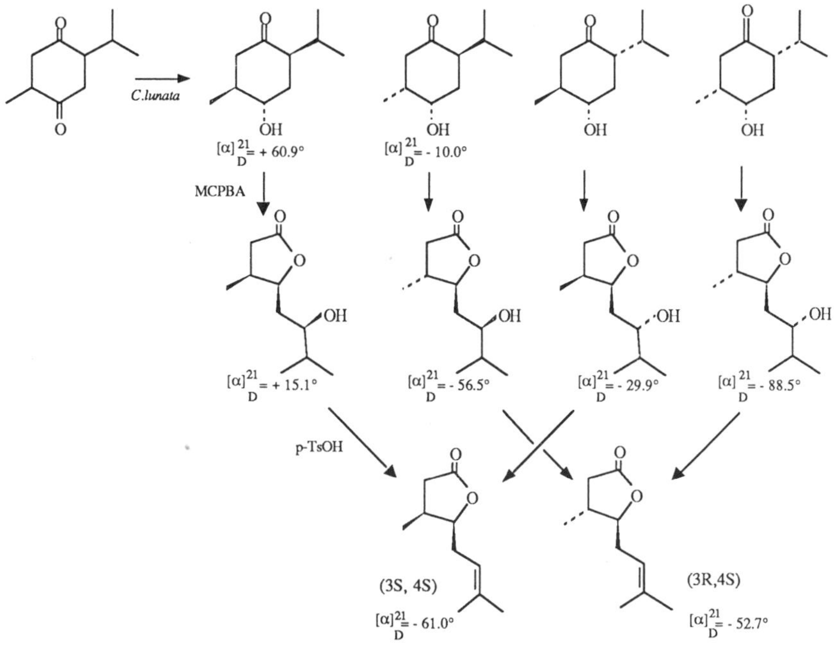

Scheme 3

For preparative purpose, MCPBA oxidation was effected on the (4S)-ketol mixture (*Scheme 4*), affording the corresponding mixture of rearranged diastereoisomeric 4(S)-lactones, which was dehydrated to the mixture of both unsaturated lactones. These lactones were easily separated by silicagel chromatography. On the other hand, reduction of the initial mixture of methyl isopropyl cyclohexanediones with Redal® afforded the complete mixture of diastereomeric 4-hydroxyketones (*Scheme 4*) which was completely resolved using the Baeyer-Villiger like oxidation by *C.lunata* , as previously shown for the tetrame-

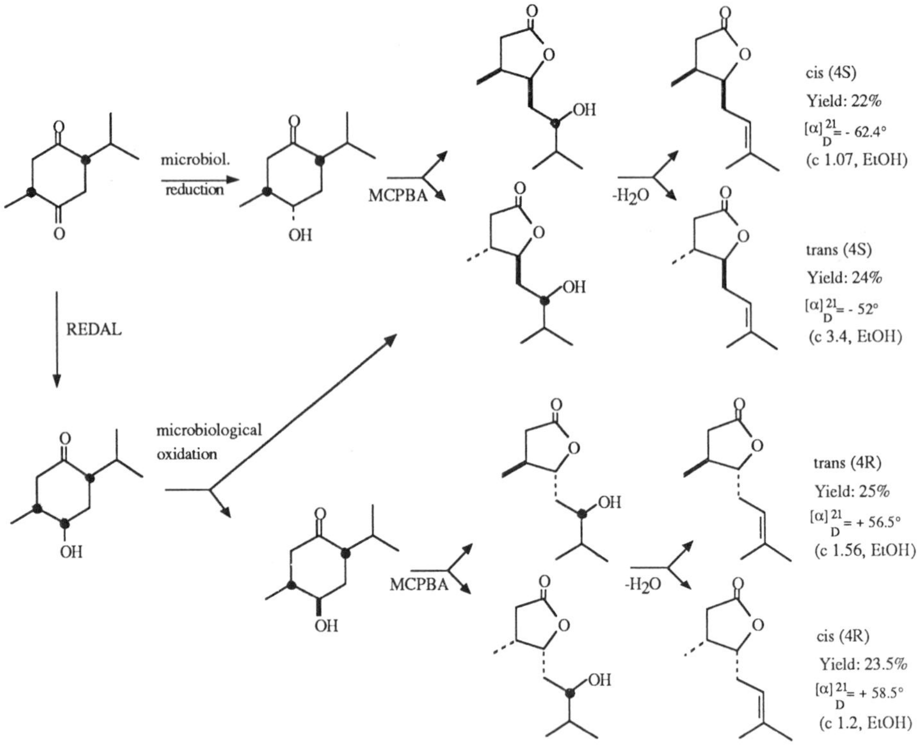

Scheme 4

thylketols [(2)].Only S-ketols were again oxidized, leaving pure diastereomeric 4(R)-ketols, the MCPBA-oxidation of which, followed by dehydration, afforded both diastereomeric rearranged lactones, one of these lactones is identical with natural 4(R)-*trans* eldanolide. About 50% total yield was obtained in the whole operation and the enantiomeric excess of all four lactone stereoisomers was higher than 95%.

Bioconversions of several other substituted 1,4-cyclohexanediones are currently investigated and preliminary results show that the combination of microbial reduction and oxidation may constitute a powerful tool for the preparation of various optically pure substituted γ-lactones and lactonic synthons.

References

1- D.Buisson, R.Azerad, G.Revial and J.D'Angelo, *Tetrahedron Lett.*, **25** (1984) 6005; J. D'Angelo, G.Revial, R.Azerad and D.Buisson, *J.Org.Chem.*, **51** (1986) 40.

2- J.Ouazzani-Chahdi, D.Buisson and R.Azerad, *Tetrahedron Lett.*, **28** (1987) 1109.

3- P.R.Atkinson, *J.Entomol.Soc.South Afr.*, **43** (1980) 171.

4- J.P.Vigneron, R.Méric, M.Larchevêque, A.Debal, G.Kunesch, P.Zagatti and M.Gallois, *Tetrahedron Lett.*, **23** (1982) 5051; T.Uematsu, T.Umemura and K.Mori, *Agric.Biol. Chem.*, **47** (1983) 597; J.P.Vigneron, R.Méric, M.Larchevêque, A.Debal, J.Y.Lallemand, G.Kunesch, P.Zagatti and M.Gallois, *Tetrahedron*, **40** (1984) 3521; K.Suzuki, T.Ohkuma, G.Tsuchihashi, *Tetrahedron Lett.*, **26** (1985) 861; H.G.Davies, S.M.Roberts, B.J. Wakefield and J.A.Winders, *J.Chem.Soc.Chem.Commun.*, (1985) 1166; H.Frauenrath and T. Philipps, *Tetrahedron*, **42** (1986) 1135; R.M.Ortuno, R.Mercé and J.Font, *Tetrahedron Lett.*, **27** (1986) 2519; D.S.Matteson, K.M.Sadhu and M.L.Petterson, *J.Am.Chem.Soc.*, **108** (1986) 810.

5-R.D.Stolow, P.M.McDonagh and M.M.Bonaventura, *J.Am.Chem.Soc.*, **86** (1964)2165.

EFFECT OF TEMPERATURE ON ENZYMATIC SYNTHESIS OF CEPHALOSPORINS

Eva M. Baldaro

Sclavo S.p.A. - Biochemical Division De.Bi.
S. S. Padana Superiore km 160
I-20060 Cassina de' Pecchi (MI)
Italy

SUMMARY

The effect of temperature on enzymatic synthesis of cephalexin (CEX) from 7-amino-3-deacetoxycephalosporanic acid (7-ADCA) and D-alpha-phenyl glycine methyl ester (PGM) using immobilized penicillin G amidohydrolase (E.C.3.5.1.11) from E. coli ATCC 9637 was investigated. It was found that decreasing temperature results in a marked enhancement in both yield in condensation product and stability of catalyst. Analogous results were obtained using 7-amino-3-chloro(deacetoxymethyl)cephalosporanic acid as substrate for enzymatic synthesis of cefaclor.

INTRODUCTION

Since Takahashi et al[1] in 1972 reported the first enzymatic synthesis of cephalosporins from 7-aminocephem compounds and organic acid esters, several groups[2-6] focused their attention on this field. The most remarkable advantages of the enzymatic process are high yield, one-step reaction in very mild conditions and relatively fast reaction rate. Nowadays the commercial availability of immobilized enzymes renders the enzymatic process a possible alternative to the chemical one.

On the other hand this process suffers of some drawbacks, the main is the side hydrolysis of aminoacid ester catalysed by the enzyme itself[4,5]. This hydrolysis makes troublesome the downstream process and, being the overall conversion controlled by the availability of the ester as an acyl donor[7], renders necessary to operate with large excess of activated substrate. Moreover, if aminoacid concentration in the reaction mixture is above its solubility, this starts to precipitate coating the catalyst.

Some authors[4,8] suggested that the presence of alcohols may be used to regenerate the ester, because of the reversibility of all the reactions involved in the mechanism. In our laboratories we found that the reaction mechanism for CEX synthesis catalysed by immobilized penicillin G amidase from E. coli is in good agreement with the one proposed by Nam, Kim and Ryu[7] (Figure 1) for enzyme from Xanthomonas citri, in which the acyl enzyme (E··PG) can be irreversibly hydrolized or can react with 7-ADCA by nucleophilic attack. We supposed that the energy involved in these two reactions could be different and the effect of temperature on enzymatic synthesis of CEX could consequently be pronounced.

Bioorganic Chemistry in Healthcare and Technology, Edited by U.K. Pandit and F.C. Alderweireldt, Plenum Press, New York, 1991

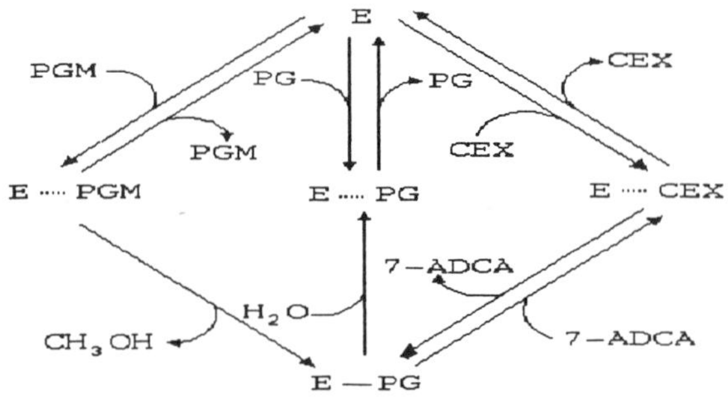

Fig. 1. Schematic diagram of the mechanism proposed by Nam et al.[7] for CEX synthesis.

Up to date most experiments in literature have been performed at about 37°C and influence of temperature has been examined by few groups[3,5,9,10]; unfortunately the reported results are not always exhaustive and consistent.

We report in this article the effect of temperature on maximum yield in condensation product, aminoacid formation and stability of the catalyst to repeated uses in CEX synthesis.

MATERIALS AND METHODS

Materials

Immobilized penicillin G amidase is industrially produced and commercialized by Sclavo S.p.A.-De.Bi. division. 7-ADCA was prepared from cephalosporin G by enzymatic hydrolysis using immobilized penicillin G amidase. PGM (as hydrochloride salt) was kindly provided by ing. Hendrickx (Deretil). 7-Amino-3-chloro(deacetoxymethyl)cephalosporanic acid was synthesized as described in ref.[11]. CEX, cefaclor and phenyl glycine (PG) from Sigma were used as standards. For experiments with the free form of penicillin amidase an enzymatic extract of about 10 U/mg of protein was used.

Experimental Methods

All runs for CEX synthesis were carried out under pHstat conditions and temperature control in a 50 ml stirred batch reactor containing the catalyst and equipped at the bottom with a sintered glass filter. The reaction started by addition of reagents dissolved in distilled water at pH 7. Reaction time 120'. For cefaclor synthesis a 5 ml reactor was used.

Analytical Procedures

Enzyme activity was determined by measuring the amount of CEX produced. One unit (U) of enzyme was defined as the amount required to produce 1 µM/min at pH 7 and 37°C (93 mM 7-ADCA, 233 mM PGM). The reaction mixture was analysed by HPLC with Waters pump 510 and revealed at 220 nm using a Spectra Physics SP8480 UV detector. The appropriately diluted sample was applied to a Hibar RP-8 Merck column and eluted with CH_3CN 14% in phosphate buffer at pH 2.3.

RESULTS AND DISCUSSION

To get an insight into the effect of temperature on the maximum yield in CEX, experiments at different concentration of 7-ADCA and at different

Table 1. Effect of temperature on CEX synthesis and PG formation by *E. coli* immobilized enzyme.

7-ADCA (mM)	PGM (mM)	Enzyme (U/10mM 7-ADCA)	Temperature (°C)	CEX yield (% 7-ADCA)	PG yield (% PGM)
93	233	45	37	57	36
93	233	90	22	77	31
93	233	340	4	86	26
93	233	1500	-5[a]	89	24
140	350	45	37	67	31
140	350	90	22	82	28
140	350	340	4	90	26
93	355	45	37	70	32
93	355	90	22	88	29
93	355	340	4	93	26

[a] experiment performed in ethylene glycol 20% in water.

7-ADCA : PGM molar ratios were performed. Naturally, enzymes show a reduced activity lowering temperature. Therefore in order to mantain the reaction time within 120 min, the amount of catalyst was gradually increased as temperature was lowered. Enzyme concentration has been demonstrated[3,12], and confirmed by us for immobilized enzyme, to influence the time taken to reach the maximum but not the maximum itself.

In Table 1 are summarized data of the effect of temperature on CEX synthesis and PG formation by immobilized enzyme; identical results were obtained using the enzymatic extract before immobilization. As can be inferred from Table 1, in all sets of experiments two effects correspond to the decrease in temperature: CEX yield rises and PG formation slows down. These results are consistent with the hypothesis of two competitive kinetically controlled reactions at a common acyl enzyme intermediate: nucleophilic attack by 7-ADCA and hydrolysis. The decrease in PG formation also permits to work at high concentrations of 7-ADCA or PGM avoiding the problem of amino acid precipitation.

The same trend reported for CEX was observed using 7-amino-3-chloro (deacetoxymethyl)cephalosporanic acid as nucleophile for the enzymatic synthesis of cefaclor (Table 2).

To test the possibility of practical application of the low temperature process for cephalosporins enzymatic synthesis, the stability of catalyst at repeated CEX synthesis at various temperatures was checked. The effect of temperature was quite pronounced: after 25 h of work at 37°C the retained activity was only 44% of the original, while at 22°C it was 87% after 25h and 78% after 60 h. At 4°C no significant loss of enzymatic activity was observed even after 60 h of work.

Table 2. Effect of temperature on cefaclor synthesis and PG formation by *E. coli* immobilized enzyme.

Nucleophile (mM)	PGM (mM)	Enzyme (U/10mM nucleophile)	Temp. (°C)	Cefaclor yield (% nucleophile)	PG yield (% PGM)
43	107	50	37	38	33
43	107	100	22	56	27
43	107	380	4	66	23

Table 3. Repeated uses of immobilized enzyme for CEX synthesis (93 mM 7-ADCA, 233 mM PGM, pH 7).

Temperature (°C)	Enzyme (U/10mM 7-ADCA)	$\frac{(U)_{T°C}}{(U)_{37°C}}$	A 50% (h)	$\frac{(A\ 50\%)_{T°C}}{(A\ 50\%)_{37°C}}$
37	45	1	22	1
22	90	2	120[a]	5.5
4	340	7.5	800[a]	36

[a] Data extrapolated after 60 h of repeated use.

In Table 3 are summarized experimental data about the half-life time (A 50 %, number of hours of work corresponding to 50 % of retained activity) and amount (U) of catalyst at different temperatures.

From this data it becomes evident that the increase in stability is achieved working at low temperature amply compensates for the larger amount of enzyme to be used.

On the base of these results presented in this article we can conclude working at a low temperature has several positive effects on the reaction. The decreased amount of activated substrate hydrolized permits to obtain higher yields in condensation product and to work at higher concentration of substrates avoiding precipitation problems; besides, the increased stability of the immobilized enzyme cuts down the incidence of catalyst cost on the product. For these reasons the use of a low temperature can render the enzymatic synthesis of cephalosporins a real alternative to the chemical process in both environmental and economic terms.

ACKNOWLEDGEMENT

I wish to thank Mrs. A. M. Bertolaso for her enthusiastic collaboration.

REFERENCES

1 T. Takahashi, Y. Yamazaki, K. Kato and M. Isono, J. Am. Chem. Soc., 1972, 94, 4035.
2 T. Fujii, K. Matsumoto and T. Watanabe, Process Biochem., 1976, 11, 21.
3 W. G. Choi, S. B. Lee and Dewey D. Y. Ryu, Biotechnol. Bioeng., 1981, 23, 361.
4 V. Kashe, Biotechnol. Lett., 1985, 7, 877.
5 Y. W. Ryu and Dewey D. Y. Ryu, Enzyme Microb. Technol., 1987, 9, 339.
6 C. K. Hyun, J. H. Kim and Y. J. Kim, Biotechnol. Lett., 1989, 11, 537.
7 D. H. Nam, C. Kim and Dewey D. Y. Ryu, Biotechnol. Bioeng., 1985, 27, 953.
8 J. Konecny, A. Schneider and M. Sieber, Biotechnol. Bioeng., 1983, 25, 451.
9 V. Kashe, Enzyme Microb. Technol., 1986, 8, 4.
10 F. Knauseder and N. Palma, '3rd European Congress on Biotechnology, Munich', Verlag Chemie, Weinheim, 1984, vol. 1, p. 431.
11 R. R. Chauvette, D. A. Pennington, J. Med. Chem., 1975, 18, 403.
12 V. Kashe, U. Haufler and L. Riechmann, Ann. N. Y. Acad. Sci., 1984, 434, 99.

STUDY OF THE ACTION OF HYDROLASES ON INSOLUBLE SUBSTRATES IN AQUEOUS AND NON-AQUEOUS SYSTEMS

Cristina Otero, Maria L.Rúa,
Carmen Cruzado and Antonio Ballesteros

Unidad de Biocatálisis, Instituto de Catálisis
C.S.I.C., 28006 Madrid, Spain

INTRODUCTION

Lipases belong to the group of enzymes acting on hydrophobic substrates, and so, while they are soluble in water, their substrates are insoluble in it. Therefore, it is of great importance for the catalysis to facilite the contact between the substrate and the active site of the protein.

p-Nitrophenyl esters are generally employed to determine, in standard reference conditions, the activity and behaviour of hydrolytic enzymes. The low solubility in water of these substrates is usually increased by addition of a low concentration of organic solvents (acetonitrile, acetone,..), surfactants, ultrasonic disintegration, etc.

We present below two methods for the determination of the activity of these enzymes with p-nitrophenyl esters: i) Addition of cyclodextrins (CDs) to the reaction mixture; ii) Formation of microemulsions.

MATERIALS AND METHODS

Candida cylindracea type VII lipase was purchased from Sigma. Partial purification of this extract yielded two isozymes, lipases A and B (1). p-nitrophenyl butyrate (p-NPC4),p-nitrophenyl laurate (p-NPC12), cyclodextrins, and maltoheptaose were also from Sigma, and sodium bis-2-ethylhexyl sulfosuccinate (AOT) was from Fluka.

Water-in-oil (w/o) microemulsions were prepared by adding the appropriate volume of aqueous buffer solution to 25 ml of 0.1 AOT in heptane. Then, a solution of p-NP ester in acetone was added (4% acetone in the final volume).

The kinetics of hydrolysis of p-Np esters was followed

Bioorganic Chemistry in Healthcare and Technology, Edited by U.K. Pandit and F.C. Alderweireldt, Plenum Press, New York, 1991

spectrophotometrically at 400 nm and 30°C in a Varian Cary spectrophotometer provided with magnetic stirring. The value of the molar absorption coefficient (ϵ) at 400 nm of p-nitrophenolate in these conditions was 13400 M^{-1} cm^{-1}.

Table 1. Kinetic Parameters (crude Candida c. lipase)[a]

	Km (mM)	kcat (s^{-1})	k_2^0 ($M^{-1}s^{-1}$)
Aqueous System (3% acetonitrile)			
p-NPC4	0.35	501	1.43×10^6
CDs System (β-CD)			
p-NPC4	0.27	395	1.46×10^6

[a]Lipase concentration; 0.05 g/l; pH 7.26, 30°C, 0.1M sodium phosphate

RESULTS AND DISCUSSION

i) Cyclodextrins (CDs). Many hydrophobic compounds are solubilized in water by adding a low amount of a water-miscible organic solvent to the reaction mixture (e.g., acetonitrile at 1-4%). Another approach could be, instead of the addition of another solvent, the use of inclusion compounds having hydrophobic cavities able to include the insoluble substrates and products. CDs enhance the solubility of complexed substrates in aqueous media, contributing to the stabilization of sensitive substrates. Taking advantage of the solubilizing properties of inclusion compounds (cyclodextrins (2)), we have proposed their use in enzymology to enhance the solubility of hydrophobic compounds in water (3).

In control experiments without enzyme, hydrolysis by the added CD was insignificant. Table 1 compares the values of the kinetic parameters obtained in the regular medium (water containing 3% acetonitrile) and in the same medium plus cyclodextrin. One can see that the specificity constant, k_2^0, is not affected. β-CD increases slightly (1.5-fold) the stability of Candida lipase at 50°C; however, a non-cyclic heptasaccharide (maltoheptaose) does not exert any effect on stabilization. On the other hand, α-CD slightly destabilizes this lipase. Table 2 presents the effect of β-CD concentration in the hydrolysis of p-NPC4. At the CD:substrate ratio used (1.5/1) there is no inhibition. Furthermore, the inhibition is not very large even with ten times more CD (3).

Several conclusions can be drawn: 1) More concentrated solutions of substrates can be prepared. 2) The enzyme parameters are not affected substantially when compared with their values in water with 3% acetonitrile. 3) The reproducibility is higher, owing to the true solution obtained.

Table 2. Inhibition by β-CD[a]

CD (mM)	% Inhibition
0.44	0
0.80	13
1.30	22
4.40	21

[a]Initial rate of p-NPC4 (0.29 mM) hydrolysis by Candida lipase (0.05 mg/mL) vs β-CD concentration

ii) Water-in-oil (w/o) microemulsions present many advantages in enzyme catalysis: they protect the protein against denaturation by the organic solvent(s) used in many synthetic processes, and facilitate mass transfer between the substrate-containing and the enzyme-containing phases. In order to partition the enzyme and the substrate in the aqueous pseudophase and in the organic phase, respectively, the same reaction (hydrolysis of p-NPC4) has been carried out in w/o microemulsions.

Table 3. Kinetic Parameters of Lipases A and B[a]

	K_m (mM)		k_{cat} (s^{-1})	
	L.A	L.B	L.A	L.B
Aqueous System (4% acetone)				
p-NPC4	0.14	0.14	842	836
Microemulsion System (R = 5):				
p-NPC4	0.19	0.31	42	1.54
p-NPC12	5.70	7.10	290	5.30

a) 26°C, pH 6.1, 50 mM Mes in aqueous system, and 1.8 mM Mes in microemulsions. [L.A] = 2.2 nM and [L.B] = 3.27 nM and 0.29 nM, respectively, in aqueous and microemulsion system.

The kinetic constants of lipases A and B in the aqueous and in the microemulsion systems appear in Table 3. In the case of p-NPC4 the specificity constant in microemulsions ($k_2^0 = k_{cat}/K_m$) is 30 times smaller for lipase A but 10^3-fold lower for lipase B. Using p-NPC12, whose hydrolysis in aqueous medium is negligible owing to its low solubility, the kinetic parameters found are higher than those of p-NPC4 (also in microemulsions).

In conclusion, as far as the specificity constant of the hydrolysis of p-nitrophenyl butyrate is concerned, its value is not altered by the addition of cyclodextrins, while in the case of microemulsions much smaller values are found.

ACKNOWLEDGEMENTS

Supported by the EEC (grant BAP.0402.E) and the Spanish CICYT (grant BI088-0241)

REFERENCES

1. M.L. Rúa, V.M. Fernández, C. Otero and A. Ballesteros, Purification and partial characterization of lipase from Candida cylindracea, in preparation (1990).

2. M.L. Bender and M. Komiyama (1978). "Cyclodextrin Chemistry", Springer-Verlag, New York.

3. C. Otero, C. Cruzado, and A. Ballesteros, Use of cyclodextrins in enzymology to enhance the solubility of hydrophobic compounds in water, Appl. Biochem. Biotechnol., accepted (1990).

ENZYMES IN ORGANIC SOLVENTS: ENANTIOSELECTIVE TRANSESTERIFICATION OF ALPHA-METHYL SUBSTITUTED PRIMARY ALCOHOLS CATALYZED BY A LIPASE

E. Santaniello, P. Ferraboschi, P. Grisenti, A. Manzocchi

Dipartimento di Chimica e Biochimica Medica
Università degli Studi di Milano, Italy

INTRODUCTION

The enzyme catalyzed transesterification of racemic alcohols has been recently introduced for the preparation of enantiomerically pure alcohols and esters of opposite configurations [1]. This procedure makes use of lipases from different sources as the biocatalysts and vinyl esters as acyl donors. The peculiarity of the method resides in the fact that the transesterification is irreversible, since one of the products, namely vinyl alcohol, rapidly tautomerizes to acetaldehyde and this prevents the back reaction. Further, the reaction can be efficiently carried out in organic solvents (Eq. 1).

$$R\text{-}CH_2OH + CH_2\text{=}CH\text{-}OCOCH_3 \xrightarrow{\text{Enzyme}} R\text{-}CH_2OCOCH_3 + CH_3\text{-}CHO \qquad (1)$$

We have used the above method for the resolution of primary alcohols which bear a chiral carbon at the alpha-position. The common structural feature of these substrates can be summarized in the formula **$R\text{-}CH(CH_3)CH_2OH$**. The lipase used was from Pseudomonas fluorescens (PFL), the organic solvents were dichloromethane or chloroform and vinyl acetate was the acylating reagent.

MATERIALS AND METHODS

Pseudomonas fluorescens lipase (PFL) and all the chemicals were purchased from Fluka (Buchs, Switzerland). The transesterification procedure was essentially identical to the method described by Wang et al. [1]. The optical purities were determined by measuring the optical rotations on a Perkin-Elmer Model 241 polarimeter and by ^{1}H-NMR analysis of the esters with (-)-MTPA chloride [2]. The ^{1}H-NMR spectra were recorded in $CDCl_3$, using $SiMe_4$ as internal standard. The 200 and 500 MHz ^{1}H-NMR spectra were recorded on Varian XL 200 and AM 500 Bruker spectrometers.

RESULTS AND DISCUSSIONS

The PFL-catalyzed transesterification was achieved with high enantioselectivity (>98% ee) when (R,S)-4-phenylthio- and 4-phenylsulphonyl-2-methylbutanol [3] were the substrates. By this procedure, the corresponding (R)-alcohol and (S)-acetate could be prepared. Analogous results were obtained using 4-phenylselenyl -2-methylbutanol as substrate [4] (Eq. 2).

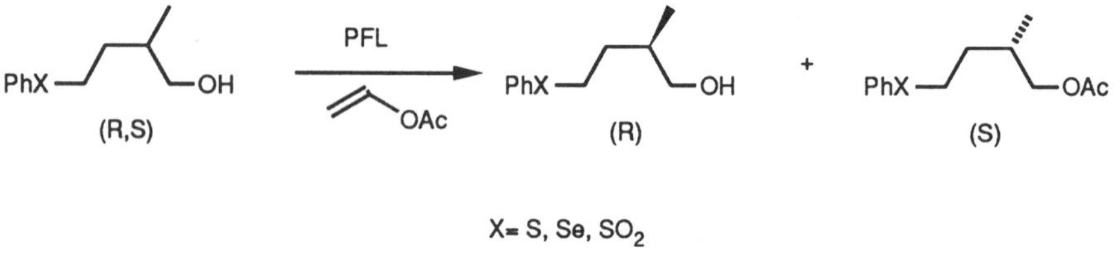

(2)

It should be mentioned that, in order to achieve a preparation of the alcohol with the highest enantiomeric excess, the transesterification has to be brought about to 60% conversion to the acetate. In this way, the unreacted (R)-alcohol (40% yield) is nearly optically pure (>98% ee), as established by ^{1}H-NMR analysis of its MTPA ester. If a >98% ee is desired for the (S)-acetate, the reaction has to be stopped at 40% conversion and the recovery of the product is essentially quantitative (38-40%).

The above examples are of special synthetic significance, since the chiral compounds so far obtained are useful chiral synthons. In fact, enantiomerically pure (R)- and (S)-4-phenylthio- or 4-phenylsulphonyl -2-methylbutanol are C-5 isoprenoids chiral building blocks [5]. We have already used the (S)-isomer for a synthesis of (25S)-26-hydroxy cholesterol [6]. Furthermore, due to the presence of the phenylselenyl moiety in the molecule, we have easily transformed (R)- and (S)-4-phenylselenyl-2-methylbutanol into two useful chiral synthons, namely, 2-methyl-3-butenol and 2-methyl-1,3-propanediol derivatives of high ee (Eq. 3).

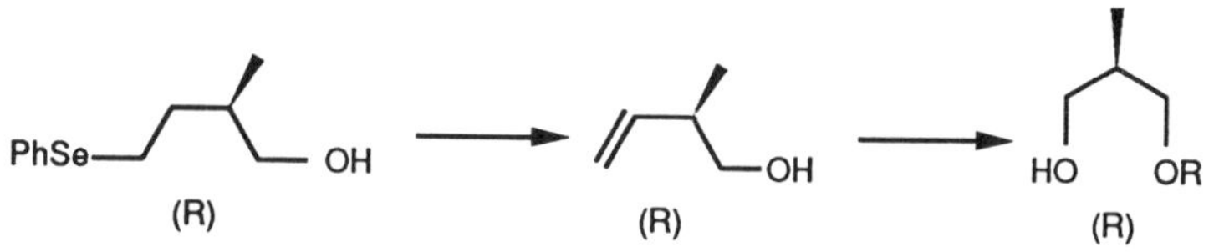

(3)

Derivatives of 2-methyl-1,3-propanediol could be prepared by an alternative approach. Thus, the enantioselective acetylation of racemic 3-diphenyl-_t_-butylsilyl ether of 2-methyl-1,3-propanediol (Eq. 4, R= $SiPh_2$ _t_-Bu) could afford a >98% ee (S)-alcohol or (R)-acetate [7].

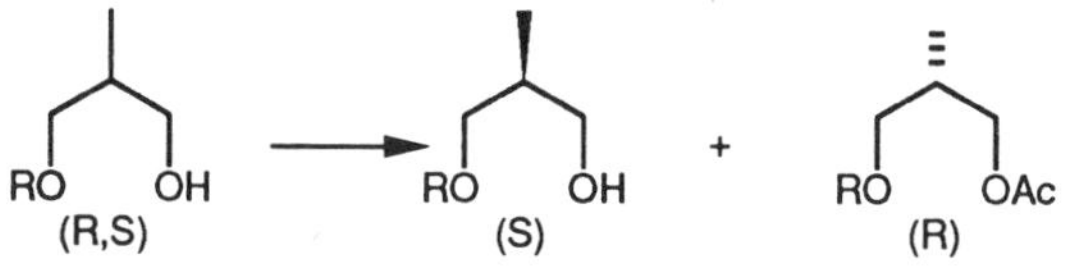

(4)

If the racemic monoacetate of 2-methyl-1,3-propanediol (Eq. 4, R= $COCH_3$) is the substrate, the unreacted (S)-monoacetate can be isolated (38%, > 98% ee) when the reaction is brought about to 60% of diacetate. When the transesterification procedure is applied to the prochiral 2-methyl-1,3-propanediol itself, the (S)-monoacetate can be obtained (52%) with 60% maximum ee if 40% of the diol is present in the mixture and 60% of the monoacetate is formed. If the reaction is carried out (Eq. 5) to the disappearance of the diol and formation of mono- and diacetate, the optically pure (S)-monoacetate (>98% ee) is obtained (40%), when 60% of the diacetate is formed.

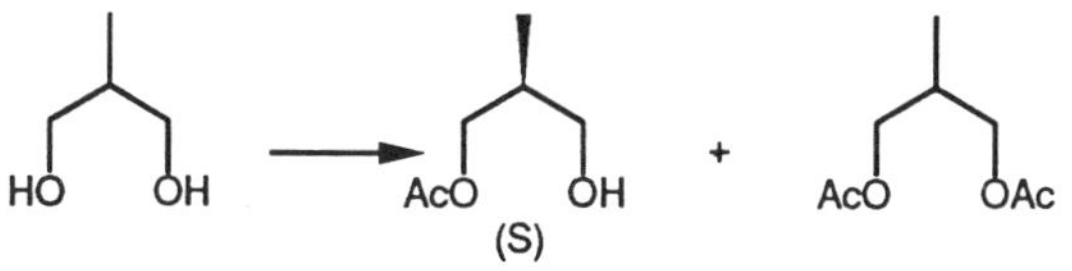

(5)

Interestingly, the aqueous hydrolysis of the 2-methyl-1,3-propanediol diacetate catalyzed by PFL leads to (R)-monoacetate (90% ee).

CONCLUSIONS

The irreversible PFL-catalyzed transesterification in organic solvents seems especially suitable for the resolution of alpha-methyl substituted primary alcohols. The compounds which can be prepared by this procedure are nearly optically pure and, for the the substrates examined by us, the stereochemical outcome of the reaction is constant.

REFERENCES

1. Wang, Y.F.; Lalonde, J.J.; Momongan, M.; Berghreiter, D.E.; Wong, G.H. J. Am. Chem. Soc., **1988**, 110, 7200.

2. Dale, J.A.; Mosher, H.S. J. Am. Chem. Soc. **1988**, 95, 512.

3. Ferraboschi, P.; Grisenti, P.; Manzocchi, A.; Santaniello, E. J. Org. Chem., **1990**, in press.

4. Ferraboschi, P.; Grisenti, P.; Santaniello, E. Synlett., **1990**, in press.

5. Fuganti, C.; Grasselli, P.; Servi, S.; Hogberg, H.E. J. Chem. Soc., Perkin Trans. 1 **1988**, 3061.

6. Ferraboschi, P.; Fiecchi, A.; Grisenti, P.; Santaniello, E. J. Chem. Soc., Perkin Trans. 1., **1987**, 1749.

7. Santaniello, E.; Ferraboschi, P.; Grisenti, P. Tetrahedron Lett., to be published.

PRODUCTION OF MANNITOL BY <u>LEUCONOSTOC</u> <u>MESENTEROIDES</u>,

IMMOBILIZED ON RETICULATED POLYURETHANE FOAM

W. SOETAERT and E.J. VANDAMME

Laboratory of General and Industrial Microbiology, Faculty of Agricultural Sciences, State University of Gent, Coupure Links, 653, 9000 Gent, Belgium

INTRODUCTION

Mannitol is a sugaralcohol with extensive use in food, chemistry, medicine and research. It is about half as sweet as sucrose. At present it is produced industrially by katalytic hydrogenation of fructose at high temperature and pressure. This hydrogenation yields mannitol as well as the isomer sorbitol in about equal amounts. The sorbitol formation is unwanted because sorbitol can also be made by hydrogenation of glucose.
Mannitol is present in several microorganisms, mostly in fungi, in which mannitol is a well known reserve metabolite, and as a primary metabolite produced by heterofermentative lactic acid bacteria.

Fructose is an alternative electron acceptor in the fosfoketolase breakdown of sugars by <u>Leuconostoc</u> <u>mesenteroides</u> and is thereby converted to mannitol. The reaction is mediated by the enzyme mannitol dehydrogenase.

$$\text{fructose} \xrightarrow[\text{NADH} \quad \text{NAD}]{} \text{mannitol}$$

In this process live <u>Leuconostoc</u> <u>mesenteroides</u> cells are used with glucose as the electron donor. The theoretical fermentation balance is :

$$\text{Glucose + 2 Fructose} \longrightarrow CO_2 \text{ + acetic acid + D(-)lactic acid + 2 mannitol}$$

RESULTS AND DISCUSSION

<u>Batch</u> <u>Fermentations</u>

Several batch fermentations were performed at different temperatures and with free pH or pH-stat culture at pH 4.5. The results indicate that working at low temperatures and low pH favours mannitol production (Fig. 1). The highest conversion efficiency was obtained working at pH 4.5 and 20 °C. Working at low temperature prolonged conversion time considerably. The overall volumetric productivity of the batch fermentations was about 1 g mannitol/l.hr.

Bioorganic Chemistry in Healthcare and Technology, Edited by U.K. Pandit and F.C. Alderweireldt, Plenum Press, New York, 1991

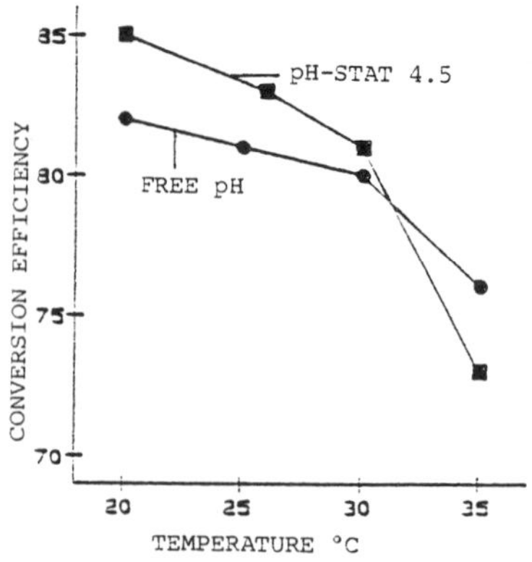

Fig. 1. Conversion efficiency as a function of temperature for free pH and pH-controlled (pH 4.5) batch fermentations.

Continuous Fermentation

For the continuous fermentation the cells were immobilised by adhesion to polyurethane foam. An important characteristic which enhances the adhesion is the hydrophobicity of the cells[1]. This cell hydrophobicity was drastically influenced by the carbon source. Cells grown on sucrose were extremely hydrophilic (hydophobicity index = 6 %), and as a consequence adhered very poorly to the foam. Cells grown on glucose and/or fructose were much more hydrophobic (hydophobicity index = 49 %) and adhered very well to the foam. The difference was mainly caused by the production of dextran when grown on sucrose. pH also had a very strong influence on the adhesion of the cells to the polyurethane foam.

Continuous fermentation was performed in a tubular reactor with reticulated polyurethane foam as the carrier material. Free cells were present in the effluent up to 100 hours after inoculation. After this the free cell content decreased drastically and a practically clear solution containing the product was continuously produced for another 150 hours, after which the fermentation was stopped. This indicated the very efficient attachment of the cells to the foam. During this period no stability problems occurred. The conversion efficiency was about 60 %, substantially lower than in batch culture. The overall residence time in the column was less than 50 minutes. The volumetric production rate was 8 g mannitol/l.hr, about 8 times as high as in batch culture. This high production rate is due to the high concentration of biomass adhered to the foam. Obtaining higher biomass and corresponding productivity is one of the main advantages of the use of immobilised cells[2].

REFERENCES

1. VAN LOOSDRECHT, M.C.M.; LEIKLEMA, J.; NORDE, W.; SCHRAA, G.; ZEHNDER, A.J.B., The role of bacterial cell wall hydrophobicity as a measure to predict the initial steps of bacterial adhesion, Appl. Env. Microbiol., 53: 1893-1897 (1987).

2. BRODELIUS, P. & VANDAMME, E.J., Immobilized cell systems, in : "Biotechnology 7A, Enzyme Technology", 405-464. KENNEDY, J.F., Ed., Publ. V.C.H. Weinheim (1987).

SYNTHESIS OF POTENTIAL INHIBITORS OF VITAMIN D HYDROXYLASES

L. Castedo. L. Sarandeses, J. Granja, J.L. Mascareñas, M.A. Maestro, and A. Mouriño

Departamento de Química Orgánica. Facultad de Química y Sección de Alcaloides del C.S.I.C. Universidad de Santiago de Compostela. Spain

Before exerting its physiological action, vitamin D_3 (**1a**, Fig. 1) must be hydroxylated in the liver to 25-hydroxyvitamin D_3 (**1b**) by the enzyme 25-hydroxylase (25-OH-ase), and then in the kidney to 1α,25-dihydroxyvitamin D_3 (**1c**) by 1α-OH-ase. This latter metabolite,whose mechanism of action is like that of steroid hormones, stimulates important physiological functions such as intestinal calcium absorption (ICA) and bone calcium mobilization (BCM).[1] Interestingly, recent studies indicate that the hormone **1c** is also involved in regulating cell proliferation and differentiation.[2] Since we are interested in both the mode of action and biosynthetic pathways of **1c**, and since research in the latter will be facilitated by inhibitors of vitamin D hydroxylases,[3] we have started work on the design of inhibitors of the 25-OH-ase. Such compounds may also prove useful for treatment of diseases related with anomalous regulation of the enzyme.

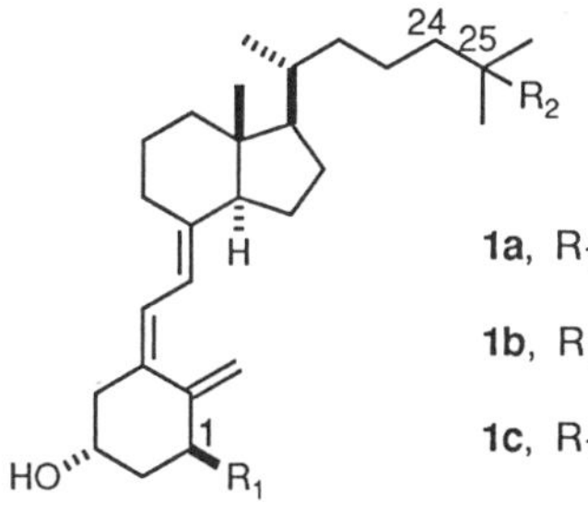

1a, $R_1 = R_2 = H$

1b, $R_1 = H$, $R_2 = OH$

1c, $R_1 = R_2 = OH$

Fig. 1

Scheme 1 describes how compounds **2a** and **2b** might work as suicide inhibitors. 25-hydroxylation of **2a** by 25-OH-ase should afford the hemi-acetal **3**, which in turn should decompose to alcohol **4** and acetone (**5**), the latter of which will feasibly react with a nucleophilic component of the enzyme such as an amino group to form a Schiff-base.[4]

Bioorganic Chemistry in Healthcare and Technology, Edited by U.K. Pandit and F.C. Alderweireldt, Plenum Press, New York, 1991

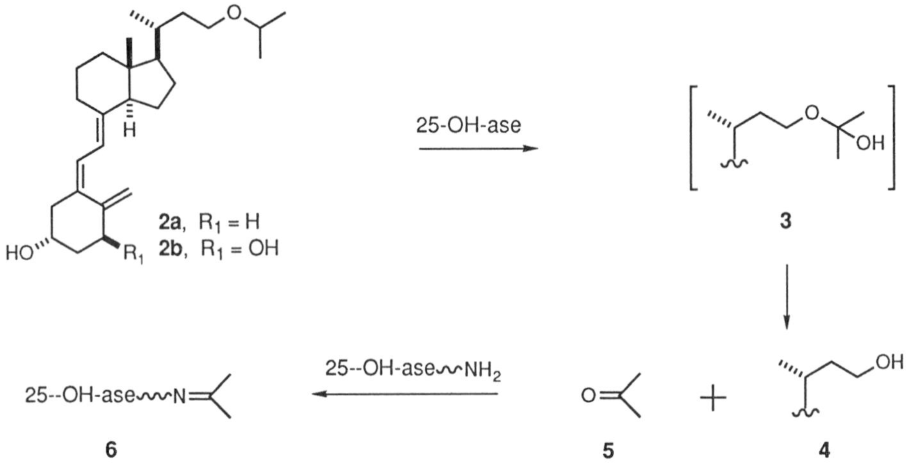

Scheme 1

We describe here the synthesis of **2a** and **2b** (Scheme 2). We started from vitamin D_2 (**7**) which was converted to alcohols **8** and **9** by known procedures.[4b,5] Protection of **8** with chloromethyl methyl ether under standard conditions followed by side-chain cleavage with ozone in methanol at -78 °C and in situ reduction with sodium borohydride (-78 to 0 °C) afforded the corresponding monoprotected diol, which was converted into its tosylate. Displacement of the tosylate with sodium cyanide afforded the required nitrile **10**. Reduction of **10** with diisobutylaluminum hydride followed by treatment with sodium hydride and subsequent reaction of the resulting alkoxide with 2-bromopropane for 12 h afforded the desired ether **11**. Deprotection of **11** with AG 50W-X4 cation-exchange resin gave the corresponding alcohol, which was oxidized with pyridinium dichromate to give ketone **12**. Enolization of **12** under kinetic conditions followed by the addition of N-phenyltrifluoromethanesulfonimide in THF at -80 °C and stirring for 10 h at RT afforded the desired vinyl triflate **13**.

The triene system of **2a** was generated by the Wittig-Horner coupling method developed by Lythgoe.[1b,5,6] The materials required for this key step are the ketone **12** and the phosphine oxide anion **14**. Coupling between **12** and **14** yielded the protected form of **2a** stereoselectively. The tert-butyldimethylsilyl group was removed by treatment of the resulting triene with tetrabutylammonium fluoride to give the desired vitamin **2a**.

The materials required for the preparation of **2b** are the vinyl triflate **13** and the enyne **15**, which had previously been synthesized in our laboratory.[7] Coupling between **13** and **15** in the presence of triethylamine in DMF at 75 °C gave the dienyne **16**.[8] Semihydrogenation over Pd-$BaSO_4$ poisoned with quinoline afforded the bis-protected previtamin **17**, which was thermolysed to the corresponding bis-protected vitamin. Removal of the tert-butyldimethylsilyl groups with tetrabutylammonium fluoride gave the desired vitamin **2b**.

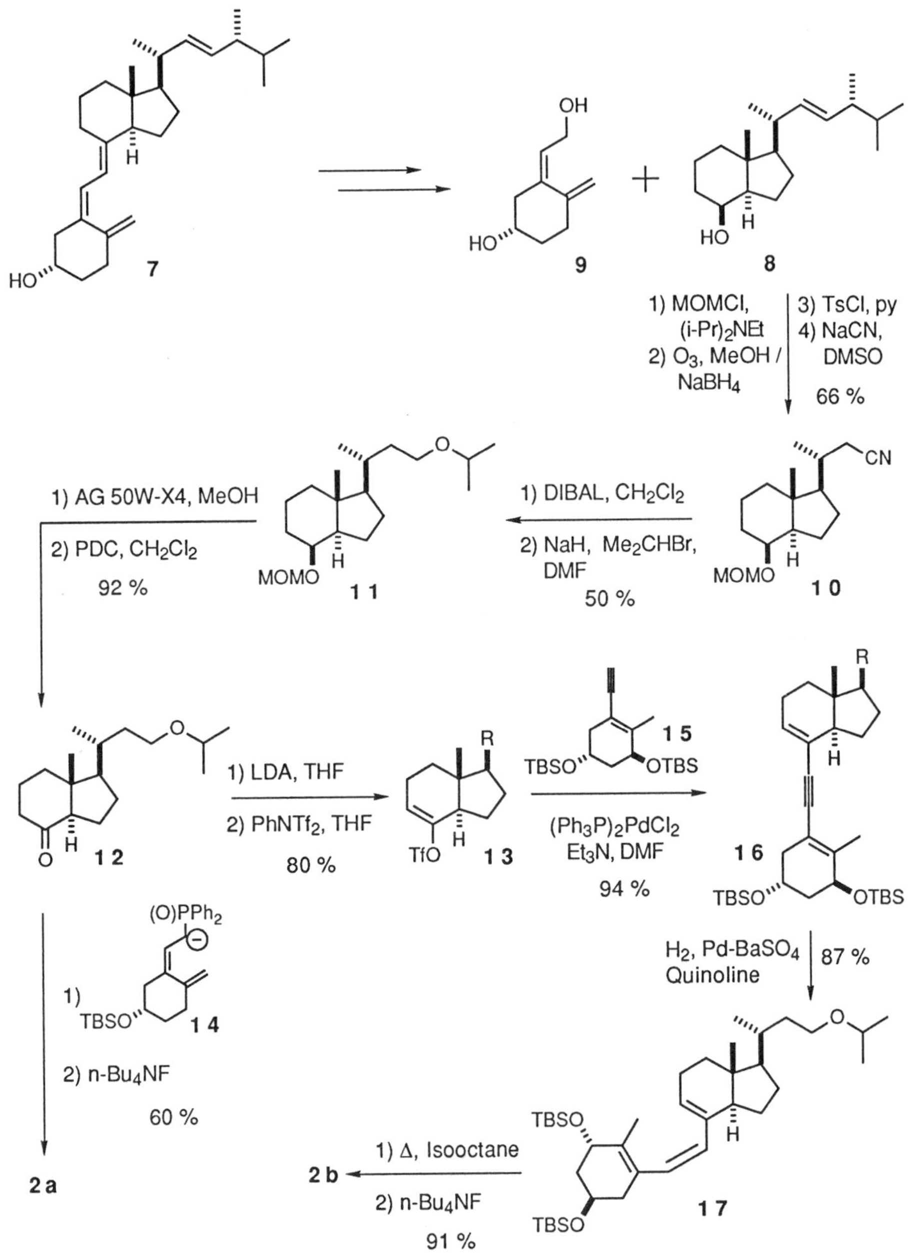

Scheme 2

Acknowledgement. We are grateful to the Dirección General de Investigación Científica y Técnica (DGICYT) for financial support of this work. We thank Duphar (Netherlands) for the generous gift of vitamin D_2.

References and Notes

1. Reviews: (a) A.W. Norman, "Vitamin D, the Calcium Homeostatic Steroid Hormone", Academic Press, NY (1979); (b) R. Pardo, R. Santelli, Bull. Chim. Soc. Fr. 98 (1985)

2. (a) C. Miyaura, E. Abe, T. Kuribayashi, H. Tanaka, K. Konno, Y. Nishii, T. Suda, Biochem. Biophys. Res. Commun. 102:937 (1981). (b) H.P. Koeffler, T. Amatrada, N. Ikekawa, Y. Kobayashi, H.F. DeLuca, Cancer Res. 44:5624 (1984).(c) J.A. MacLaughlin, W. Gange, D. Taylor, E. Emith, M.F. Holick, Proc. Natl. Acad. Sci. USA. 82: 5409 (1985).

3. (a) B. L. Onisko, H. K. Schnoes, H. F. Deluca, J. Biol. Chem. 254:3493 (1989). (b) B. L. Onisko, H. K. Schnoes, H. F. Deluca, Bioorg. Chem. 9:187 (1980).

4. A vitamin D_3 analogue containing an A-ring cyclic ether has recently been designed and synthesized to inhibit 1α-OH-ase. This approach is based on the reaction of nucleophilic components of the enzyme with α,β-unsaturated aldehydes: (a) S.A. Barrack, R.A. Gibbs, W.H. Okamura, J. Org. Chem. 53:1790 (1988). Reports on oxa-vitamin D analogues have also appeared: (b) H.T. Toh, W.H. Okamura, J. Org. Chem. 48:1414 (1983); (c) N. Kubodera, K. Miyamoto, K. Ochi, I. Matsunga, Chem. Pharm. Bull. 34:2286 (1986); (d) J. Abe, M. Morikawa, K. Miyamoto, S.-i. Kaiho, M. Fukushima, C. Miyaura, E. Abe, T. Suda, Y. Nishii, FEBS Letters 226:28 (1987().

5. J.L. Mascareñas, A. Mouriño, and L. Castedo, J. Org. Chem. 1:1269 (1986).

6. F.J. Sardina, A. Mouriño, and L. Castedo, J. Org. Chem. 51:1264 (1986).

7. L. Castedo, J.L. Mascareñas, and A. Mouriño, Tetrahedron Lett. 28:2099 (1987).

8. For original work on the coupling between vinyl triflates and enynes, see: L. Castedo, A. Mouriño, and L.A. Sarandeses, Tetrahedron Lett. 27:1523 (1986).

INHIBITORS OF ADENYLOSUCCINATE LYASE

W.A. Hartgers, M.J. Wanner and G.J. Koomen

Lab. of Organic Chemistry, University of Amsterdam
Nieuwe Achtergracht 129, 1018 WS The Netherlands

Adenylosuccinate lyase is an essential enzyme in two steps of the biosynthesis of adenosine nucleotides. In the conversion IMP → AMP it catalyzes the elimination of fumaric acid from adenylosuccinate:

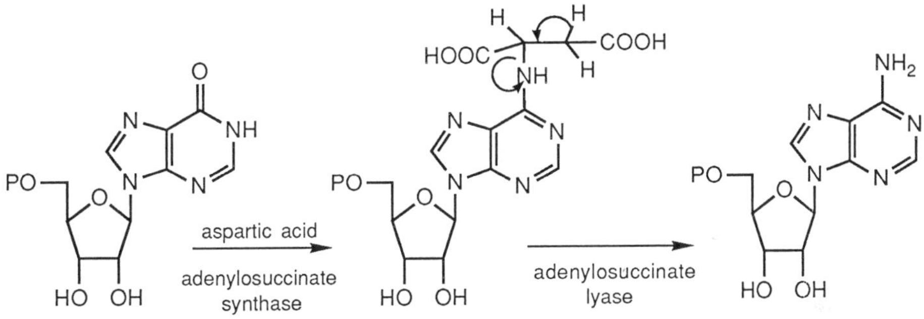

We have been involved in the synthesis of adenylosuccinate analogs of different types. In these analogs the elimination of fumaric acid was inhibited by either substituting the proton involved by a methyl group or a fluorine atom i.e. **1**. (Wanner 1978, 1980), or by replacing the amino group of adenine by a methylene group. (**2**, Odijk 1985).

Our interest in the preparation and study of inhibitors of this enzyme was recently stimulated by the following developments:

1. The activity of the enzyme was determined by Reed (1987) in human tissues of different nature. It was found that the enzyme can be used as a malignancy marker, showing significantly higher activity in human mammacarcinoma than in benign breast tumors. Effective inhibition of the enzyme thus could lead to a selective anticancer effect.
2. The enzyme from yeast in purified form recently became commercially available.
3. Methods for determining enzyme activity are well documented, both on the isolated enzyme as well as with homogenized tissue (Reed, 1987)

Bioorganic Chemistry in Healthcare and Technology, Edited by U.K. Pandit and F.C. Alderweireldt, Plenum Press, New York, 1991

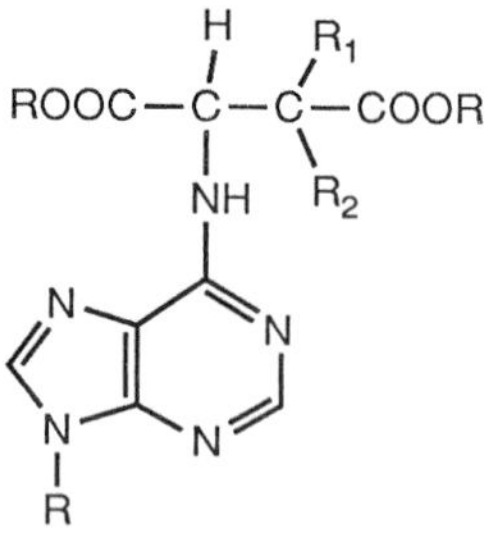

R_1, R_2 = H, F, CH_3

1

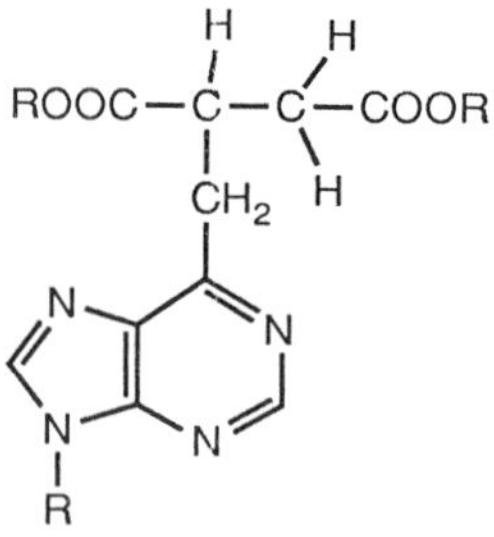

2

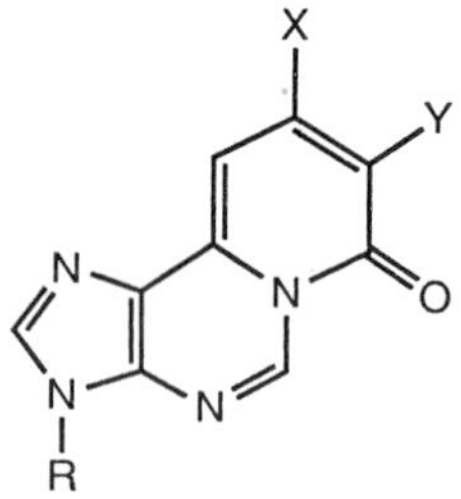

3

Carba analogs **2** easily cyclized to pyrido-purine systems of general structure **3** with highly fluorescent properties in the visible region. (Odijk, 1986, 1987). In order to prepare more effective inhibitors, wedirected our attention to the preparation of inhibitors of the transition state and suïcide type. In the last case it is not simple to devise a molecule with potential suïcide properties without major structural modifications. Since the enzyme catalyzes reactions with rather different substrates, we assumed that it would not be very selective.

In first instance we decided to prepare derivative **4a**, which, when accepted as a substrate despite its structural changes, would produce a very reactive ketene within the active site.

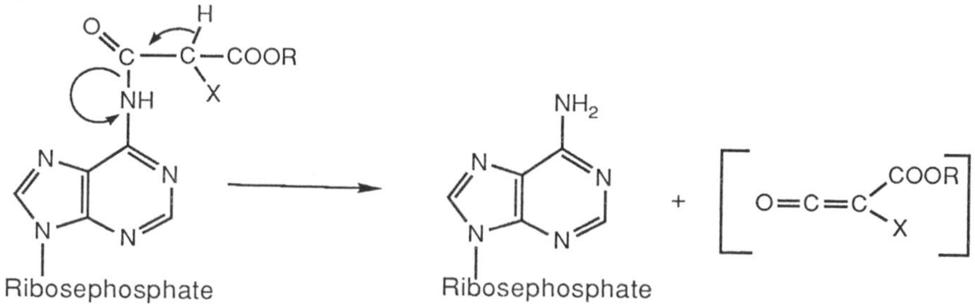

4a X = H
4b X = F

5a,b

The synthesis of **4a** was accomplished via the scheme outlined below:

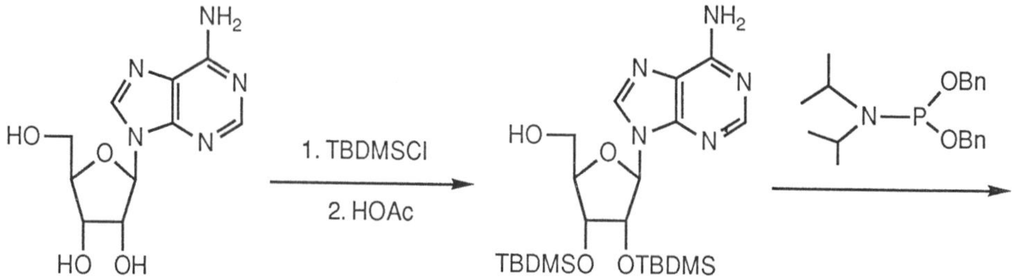

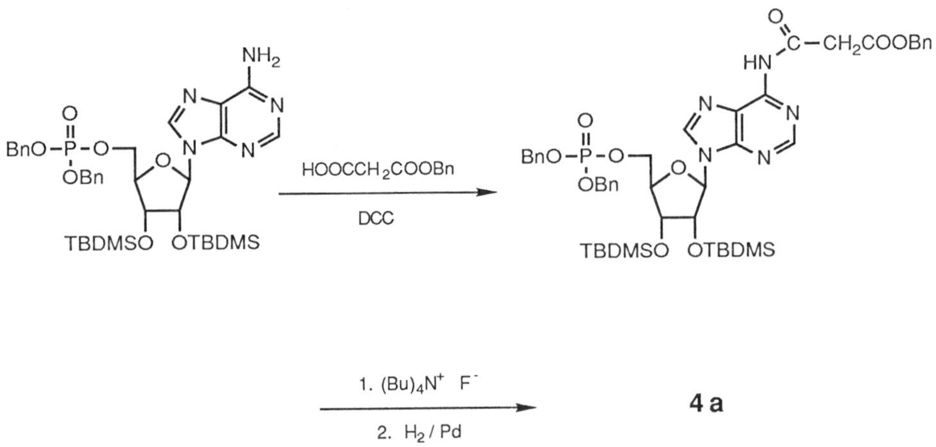

4a

Due to the good leaving group properties of the adenine ring, the amide bond showed very low stability under hydrolytic conditions. This was the reason, why phosphorylating the 5' OH function, prior to substituting the amino group appeared to be the most effective approach for the synthesis of **4a**.

The activity of **4a** as an enzyme inhibitor was determined on adenylosuccinate lyase from yeast. (Sigma). A double reciprocal (Lineweaver-Burk) plot of 1/v vs 1/S showed non competitive inhibition ($K_i = 65\,\mu M$). Studies to confirm the suïcide nature of the inhibition as well as antileukemic tests with the non-phosphorylated riboside are currently underway.

A second approach to the development of inhibitors of the adenylosuccinate system consists of the design of transition state analogs of the enzyme, for instance **6** or **7**.

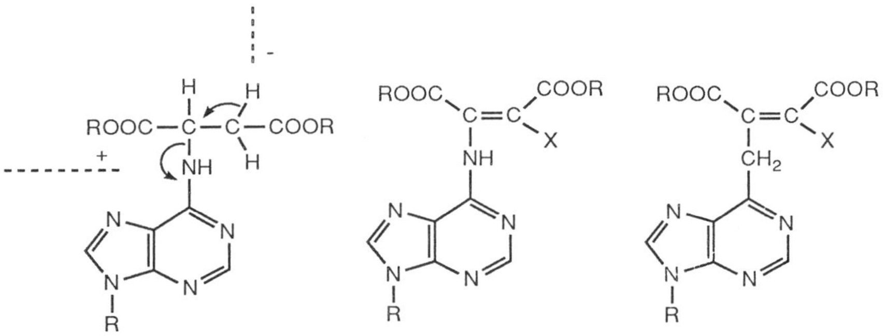

trans. state | **6** | **7**

C=C : Z and E

X = H, F

Efforts to synthesize derivatives of type **6** via addition reactions to acetylenic esters produced cyclized products (**8**) under the reaction conditions employed, even with t-butyl esters

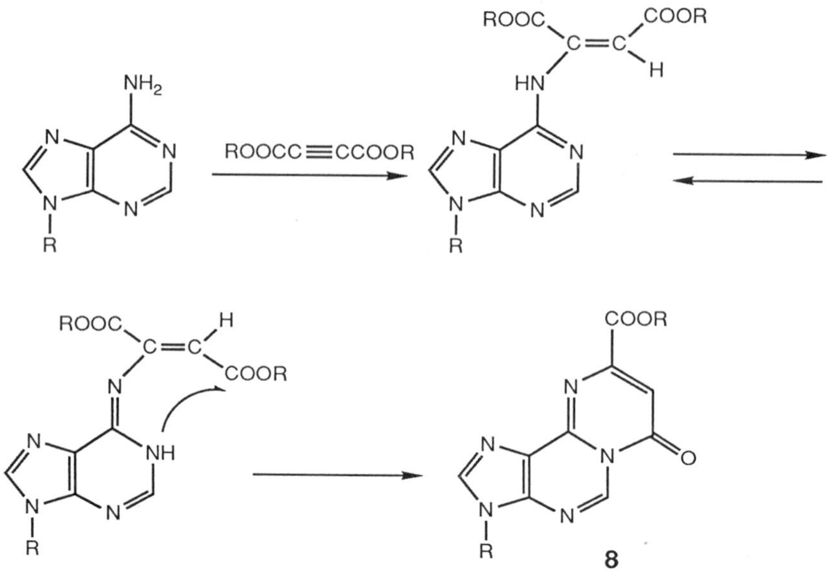

In the present synthetic approach the nucleophilicity of nitrogen N-1 is blocked via conversion into **9**. After the Michael addition, **10** will be converted into **5** (X = H) under mild conditions in order to prevent cyclization.

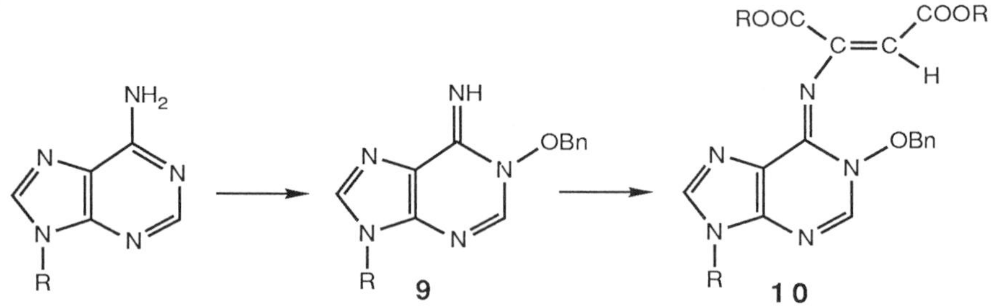

References

V.L. Reed, D.O. Mack and L.D. Smith. Clin. Biochem. 20, 349 (1987).

W.M. Odijk and G.J. Koomen. Tetrahedron 41, 1893 (1985).

W.M. Odijk and G.J. Koomen. J.Chem. Soc. Perk. II, 1561 (1986).

W.M. Odijk and G.J. Koomen. J. Histochem. Cytochem. 35, 1161 (1987).

M.J. Wanner, J.J.M. Hageman, G.J. Koomen and U.K. Pandit. Rec. Trav. Chim. 97, 211 (1978).

M.J. Wanner, E.M. van Wijk, G.J. Koomen and U.K. Pandit. Rec. Trav. Chim. 99, 20 (1980).

MODEL STUDIES OF THE THYMIDYLATE SYNTHASE REACTION. DESIGN OF MECHANISM-BASED POTENTIAL INHIBITORS OF DNA BIOSYNTHESIS

U.K. Pandit, E. Vega, J.W.G. Meissner and E. R. de Waard

Laboratory of Organic Chemistry, University of Amsterdam, Nieuwe Achtergracht 129 1018 WS Amsterdam, The Netherlands

The enzyme thymidylate synthase (TS) mediates the conversion of dUMP to dTMP which is utilized in the biosynthesis of DNA. The methyl group transferred to the dUMP substrate in this reaction is provided by the cofactor 5,10-methylene-tetrahydrofolate (5,10-CH_2-H_4-folate) in a multi-step process. The proposed mechanism of the transformation is outlined in scheme I [1].

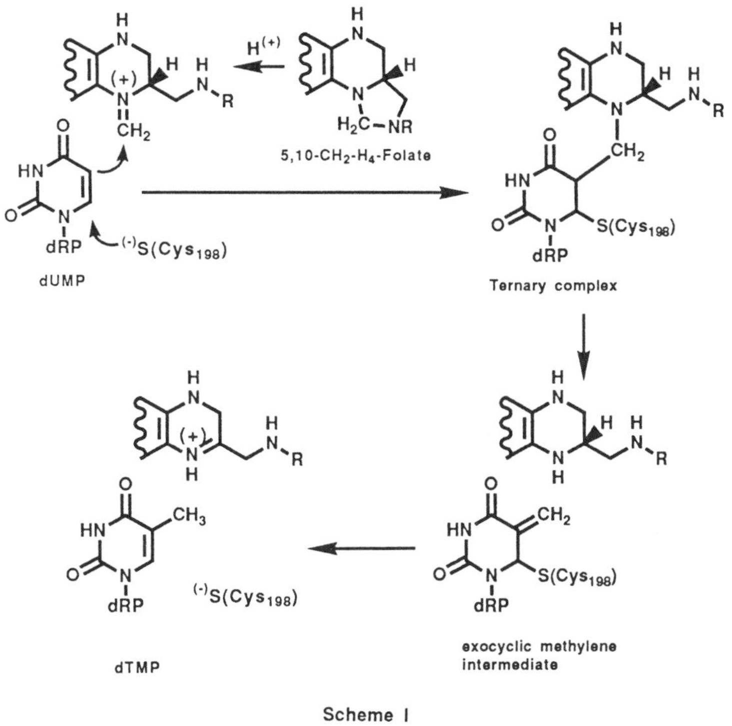

Scheme I

Bioorganic Chemistry in Healthcare and Technology, Edited by U.K. Pandit and F.C. Alderweireldt, Plenum Press, New York, 1991

It has been shown in our laboratory that the combined CH_2-transfer and reduction steps, mediated by 5,10-CH_2-H_4-folate, can be mimiced by the reaction of cofactor model 1 with a C(5)-nucleophilically activated uracil derivative 2 [2]. The deuterium label in 1 was found in the C(5)-methyl group of the corresponding thymine product (scheme II).

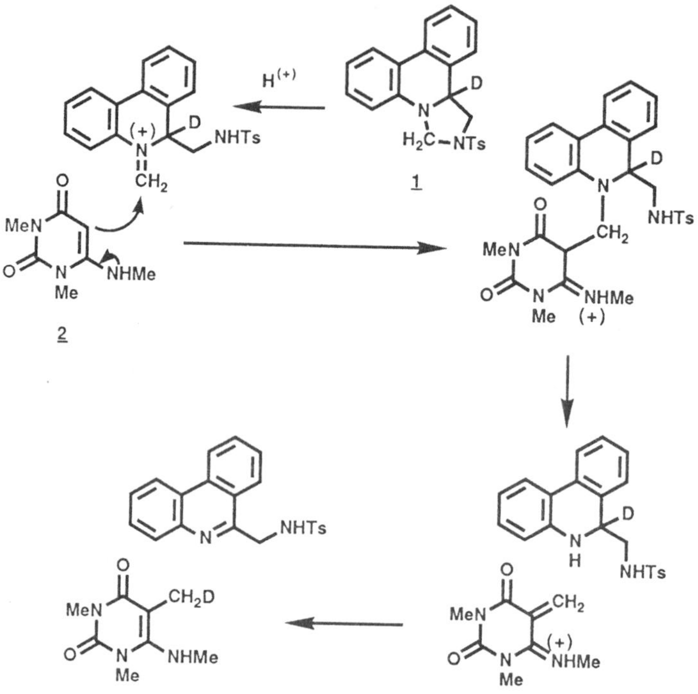

Scheme II

We have studied the reaction of 1,3-dimethyl-5-(1-pyridiniummethylene)uracil salt (3) with thiols in an effort to generate the 1,3-dimethyl-5-methyleneuracil-6-thio ether intermediate (4). When 3 was heated with RSH (R = p(NO_2)$PhCH_2$, $PhCH_2$, t-Butyl) in benzene or toluene, the reaction product was 1,3-dimethylthymine 5 [3]. The conversion of uracilyl pyridinium salt 3 to thymine derivative 5 can be rationalized on the basis of the formation of exocyclic methylene intermediate 4 (analogous to that formed in the enzymic reaction) and its subsequent reduction by thiols, presumably via a radical mechanism (Scheme III).

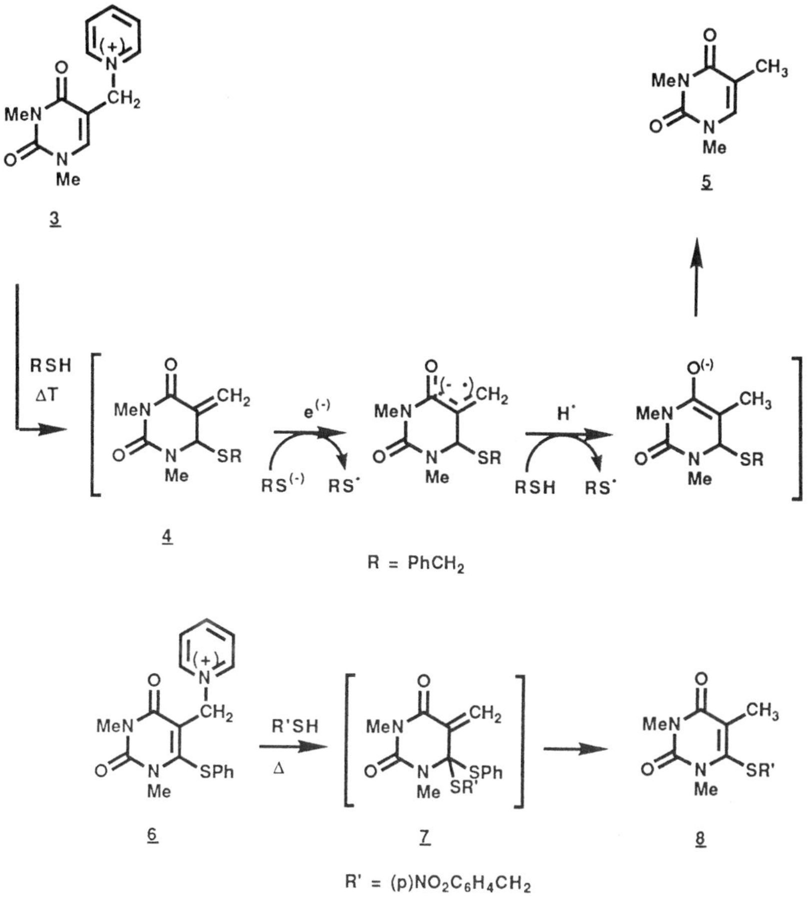

Scheme III

Evidence for the proposed mechanism is derived from the following observations.

(a) The rate of formation of 5 is strongly influenced by the nature of substituents both in the thiol and the C(6) position of the substrate (3)[3]. Sterically large groups, e.g. t-Butyl in the thiol and CH_3 instead of H at C(6) in 3 cause reduction of the rate.

(b) When salt 6 was allowed to react with thiol R'SH (R' = $p(NO_2)PhCH_2$), the thymine product was the one containing the R'S residue at position C(6). This is the expected result of the aforementioned mechanism. The exocyclic methylene intermediate 7 undergoes reduction with expulsion of the more stable $PhS^{(-)}$ ion in the final step.

(c) Addition of galvinoxyl radicals 9 to the reaction mixture of substrate 3 and $PhCH_2SH$ causes a decrease in the rate of formation of product 5. This decrease is dependent upon the concentration of the scavenger radical 9 (fig.1). The last mentioned result strongly support the proposed radical mechanism.

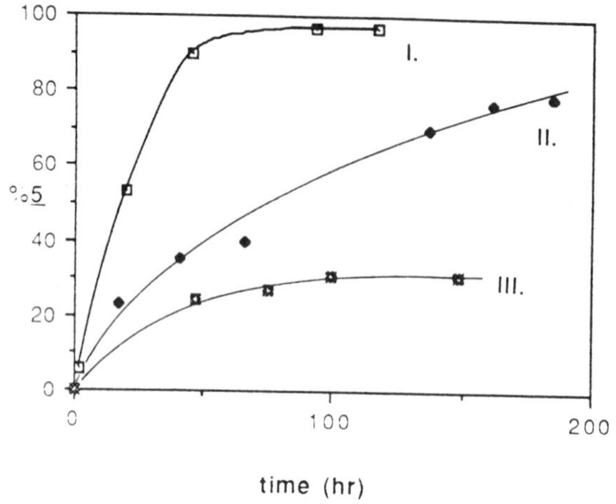

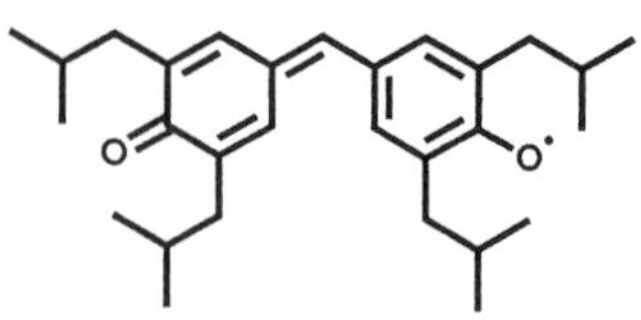

9

Figure 1. Formation of 5 upon heating 3 with $PhCH_2SH$ in the presence of galvinoxyl radicals (9). I. 3 (1 eq.), RSH (2.5 eq.); II. 3 (1 eq.), RSH (2.5 eq.), 9 (1 eq.); III. 3 (1 eq.), RSH (2.5 eq.), 9 (2 eq.).

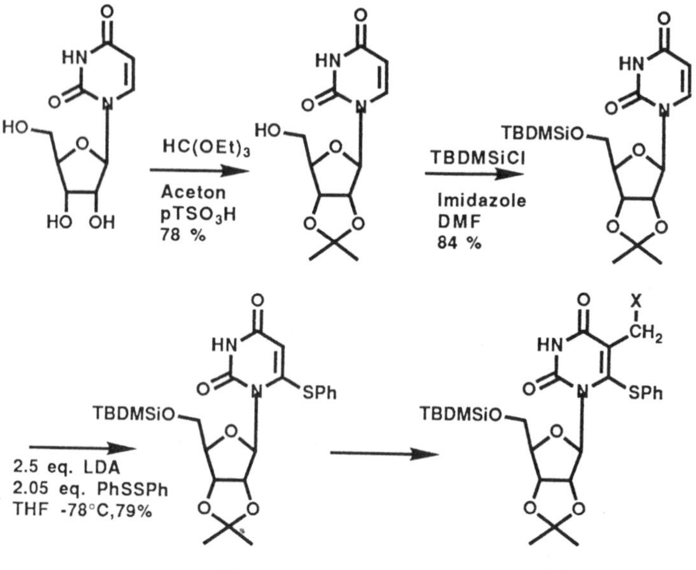

10 11

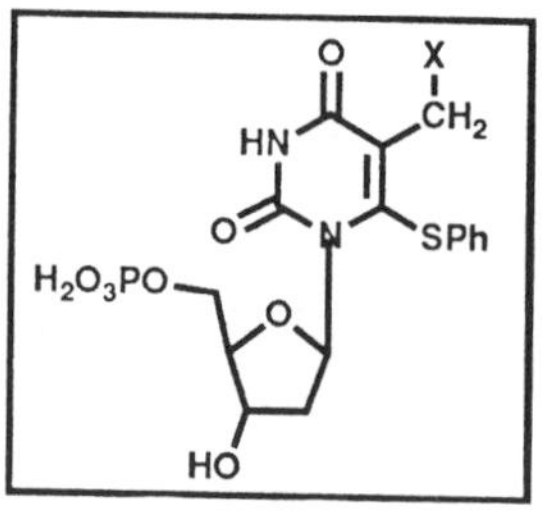

12

Scheme IV

DESIGN OF MECHANISM-BASED TS INHIBITORS

The exchange of the thiol residue during the conversion of 6 to 8, according to the proposed mechanism, projects the opportunity of developing TS inhibitors which could operate via the formation of (apo)enzyme-bound analogues of exocyclic methylene intermediate 7.Potential TS inhibitors, based on this mechanistic concept, are exemplified by compounds of type 12 (e.g. X = OR, NR_1R_2 etc.). The synthetic strategy leading to 12 is described for the corresponding ribose analoge (11) in scheme IV. In this context, the pivotal intermediate 10 has been synthesized in three steps from uridine in good overal yield (52%). Transformation of 10 into 11 possessing various leaving groeps X, is being actively pursued.

REFERENCES

1. D.V.Santi and P.V.Danenberg, in : R.L.Blakley and S.J.Benkovic,"Folates and Pteridines, vol I", Wiley, New York,345(1984).
2. P.F.C.v.d.Meij, T.B.R.A.Chen, E.Hilhorst, E.R.de Waard and U.K.Pandit, *Tetrahedron*, 43, 4015 (1987).
3. E.Vega, E.R.de Waard, U.K.Pandit, *Recl.Trav.Chim.,Pays-Bas*, 1990,109, 131.

DESIGN OF PLANT STEROL BIOSYNTHESIS INHIBITORS - INHIBITION OF THE STEPS INVOLVED IN THE REMOVAL OF THE 14α-METHYL GROUP

Maryse Taton, Pierre Benveniste and Alain Rahier

Département d'Enzymologie Moléculaire et Cellulaire de l'IBMP (UPR CNRS 406- Institut de Botanique, 28 rue Goethe 67083 - STRASBOURG Cédex

INTRODUCTION

Over the past 20 years, an increasing number of sterol biosynthesis inhibitors (SBI's) have been developed as fungicides widely used in crop protection or medicine, but also as potential hypocholesterolemic agents in mammals. In common with most eukaryotic organisms, higher plants also possess sterols and here too interference with their biosynthesis have profound biological effects which may lead to the development of new herbicides. Moreover, little is known about the impact of agricultural fungicides on plant sterol biosynthesis and the understanding of the mechanisms controlling their selective toxicity towards the fungus and relative safeness to the plant. Until recently most of SBI's have been discovered from extensive screening programmes and only a few inhibitors have been rationally designed, particularly in the case of enzymes involved in plant sterol biosynthesis |1,2|.

The biosynthetic pathway, leading from the cyclization product of squalene oxide to the terminal sterol structure, involves the removal of the 14α-methyl group. In higher plants this elimination is carried out by the cytochrome P-450 dependent obtusifoliol (1) 14α-demethylase (P450-OBT. 14DM) |3| which converts (1) to the corresponding Δ8,14-sterol diene (2) subsequently reduced to the Δ8-sterol (3) by a distinct NADPH-dependent Δ8,14-sterol 14-reductase (14-RED) |4|.

The present communication describes the potent *in vitro* and *in vivo* inhibition of these both enzymes. The strategies leading to the inhibitors were twofold. Firstly, biochemical *in vitro* screening of a series of potential ligands of cyt-P450 previously described as fungicides or plant growth regulators led to potent inhibitors of OBT.14DM. Secondly, rational design of synthetic high energy intermediate analogs led to an extremely potent inhibition of the 14-RED.

INHIBITION OF OBTUSIFOLIOL 14α-METHYL-DEMETHYLASE BY NITROGEN-BASED LIGANDS

We recently described a microsomal system isolated from *Zea mays* embryos which is capable of efficiently demethylating the 14α-methyl group of (1) |3|. Several lines of evidence strongly support the involvement of cyt-P450 in this reaction and this enzyme constituted a novel member of the increasing but still limited number of endobiotic cyt-P450s and associated monooxygenase activities identified in higher plants |3|.

Azoles, pyridines and pyrimidines belong to a large group of fungicides

Bioorganic Chemistry in Healthcare and Technology, Edited by U.K. Pandit and F.C. Alderweireldt, Plenum Press, New York, 1991

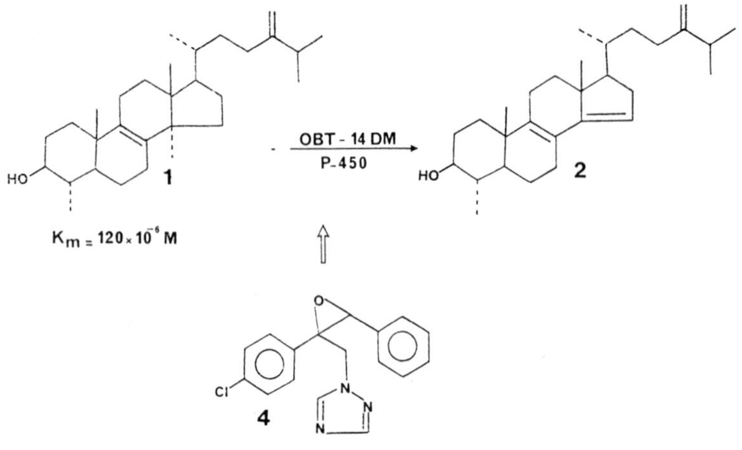

$K_m = 120 \times 10^{-6}$ M

$IC_{50} = 0.05 \times 10^{-6}$ M

used for the control of fungal diseases in agriculture and medicine. Structurally related compounds display potent plant growth regulator (PGR) properties. Recently, a triazole derivative with herbicidal properties has been described. The main inhibitory site of these fungicides appears to be the lanosterol-14α-methyl demethylase from fungi and evidences support the assumption that the molecular basis of the observed inhibition of the 14-demethylase is due to the strong binding of these chemicals to the ferric protohaem iron of cyt-P450. In contrast to the extensive studies performed with fungi, much less attention has been paid to the effects of such compounds in plants.

We recently demonstrated the *in vitro* and *in vivo* inhibition of the P450-OBT.14DM by a series of such heterocyclic derivatives containing a sp^2 nitrogen atom |5|. The results clearly indicated that many of these different azole derivatives are potent inhibitors of the P450-OBT.14DM. In particular, LAB 170250F (4), a fungicide, revealed to be a potent inhibitor of the OBT.14DM (IC50 = 50 nM ; IC50/Km = 10^{-3}), whereas most of the PGR's tested displayed low activity (IC50 >10 μM). In addition, the *in vivo* effects of this series of compounds on the sterol profile of maize seedlings were in complete accordance with the IC50 values measured *in vitro*. Moreover we could demonstrate the direct binding of these derivatives to the total cyt-P450 present in maize microsomes by difference spectroscopy. However, the measured Kd values did not match the IC50 values found for the same inhibitors underlining the probable presence of multiple forms of P450 and associated monooxygenase activities in the microsomal preparation used.

As a consequence of these results, the lack of selectivity of a number of azole inhibitors used as fungicides in plant protection clearly appears since we showed they affected strongly the plant host. Among other possibilities, a better selectivity for the pathogenic fungus versus the plant or human host would imply that the azole inhibitor should bind more tightly to the fungal 14α-demethylase than to the host 14α-demethylase. Likewise, azole inhibitors should bind more efficiently the plant 14α-demethylase than the animal demethylase when used as herbicides. The selectivity of azole inhibitors will result in part from the structural features necessary for binding to any of the three main 14α-demethylases found in nature, namely the lanosterol-14α-demethylase in mammals and *Saccharomyces cerevisiae*, the eburicol-14α-demethylase in most pathogenic fungi and the OBT.14DM in higher plants. The surprising difference in substrate specificity between 14α-demethylase enzymes from different sources may imply different structure activity relationships for the binding of azole inhibitors as far as the substrate and inhibitors binding domains are overlapping. The purification of this new P450 dependent monooxygenase and the precise knowledge of its hydrophobic substrate- and azole inhibitor-binding active site pocket could be of great value in the design of specific inhibitors of this enzyme. Work to this end is currently in progress in our laboratory.

INHIBITION OF Δ8,14-STEROL 14-REDUCTASE

The last step in the 14-demethylation process is the reduction of the Δ14-double bond. This enzymatic double bond reduction is thought to proceed through an electrophilic addition mechanism, involving a C14 putative carbonium ion high energy intermediate (HEI). It has been previously shown in our laboratory |1,2| and discussed in this issue by Taton *et al.* that ammonium analogues of carbocationic HEI involved in enzymic reaction lead to potent inhibitors.

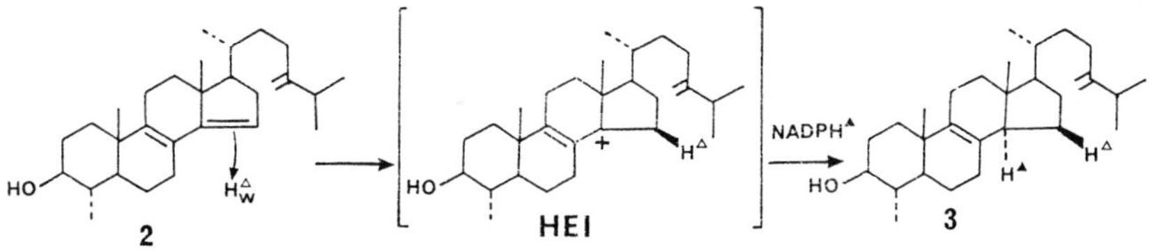

Inhibition of Δ-14-reductase by ammoniun containing species.

Compound	$I_{50}\times10^{-6}$ M	Compound	$I_{50}\times10^{-6}$ M
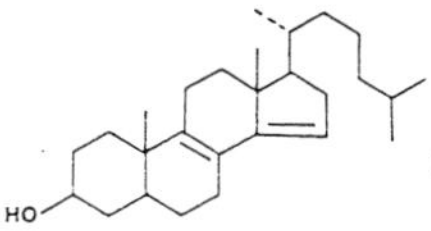	$K_m = 450\times10^{-6}$ M	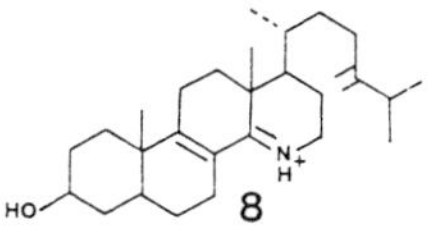 8	0.03
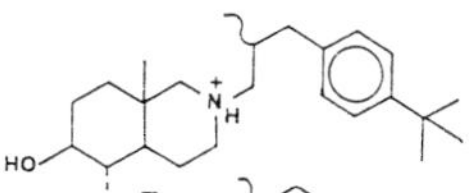 5	0.07	9 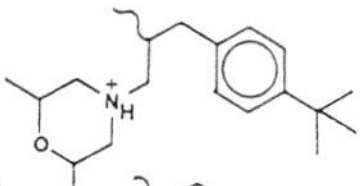	0.8
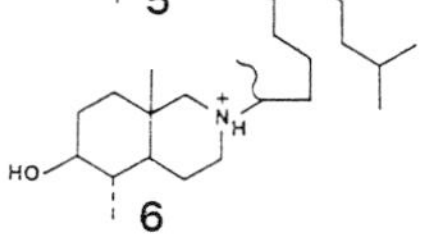 6	0.3	10 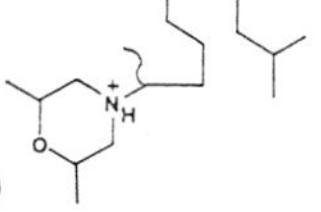	25
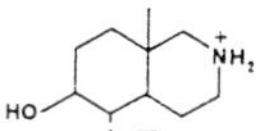 7	4.5	11 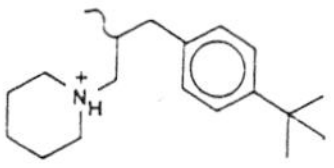	3
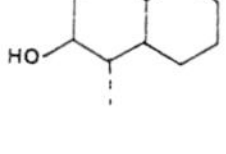	N.I.		

Indeed, synthetic ammonium and imminium analogues of the C14 cationic HEI were shown to be potent inhibitors of the reduction reaction |4|.

Thus, derivatives of the N-alkyl-8-aza-4α,10-dimethyl-*trans*-decal-3β-ol series (5-7) strongly inhibited sterol reductase (IC50 = 0.07-4 μM) (IC50/Km = 10^{-4}-10^{-3}), as did the antimycotic agent 15-azasterol (8) (IC50 = 0.03 μM). Moreover, the *in vitro* inhibition of the plant sterol reductase by a variety of ammonium-ion containing fungicides (9-11) was demonstrated and suggested these compounds to act also as reaction-intermediate analogues of the proposed C14 carbonium HEI.

To conclude, these results show that tetrahedral species such as ammonium are able to mimick a planar carbonium ion. This fact stresses the overwhelming importance of electrostatic interactions during the attachment of such inhibitor to the enzymatic site. Moreover the results clearly show that low molecular weight model compounds possessing key structural features of a reaction intermediate, such as compounds 5-7, lead to an affinity similar to that obtained with a complete analogue, such as compound 8.

Besides their value as potential fungicides, herbicides or hypocholesterolemic drugs |6|, the inhibitors described herein are powerful tools to study the physiological and biochemical roles of sterols in higher plants |7|, and in this respect, constitute an alternative to the use of sterol auxotrophs and genetic mutants blocked at specific biochemical stages, which have been widely studied in mammals or fungi.

REFERENCES

(1) M. Taton, P. Benveniste and A. Rahier (1987), Pure & Appl. Chem. 59, (3), 287-294.

(2) A. Rahier, M. Taton and P. Benveniste (1990), in Biochemistry of Cell Walls and Membranes in Fungi, P.J. Kuhn *et al.* Eds, London, 205-222.

(3) A. Rahier and M. Taton (1986), Biochem. Biophys. Res. Commun. 140, (3), 1064-1072.

(4) M. Taton, P. Benveniste and A. Rahier (1989), Eur. J. Biochem. 185, 605-614.

(5) M. Taton, P. Ullmann, P. Benveniste and A. Rahier (1988), Pestic. Biochem. Physiol. 30, 178-189.

(6) N. Gerst, A. Duriatti, F. Schuber, M. Taton, P. Benveniste and A. Rahier (1988), Biochem. Pharmacol. 37, (10), 1955-1964.

(7) P. Benveniste and A. Rahier (1990), CRC

BIOSYNTHESIS OF MELANIN BY *PYRICULARIA ORYZAE* :

MECHANISM OF REDUCTION OF POLYHYDROXYNAPHTHALENES

Fabrice Viviani, Michel Gaudry, Andrée Marquet

Laboratoire de Chimie Organique Biologique - URA CNRS 493
Université Paris 6 - 4, place Jussieu - 75005 PARIS

INTRODUCTION

The dehydroxylation of polyphenols, a rather unusual reaction, is involved in the biosynthesis of aflatoxin B1[1], during the transformation of emodin into chrysophanol[2] or during the biosynthesis of melanin[3], a dark high molecular weight pigment produced by pathogenic fungi (Scheme 1).

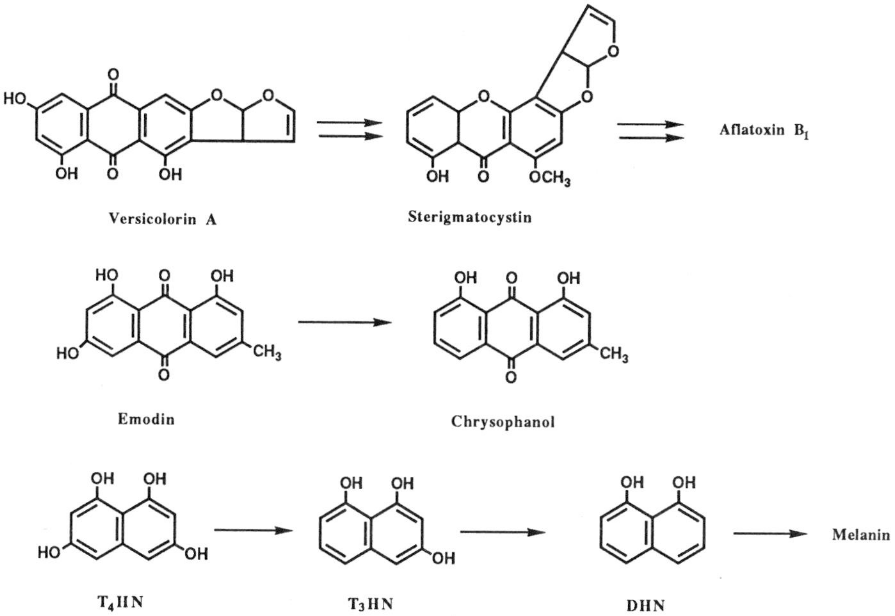

Scheme 1 Biochemical dehydroxylations of polyphenols

This reaction involves two steps : firstly the reduction of the paradiphenol followed by the dehydration of the keto alcohol and rearomatization. Two such sequences take place during the transformation of 1,3,6,8 tetrahydroxynaphthalene (T_4HN) into

Bioorganic Chemistry in Healthcare and Technology, Edited by U.K. Pandit and F.C. Alderweireldt, Plenum Press, New York, 1991

1,8-dihydroxynaphthalene (DHN), the last identified precursor of fungal melanin of *Pyricularia oryzae* or *Verticillium dahliae* : T4HN is first reduced to scytalone which is dehydrated to 1,3,8-trihydroxynaphthalen (T3HN). Reduction of T3HN to vermelone and dehydration yields DHN (Scheme 2).

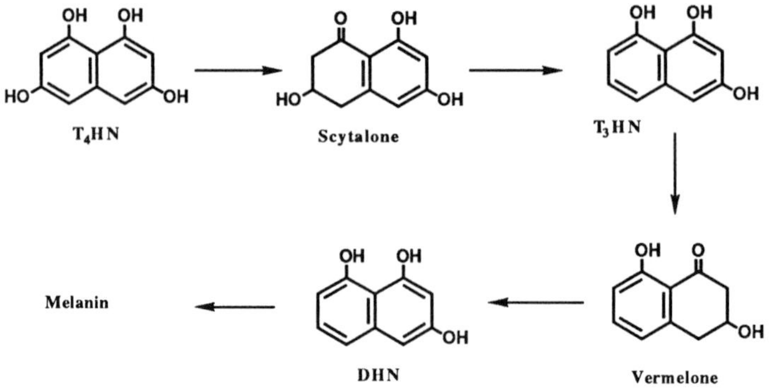

Scheme 2 Biosynthesis of fungal melanin.

The biosynthesis of fungal melanins is efficiently inhibited at the reduction steps by tricyclazole[4], a fungicide commonly use to contral rice blast disease. In order to obtain more information concerning this inhibition, we have examined the mechanism of the reduction of T4HN and T3HN.

MECHANISM AND STEREOCHEMISTRY OF THE REDUCTION

Using cell free extracts, Wheeler noted that the reduction of T4HN was NADPH dependent[5] whereas Anderson *et al.* observed an hydride transfer between labeled NADPH and emodin during its transformation in chrysophanol[6], which suggested that in that case the reduction step was achieved by a NADPH dependent reductase.

We have studied the reductases from *V. dahliae* and *P. oryzae* using specifically deuterated NADPH : [4(R)-^{2}H] and [4(S)-^{2}H] NADPH.

At pH 5.5, where the reduction rate is maximum, we observed that deuterium was incorporated in scytalone (when T4HN was the substrate) or in vermelone (when T3HN was the substrate) only when [4S-^{2}H] NADPH was used, whereas, no label transfer could be detected with [4R-^{2}H] NADPH (Table 1)[7].

Table 1. Reduction of T4HN and T3HN by the naphthol reductase from *P. oryzae* with 4R and 4S [4-^{2}H] NADPH (1 equivalent)

Substrat	NADPH	Incubation Time (hr)	% Bioconversion	% D*
T4HN	4R	3	60-70	0
	4S	"	40-50	65
T3HN	4R	4	70-80	0
	4S	"	50-60	55

* : Estimated by ^{1}H NMR and mass spectrometry

The same feature was observed with the reductase from *V. dahliae*. Both reductases belong thus to class B.

The absolute configurations of (+) scytalone and (-) vermelone which are produced enzymatically were not known. In order to achieve the elucidation of the stereochemistry of the reduction we have determined their absolute configurations by chemical correlation[8].

The R configuration of (+) scytalone was established by correlation with (-)(S)β tetralol according to Scheme 3.

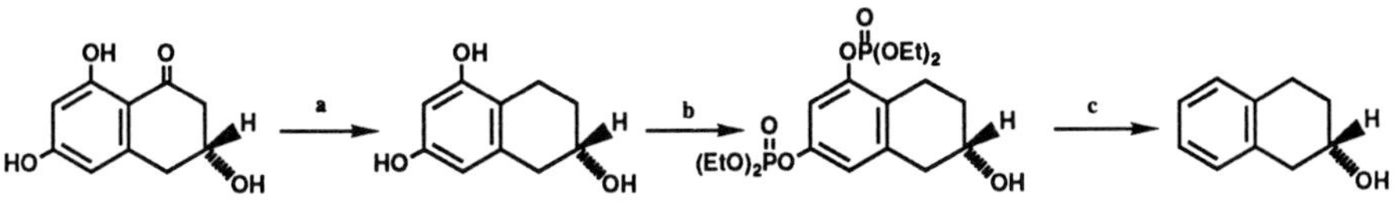

(+) Scytalone a: $NaBH_4$, $AlCl_3$/THF b: NaH, $(EtO)_2POCl$ c:Na/NH_3 (-) (S)β-tetralol

Scheme 3 determination of the absolute configuration of (+) scytalone

The configuration of (-)vermelone was established by correlation with (+)(R)scytalone according to Scheme 4[7].

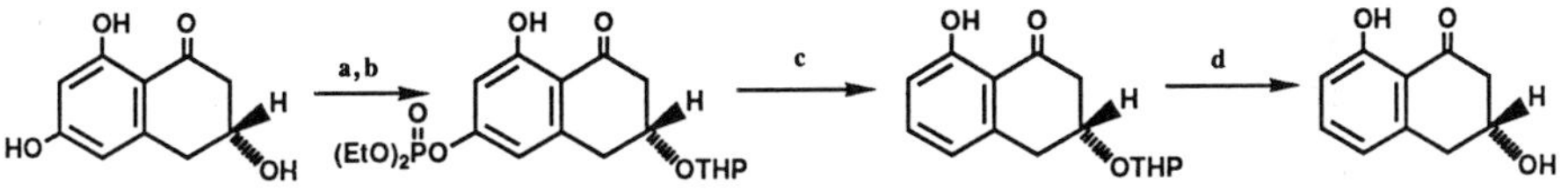

(+) (R) Scytalone (-) Vermelone

a: NaH, $(EtO)_2POCl$ b: dihydropyran, PPTS c: Na/NH_3 d: PPTS/EtOH

Scheme 4 determination of the absolute configuration of (-) vermelone

Thus the reductase transfers of proS hydrogen of NADPH to the si face of the carbonyl group of the tautomeric form of the metadiphenol (Scheme 5).

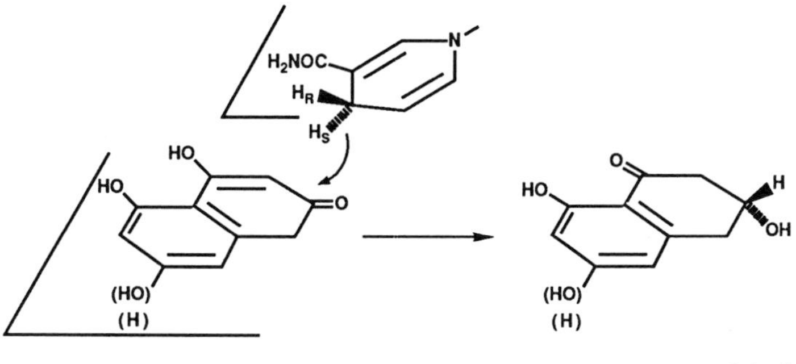

Scheme 5 Stereochemistry of the NADPH dependent reduction of polyhydroxynaphthalenes by *V. dahliae* and *P. oryzae*

KINETIC BEHAVIOUR OF THE *P. ORYZAE* REDUCTASE

The reduction could be monitored spectrophotometrically and the constants have been determined. The enzyme reduced T_4HN and T_3HN with very similar V_{max}, but the affinity of T_3HN (K_M = 7.5 μM) was higher that the affinity for T_4HN (K_M = 80 μM). The K_M for NADPH was determined as 9 μM at pH 5.5.

The reaction catalyzed by the reductase is reversible but the optimal pH for the oxidation is 7.2. At that pH the K_M for scytalone is 0.22 mM and it can be deduced from the kcat/K_M values that the equilibrium is almost totally displaced toward the reduction.

Preliminary results revealed that tricyclazole is a very efficient inhibitor of the reduction with IC_{50} in the 20 nM range.

REFERENCES

1. Hsieh, D.P.H., Lin, M.T., Yao, R.C. Biochem. Biophys. Res. Commun. 1964, 52, 992 ; Townoend, C.A., Christensen, S.B., Davis, S.G. J. Amer. Chem. Soc. 1974, 104, 6152 : Simpson, T.J., Stenzel, D.J.J. Chem. Soc. 1983, 338.
2. Anderson, J.A. Phytochem. 1986, 25, 103.
3. Bell, A.A., Wheeler, M.H. Ann. Rev. Phytopathol. 1986, 24, 411.
4. Troyd, J.D., Paget, C.J., Guse, L.R., Dreikorn, B.A. Pafford, J.L. Phytopathology 1976, 66, 1135.
5. Wheeler, M.H. Exp. Mycol. 1982, 6, 171.
6. Anderson, J.A., Bor-Kang, L., Williams, H.J., Scott, A.I. J. Amer. Chem. Soc. 1988, 110, 1623.
7. Viviani, F., Gaudry, M., Marquet, A. J. Chem. Soc. Perkin Trans I 1990, 46, 2827.
8. Viviani, F., Gaudry, M., Marquet, A. unpublished results.

THE PREDICTIVE VALUE OF HISTORICAL AND FUNCTIONAL MODELS: COENZYME STEREOSPECIFICITY OF GLUCOSE DEHYDROGENASE

Helga Schneider-Bernlöhr, Hans Werner Adolph and Michael Zeppezauer

Fachrichtung Biochemie, Universität des Saarlandes
D-6600 Saarbrücken 11, FRG

INTRODUCTION

NAD(P)-dependent dehydrogenases catalyze the stereospecific transfer of hydrogen between substrates and the C-4 position of pyridine nucleotides. With respect to the coenzyme about half of the investigated enzymes show pro-R, the other half pro-S stereospecificity[1]. Historical and functional models have been proposed to explain the stereochemical diversity[2,3]. We have pointed out that neither the historical nor the functional model is able to predict the coenzyme stereospecificity of the ubiquitiously distributed alcohol/polyol dehydrogenases and we have introduced a new hypothesis: **Zn-containing "long" forms of alcohol/polyol dehydrogenases are pro-R, "short" forms without zinc are pro-S specific**[4]. We report here that our hypothesis is also in agreement with new findings shown for a sugar dehydrogenase.

RESULTS AND DISCUSSION

Glucose dehydrogenase (E.C. 1.1.1.47) catalyzes the following reaction:

$$\beta\text{-D-glucose} + \text{NAD(P)} \rightleftharpoons \text{D-gluconolacton} + \text{NAD(P)H}$$

The enzyme from beef liver has been found earlier to be pro-S specific[5]. No structural or mechanistic analysis of this enzyme (MW 230,000), however, is available until now. Therefore it is not feasible to test whether our hypothesis can be extended to sugar dehydrogenases.

A recent structural analysis of the tetrameric glucose dehydrogenase from Bacillus megaterium (MW 118,000) has shown similarities to the "short" non-metallic alcohol/polyol dehydrogenases[6]. We investigated the stereospecificity of the hydride transfer from D-glucose-1-d_1 (96.1% D) to either NAD or NADP. 4±1% [4(S)-^{1}H]- and 96±1% [4(R)-^{1}H] NADD or NADPD, respectively, were detected by ^{1}H-NMR, which means that the hydride attacks the pyridinium ring at the si-face (Fig. 1).

Bioorganic Chemistry in Healthcare and Technology, Edited by U.K. Pandit and F.C. Alderweireldt, Plenum Press, New York, 1991

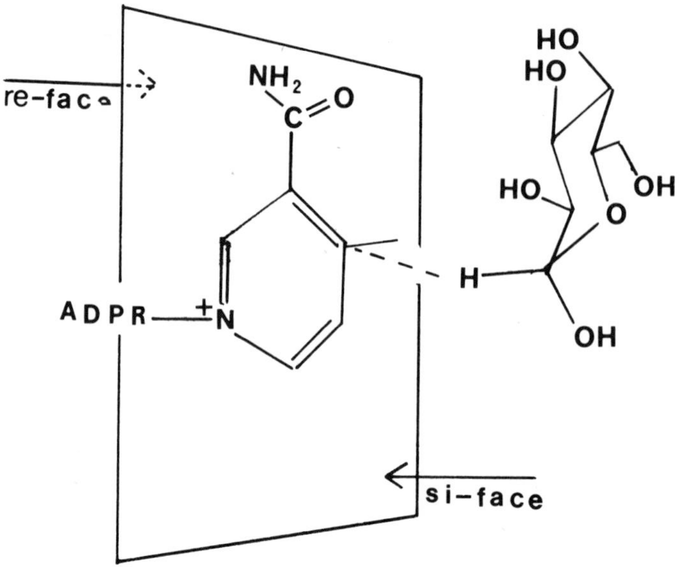

Fig. 1. Stereospecific pathway of the hydride transfer from ß-D-glucose to NAD(P). ADPR = adenosine diphosphoribose.

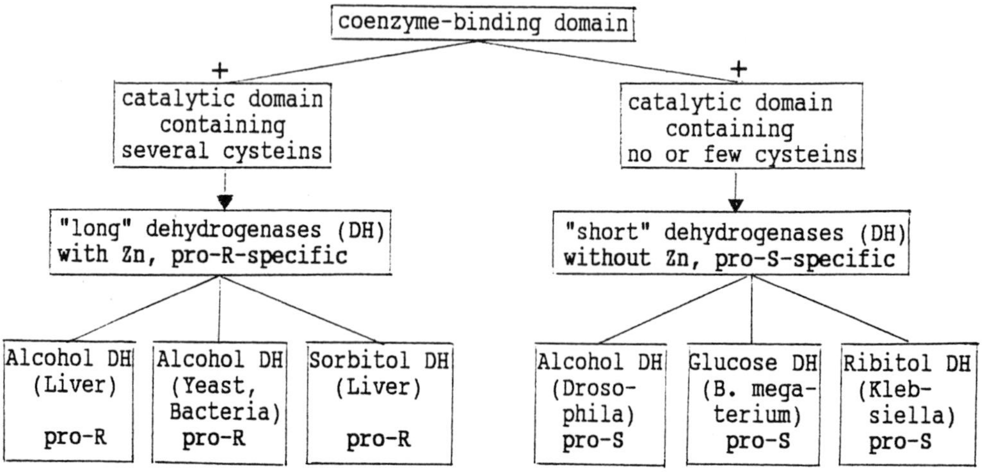

Scheme 1. Relationship between the pro-R and pro-S specific alcohol/polyol/sugar dehydrogenases.

CONCLUSION

The expected pro-S stereospecificity shows that our hypothesis can be extended to a larger family of enzymes. In addition it supports the idea that coenzyme stereospecificity is connected to alternative catalytic mechanisms. In one case substrate binding and activation depend on zinc. In the other case activation by protons could play a role. Within one family of dehydrogenases, however, coenzyme stereospecificity is conserved as shown in scheme 1.

References

1. K.-s. You, Stereospecificity for Nicotinamide Nucleotides in Enzymatic and Chemical Hydride Transfer Reactions, in: "CRC Critical Reviews in Biochemistry", vol. 17, G. D. Fasman, ed., CRC Press, Boca Raton (1984).
2. R. Bentley, Dehydrogenation and Related Processes, in: "Molecular Asymmetry in Biology", vol. II, B. Horecker, J. Marmur, N. O. Kaplan, H. Scheraga, eds., Academic Press, New York, London (1970).
3. K. P. Nambiar, D. M. Stauffer, P. A. Kolodziej and S. A. Benner, A Mechanistic Basis for the Stereoselectivity of Enzymatic Transfer of Hydrogen from Nicotinamide Cofactors, J. Am. Chem. Soc., 105:5886 (1983).
4. H. Schneider-Bernlöhr, H. W. Adolph and M. Zeppezauer, Coenzyme Stereospecificity of Alcohol/Polyol Dehydrogenases: Conservation of Protein Types vs. Functional Constraints, J. Am. Chem. Soc. 108:5573 (1986).
5. H. R. Levy, F. A. Loewus and B. Vennesland, The Enzymatic Transfer of Hydrogen V. The Reaction Catalyzed by Glucose Dehydrogenase, J. Biol. Chem. 222:685 (1956).
6. H. Jörnvall, H. von Bahr-Lindström, K.-D. Jany, W. Ulmer and M. Fröschle, Extended Superfamily of Short Alcohol-Polyol-Sugar Dehydrogenases: Structural Similarities between Glucose and Ribitol Dehydrogenases, FEBS Lett. 165:190 (1984).

MODULAR ABIOTIC AMINO ACID RECEPTORS

F. P. Schmidtchen

Lehrstuhl für Org. Chemie und Biochemie
Technische Universität München
D-8046 Garching, Fed. Rep. Germany

Natural receptors (and enzymes) bind and process their guest substrates by virtue of a peculiar array of anchor groups capable to interact with structural epitopes of the guest species. The unsurpassed selectivity in molecular recognition of biological host compounds arises from a folding process of an *open-chain modular polymer*, which places these anchor groups in a secured topology and mutual orientation in space.

This obviously successful strategy can be adapted for the construction of artificial receptors [1], if suitable binding modules can be found, which interact in dedicated fashion with complementary substructures of the guest by non-covalent and/or reversibly covalent bonds. Covalent linear connection of these modules would furnish polytopic abiotic hosts with selectivity features dictated by the type, the number and the sequence of the parent anchor groups along with the spacing and the flexibility of their juncture. Compared to macrocyclic artificial receptors open-chain polymodular hosts are less likely to suffer from slow host guest exchange and are more readily synthesized and modified.

To test this concept of open chain modular receptors we selected amino acids as guest substrates, which provide a number of attractive features: Above all the carboxylate as well as the α-amino group may be selectively recognized by known artificial anchor groups (bicyclic guanidines for binding carboxylates [2], azacrown ethers or pyridoxal analogues for binding amino groups [3]). The interactions involved are enthalpic in origin and are quite strong and thus may suffice to produce stable host-guest complexes under competitive conditions in protic solvents. Making use of pyridoxamine coenzyme derivatives would allow to model a variety of enzyme reactions (viz. transamination, racemization, decarboxylation or β- or γ-eliminations of suitable amino acid substrates), which could benefit from the wealth of published structural and kinetic data on vitamin B_6 enzyme models [4]. Contrary to the binding pattern in most of these models, the simultaneous recognition of the α-amino carboxylate moiety would conserve the option to exploit the recognition process for a directed chemical attack on amino acid side chains. Rather than for the demonstration of substrate selectivity the prospected modular pyridoxamine host compound could be elaborated into different chemoselective catalysts, which would mimick the various reaction modes of vitamin B_6.

Bioorganic Chemistry in Healthcare and Technology, Edited by U.K. Pandit and F.C. Alderweireldt, Plenum Press, New York, 1991

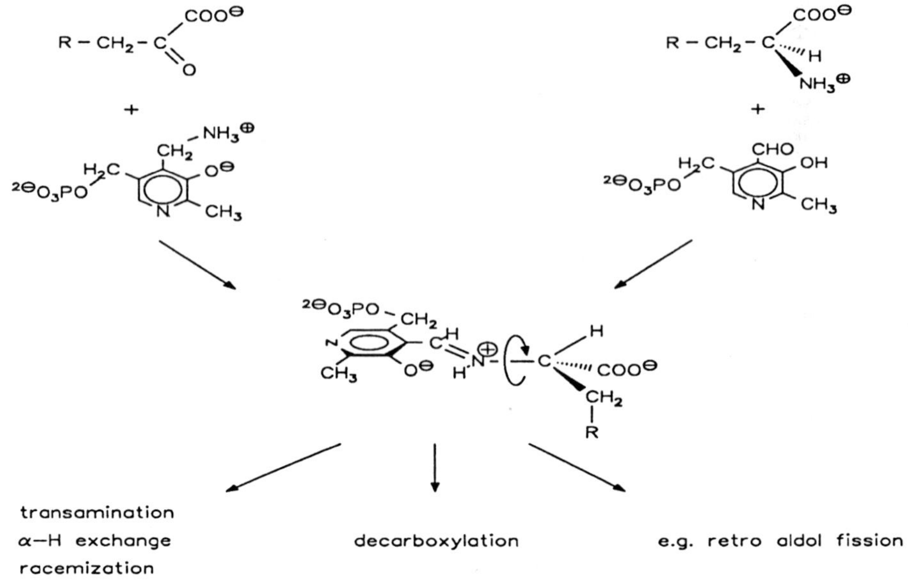

scheme 1

Our prime aim is the construction of a modular amino acid receptor, which would serve to verify the hypothesis of Dunathan [5] that B_6-dependent enzymes exhibit their different reaction modes by virtue of their ability to control the conformation of the C_α-N bond in the aldimine intermediate of the enzyme mechanism. The linear combination of

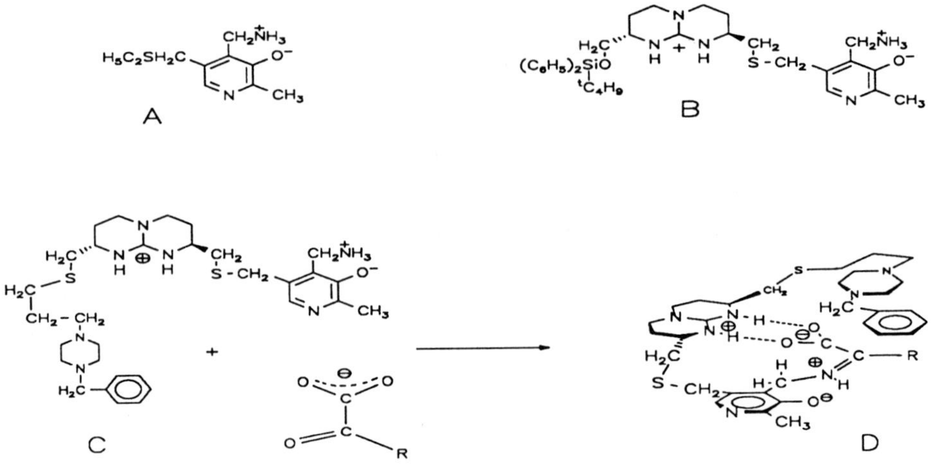

scheme 2

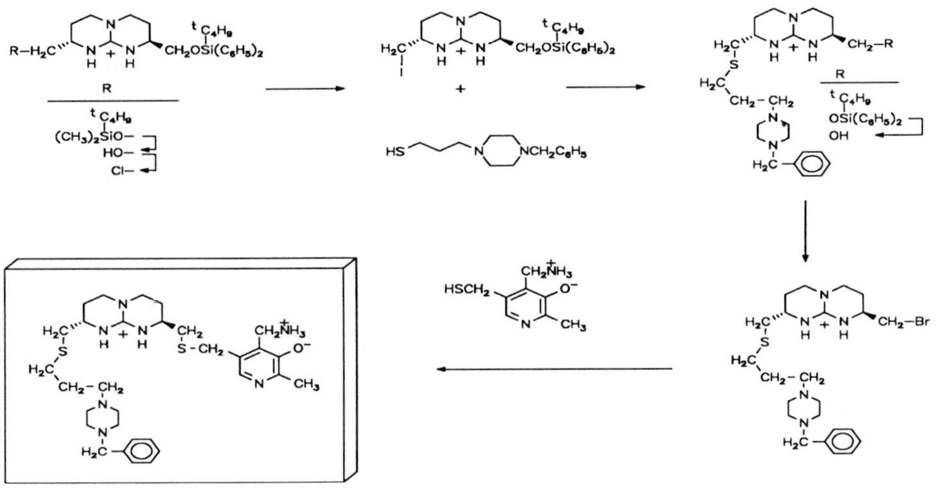

scheme 3

pyridoxamine, an artificial chiral guanidinium bicycle as carboxylate binding function and a piperazine base as in catalyst C was considered a promising setup to form an adduct with α-keto carboxylic acids, distinguished by a unique spatial structure (scheme 2).

The synthesis of the modular catalyst C made use of readily available building blocks [6, 4a] to furnish in a convergent strategy the target host using thioether module connections (scheme 3).

This catalyst proved active in transamination reactions (scheme 4) in the absence of additional metal ions, which were followed by UV-monitoring and stereochemical analysis of dansyl derivatives of the amino acids formed. The kinetics adhered to a first order rate equation from the beginning to >80% conversion. This absorption change most likely reflects the 1,3-H+ transposition in the formation of the more stable aldimine from the initially formed ketimine intermediate. In comparison to a pyridoxamine analog lacking the base (B) or the base and the carboxylate anchor group altogether (A) the modular catalyst C displayed the higher rate accelerations (2-5 fold with respect to A) and a distinct optical induction in the amino acid product with preference to the S-enantiomer. Though the asymmetric induction observed is only modest (20% ee) it is almost independent of the side chain structure and its mere existence indicates that the chirality of the carboxylate receptor function transmits stereochemical information over the respectable distance of 8 heavy atom centers (see scheme 1). Thus the involvement of a distinguished conformation of the flexible host-guest complex similar to that shown in scheme 1 is supported by the data. It is, however, very obvious that neither the reaction conditions nor the selected spacing of modules in this first generation modular receptor are optimal to bring about high asymmetric inductions. Further improvements appear very likely, if the dependencies on temperature, solvent, pH-value, ionic strength and molecular structure as elucidated by molecular modelling are incorporated. These issues are under study.

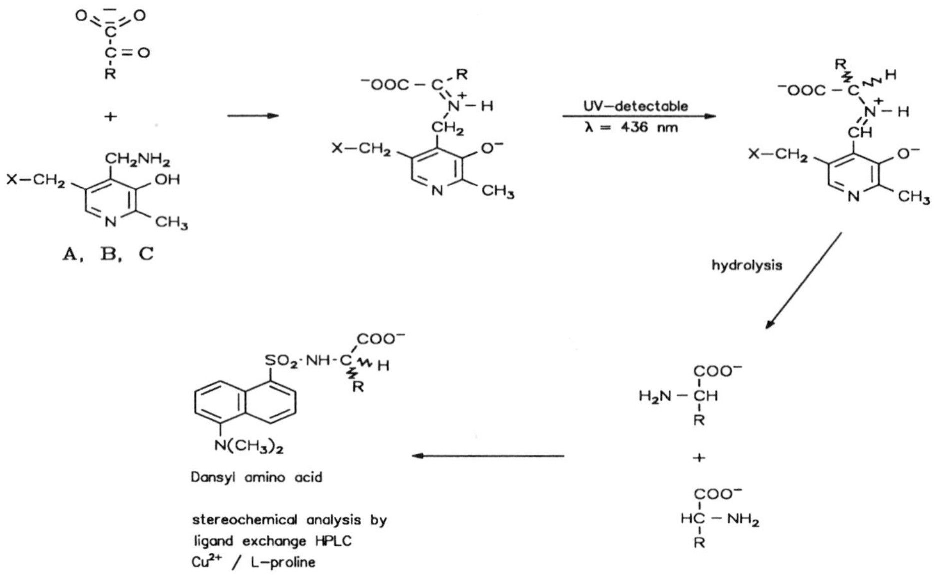

scheme 4

References

1) F. P. Schmidtchen J. Am. Chem. Soc. 1986, 108, 8249.
2) F. P. Schmidtchen, G. Müller and J. Riede Angew. Chem. Int. Ed. Engl. 1988, 27, 1516.
3) a) J. M. Lehn and P. Vierling Tetrahedron Lett. 1980, 21, 1323; b) A. E. Martell Acc. Chem. Res. 1989, 22, 115.
4) leading references include: a) R. Breslow, J. W. Canary, M. Varney, S. T. Wadell, D. Yang J. Am. Chem. Soc. 1990, 112, 5212; b) M. Ando, H. Kuzuhara, J. Watanabe Bull. Chem. Soc. Jpn. 1990, 63, 88; c) I. Tabushi Pure Appl. Chem. 1986, 58, 1529; d) Y. Murakami, J.-I. Kikuchi, N. Shiratori Bull. Chem. Soc. Jpn. 1989, 62, 2045.
5) H. C. Dunathan Proc. Natl. Acad. Sci. USA 1966, 55, 713.
6) F. P. Schmidtchen, H. Kurzmeier J. Org. Chem. 1990, 55, 3749.

NEW ALKALOIDS OF SOME TURKISH MEDICINAL PLANTS

Bilge Şener, Hülya Temizer

Department of Pharmacognosy
Faculty of Pharmacy, Gazi University
06560 Ankara-TURKEY

INTRODUCTION

Many of the Turkish medicinal plants have useful medicinal aspects and contain alkaloids of potential medicinal uses.Continuing our researches carried out on various medicinal plants have resulted in the isolation and structure elucidation of new and known alkaloids.During the last ten years, the isolation and structure elucidation of several new alkaloids from *Fumaria*,*Corydalis*,*Ruta*,*Haplophyllum*,*Galium*,*Leucojum*,*Buxus*, *Thermopsis*,*Senecio*,*Consolida* and *Veratrum* species growing in Turkey have been completed.

In recent years considerable attention has been directed towards the study of isoquinoline alkaloids.Mainly *Fumaria* and *Corydalis* species are invaluable source of isoquinoline derivatives many which possess a wide diversity of structures and pharmacological effects. Here, we describe the isolation and identification of novel alkaloids from Turkish *Fumaria* and *Corydalis* species.Their structures were established with the help of modern spectroscopic techniques.

ALKALOIDS FROM *Fumaria* L. SPECIES

The genus *Fumaria* is represented by 13 species in Turkey[1].They are small annual herbs.These species are used in folk medicine in the treatment of eczema, rheumatism,stomachache and dysentery. In Anatolia one of them,*F.vaillantii* is widespread where its extracts are used in folk medicine as a blood purifier in the treatment of skin diseases.The aerial parts of Turkish *Fumaria* species yielded 34 isoquinoline alkaloids[2-10].Of these,25 alkaloids had been isolated from other *Fumaria* species previously[11-13]. Five alkaloids (+)-juziphine, (+)-isocorydine,(-)-corledine,dihydrosanguinarine and norsanguinarine previously found in other genera of the Fumarioideae were reported for the first time from the genus *Fumaria*[14]. The remaining 4 alkaloids were characterized as new isoquinoline derivatives.

One of which was recognized as a member of indenobenzazepine group.Diazomethane o-methylation of this alkaloid(1) yielded o-methylfumarofine(2). Additionally its structure was also confirmed with the o-acetylation product(3). As a result of spectroscopic and chemical studies on this new alkaloid, it can be considered fumarofine(1) which is the reduced indenobenzazepine alkaloid[15,16].

We have also characterized E-fumaramine(4) belonged to the secophthalide-isoquinoline group. This is a geometric isomer of Z-fumaramine previously known.These isomers differ significantly in their UV and ^{1}H-NMR spectra. The assignments of protons are given on the structure[17].

The third new alkaloid had a secophthalideisoquinoline skeleton as may be deduced from its ^{1}H-NMR and mass spectra.In the light of spectral data, this was identified as microcarpine(5)[18].

The last new alkaloid isolated from Turkish *Fumaria* species was determined as a member of secospirobenzylisoquinoline group. On the basis of spectroscopic findings[19] it was characterized as secodensiflorine(6).

During pharmacological studies ethanolic extracts of *Fumaria* species showed no toxic effects. These extracts as well as the alkaloid protopine exhibited a significant spasmolytic,antiarrhytmic hypotensive and antipyretic effects. Protopine is also an anti-thrombic drug. In addition pharma cological investigations with fumariline indicated that the alkaloid is a central nervous system depressant.The *in vitro* cytotoxicity assays of the new alkaloids were carried out according to NCI protocol and found to be inactive.Our works are in progress regarding the pharmacological studies on the depurative profile of Turkish *Fumaria* species.

ALKALOIDS FROM *Corydalis* Medik. SPECIES

We had also occasion to investigate the alkaloidal contents of Turkish *Corydalis* species. Eight species are known to grow in Turkey[1].They are usually found on rocky slopes.These species are perennial plants with tubers. *Corydalis* species are common plants to central and south-east Asia.Some of them such as *C.ambigua*, *C.yanhusuo*, *C.bulleyana* and *C.govaniana* are used in folk medicine as a febrifuge,antidote or analgesic in China and India for a long time.

The tubers of 6 Turkish *Corydalis* species have led to the isolation of 61 alkaloids. Forty-six of them had been isolated from the genus *Corydalis*[20-25]. All alkaloids were identified on the basis of their physical and spectral properties.

Twelve alkaloids (-)-norjuziphine,13-methylcolumbamine,norsanguinarine, (±)-chelidimerine,oxosarcocapnidine,oxosarcocapnine,(+)-fumarophycine, (+)-fumariline,(-)-fumaritine,(+)-parfumine,(+)-fumaritine and (-)-africanine are reported for the first time from the genus *Corydalis*. Of these(+)-fumarophycine,(+)-fumaritine and (-)-africanine which are the first naturally occurring enantiomer of the known (-)-fumarophycine (-)-fumaritine and (+)-africanine isolated from *Fumaria*, *Corydalis* and *Rupicapnos* species[26-28]. New alkaloids for the genus *Corydalis* are (-)-corybrachylobine(7), (-)-corytaurine(8) and (+)-corytauricoline(9). (-)-Corybrachylobine(7) is an unusual tetrahydroprotoberberine alkaloid that bears 13,13-dimethylated substitution pattern. All spectral data confirmed the structure suggested for (-)-corybrachylobine.

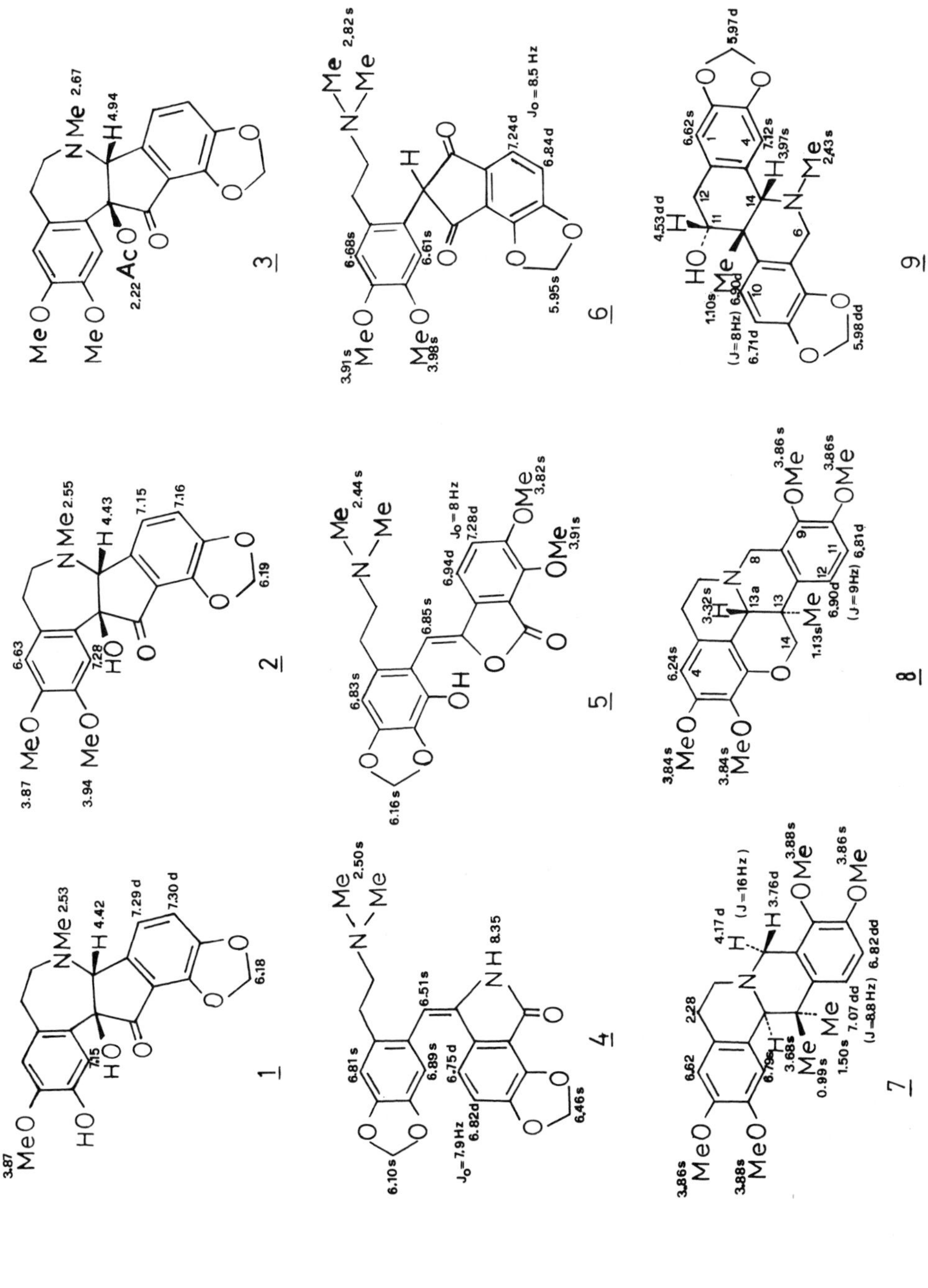
1
2
3
4
5
6
7
8
9

(-)-Corytaurine(8) gave a UV spectrum which showed no shift upon addition of NaOH determined the presence of a nonphenolic nucleus in this alkaloid. The ^{1}H-NMR spectrum of this alkaloid contained four methoxy groups and a three proton singlet at 1.13 ppm due to a tertiary methyl group at C-13.Three aromatic proton assignments were described to H-4,H-11 and H-12.The C-8 methylene protons appeared at 3.35 and 4.32 ppm while those for the C-14 methylene protons were at 4.17 and 4.67 ppm.A singlet for H-13a occurred at 3.32 ppm. The IR spectrum of this alkaloid contained a Bohlmann band at 2840 cm^{-1} characteristic of a trans-quinolizidine ring.The presence of an oxymethylene bridge between C-1 and C-13 was supported by the appearance of fragments at 206 and 191 in the mass spectrum.

(+)-Corytauricoline(9) showed a UV spectrum typical for corynoline-type alkaloids.The mass spectrum with a molecular ion at 367(base peak) also revealed a characteristic for the corynoline-group alkaloids. Studies on the ^{1}H-NMR spectrum coupled with a COSY experiments of this alkaloid gave us the proposed structure. Two allylic type long-range couplings were observed between H-1,H-12 and H-4,H-14. One long-range coupling between H-6 and H-10 was also observed.Acetylation of this alkaloid with acetic anhydride/pyridine gave the acetylated product.The mass spectrum and fragmentation pattern of this derivative were also similar to corynoline-type alkaloids.On the basis of these experiments the described structure was showed to this a new benzophenanthridine base.

Finally it is interesting to note the occurrence of corydaldine and N-methylcorydaldine belonging to the isoquinolone group is reported for the first time for the Fumarioideae. It should be pointed out here that isoquinolones are end products of the catabolism of isoquinooline alkaloids and will be found wherever isoquinoline alkaloids are present.

REFERENCES

1. P.H.DAVIS, "Flora of Turkey and the East Aegean Islands",Vol.1,University Press,Edinburgh (1965).
2. B.ŞENER, "Turkish Species of Fumaria and Their Alkaloids",Dissertation thesis for Assoc.Prof.,Faculty of Pharmacy,University of Ankara, Ankara (1981).
3. B.ŞENER, Turkish Species of Fumaria and Their Alkaloids II.Alkaloids of F.gaillardotii, Int.J.Crude Drug Res.,21(3):135 (1983).
4. B.ŞENER, Turkish Species of Fumaria and Their Alkaloids III.Alkaloids of F.judaica, J.Fac.Pharm.Gazi,1(1):15 (1984).
5. B.ŞENER, Turkish Species of Fumaria and Their Alkaloids IV.Alkaloids of F.macrocarpa, Int.J.Crude Drug Res.,22(4):185 (1984).
6. B.ŞENER, Turkish Species of Fumaria and Their Alkaloids V.Alkaloids of F.capreolata and F.asepala, J.Nat.Prod.,48(4):670 (1985).
7. B.ŞENER, Turkish species of Fumaria and Their Alkaloids VI.Alkaloids of F.capreolata, Int.J.Crude Drug Res.,23(4):161 (1985).
8. B.ŞENER, Turkish Species of Fumaria and Their Alkaloids VII.Alkaloids of F.officinalis and F.cilicica, J.Fac.Pharm.Gazi, 2(1):45 (1985).
9. B.ŞENER, Turkish Species of Fumaria and Their Alkaloids VIII.Alkaloids of F.asepala, Int.J.Crude Drug Res.,24(2):105 (1986).

10. B. ŞENER, Turkish Species of Fumaria and Their Alkaloids IX.Alkaloids of F.parviflora, F.petteri subsp.thuretii and F.kralikii,Int.J. Crude Drug Res.,26(1):61 (1988).
11. T. F. PLATONOVA, P. S. MASSAGETOV, A. D. KUZOVKOV and L. M. UTKIN, J.Gen.Chem.USSR,26:181 (1956).
12. G. P. SHEVELEVA, D. SARGAZAKOV, N. V. AKTANOVA and A. Sh. ALDASHEVA, Fiziol.Akt.Soedin.Rast.Kirg.,41:28 (1970).
13. I. A. ISRAILOV, M. S. YUNUSOV and S. Yu. YUNUSOV, Khim.Prir.Soedin, 194:32 (1971).
14. B. ŞENER, B. GÖZLER, R. D. MINARD, M. SHAMMA, Alkaloids of F.vaillantii, Phytochem.,22(9):2073 (1983).
15. G. BLASKO, N. MURUGESAN, S. F. HUSSAIN, R. D. MINARD, M. SHAMMA, B. ŞENER, M. TANKER, Revised Structure for Fumarofine,An Indenobenzazepine Type Alkaloid, Tetrahedron Lett., 22(33):3135 (1981).
16. G. BLASKO, N. MURUGESAN, A. J. FREYER, D. J. GULA, M. SHAMMA, B. ŞENER, The Indenobenzazepine-Spirobenzylisoquinoline Rearrangement,Stereocontrolled.Synthesis of (±)-Raddeanine and (±)-Yenhusomine,Tetrahedron Lett., 22(33):3139 (1981).
17. B. ŞENER, Secodensiflorine, A New Alkaloid from F.densiflora, Int.J. Crude Drug,Res.,22(2):79 (1984).
18. B. ŞENER, Microcarpine, A New Alkaloid from F.microcarpa, Int.J.Crude Drug Res.,22(1):45 (1984).
19. G. BLASKO, V. ELANGO, B. ŞENER, A. J. FREYER, M. SHAMMA, Secophthalideisoquinolines, J.Org.Chem., 47:880 (1982).
20. B. ŞENER, Alkaloids from Turkish Corydalis Species in "Perspectives in Natural Products Chemistry" ATTA-UR-RAHMAN, P. W. LE QUESNE, ed., Springer-Verlag, West Germany (1988).
21. B. ŞENER, Alkaloids from Corydalis rutifolia subsp.kurdica of Turkish Origin, J.Fac.Pharm.Gazi,3(1):13 (1986).
22. B. ŞENER, M. KOYUNCU, H. TEMİZER, Alkaloids from Turkish Corydalis solida subsp. solida in " Proceedings of the VI.Symposium on Plant Originated Crude Drugs", B. ŞENER, ed., University Press,Ankara (1987).
23. B. ŞENER, Minor Alkaloids of Corydalis rutifolia subsp.kurdica of Turkish Origin, Int.J.Crude Drug Res.,26(3):155 (1988).
24. B. ŞENER, H. TEMİZER, Pharmacognosic Investigations on Corydalis solida subsp. brachyloba II.Alkaloids of C.solida subsp.brachyloba, J.Fac. Pharm.Gazi,5(1):9 (1988).
25. B. ŞENER, Spirobenzylisoquinoline Alkaloids from Corydalis caucasica, Int.J.Crude Drug Res.,27(3):161 (1989).
26. L. CASTEDO, D. DOMINGUEZ, J. M. NOVO, Two New Spirobenzylisoquinoline Alkaloids from Rupicapnos africana, Heterocycles ,24(10):2781 (1986).
27. R. M. PREISNER, M. SHAMMA, The Spirobenzylisoquinoline Alkaloids,J.Nat. Prod.,43:305 (1980).
28. B. ŞENER, H. TEMİZER, Chemical Studies on the Minor Alkaloids from Corydalis solida subsp. brachyloba, presented at the 17th IUPAC International Symposium on the Chemistry of Natural Products,New Delhi,Feb.4-9 (1990).

TOWARDS CHEMICAL CONTROL OF GENE EXPRESSION: COPPER(II) AND THIOLS, INCLUDING GLUTATHIONE, ARE POWERFUL REAGENTS FOR CLEAVAGE OF DNA

Kenneth T. Douglas[1], David C.A. John[1] and Celia J. Reed[2]

[1]Department of Pharmacy
University of Manchester
Oxford Road
Manchester M13 9PL

[2]School of Natural Sciences
The Liverpool Polytechnic
Byrom Street
Liverpool L3 3AF

INTRODUCTION

Sequence-specific cleavage of DNA has many applications in molecular biology, but is limited by the specificities and accessibility of natural restriction endonucleases[1]. One approach to overcoming this has been to chemically modify a DNA-recognising molecule with a reagent capable of chemical cleavage of DNA. The most commonly used reagent for such chemical cleavage is probably the EDTA:Fe(II) system (1) which has been attached to oligonucleotides[1,2], intercalators[3], and to a combination of these binding-species[4]. Minor-groove directed drugs[5] and antisense oligodeoxyribonucleoside methylphosphonates[6] have joined the catalogue. The other commonly used chemical cleavage systems include bis (1,10-phenanthroline):Cu(I) (2)[7-9] porphyrin metal complexes (3)[10-13], and rhodium complexes[14]. Photochemical cleaving systems for DNA have also been described[15-19].

The metal-complex systems above all require activation to induce strand scission, frequently achieved by means of light or of a reducing agent but also by oxidising agents in some situations. There have been reports of DNA cleavage by various transition-metal ions, but these still often act by activation using hydrogen peroxide[20,21]. Sequence-directed cleavage of single-stranded DNA has been achieved recently by means of complementary oligonucleotides bearing terminal alkylating agents (4)[22].

One use for DNA and RNA sequence-specific reagents capable of controlled chemical cleavage is as molecular biological tools either in the sense of standard artificial restriction endonucleases or for larger scale genome mapping. In addition the genetic basis of several major diseases has been recognised lately and it seems a realistic goal to target aberrant DNA base sequences for occlusion by direct binding (either at DNA or, more likely, at mRNA level) or by chemical excision.

With this in mind, it is important to develop not only novel DNA-sequence reading systems but also chemical cleavage systems which are easily amenable to chemical synthetic manipulation and have suitable biocompatability (e.g. stability, cellular penetration, recycling). Thiols have long been known to autooxidise, especially in the presence of certain transition-metal ions. Thus, we felt that thiol:metal ion systems might be suitable reagents for DNA cleavage. This paper gives details of the cleavage of plasmid DNA by an efficient new combination, the Cu(II):thiol system.

The use of thiols for DNA cleavage offers a number of potential special advantages. If cellular thiols, such as glutathione, lipoate or cysteine, can be induced to cleave DNA by a redox process then *in vivo*

Bioorganic Chemistry in Healthcare and Technology, Edited by U.K. Pandit and F.C. Alderweireldt, Plenum Press, New York, 1991

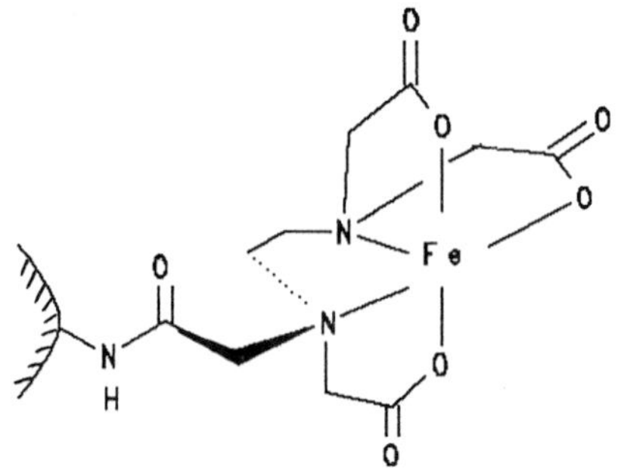

1

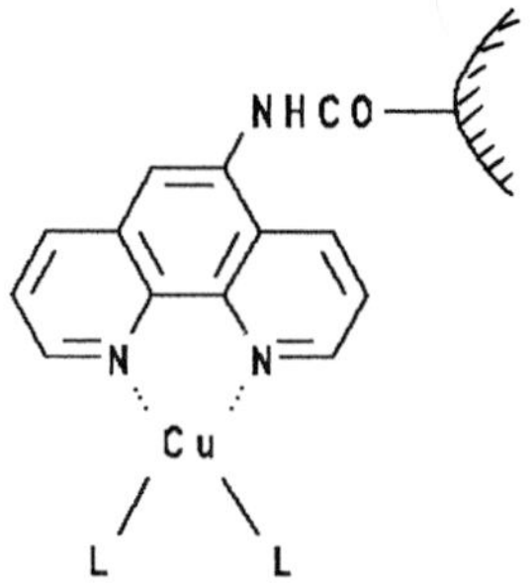

2

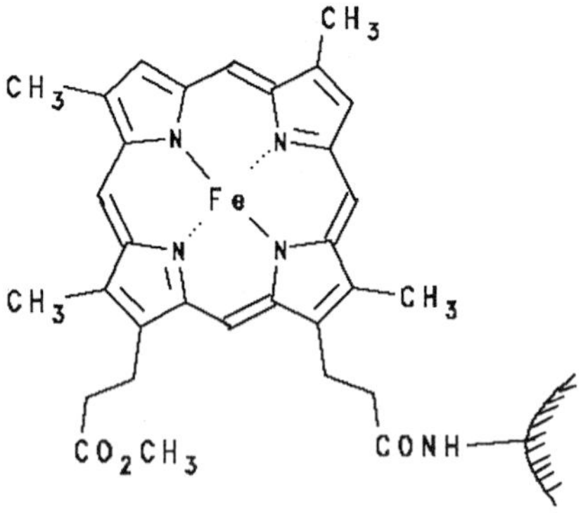

3

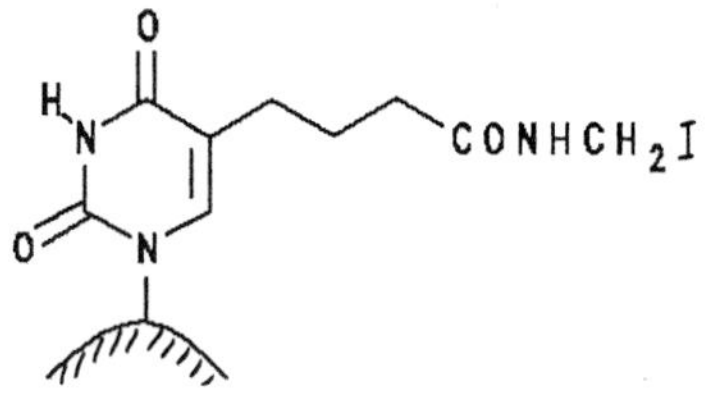

4

there is an intrinsic recycling system available in principle, e.g. for glutathione there is the glutathione reductase system to reform reduced glutathione, as in equation (1).

$$2GSH \underset{\substack{\text{glutathione}\\ \text{reductase}\\ (\text{NADPH, } H^{+}).}}{\overset{\text{oxidation}}{\rightleftharpoons}} GSSG \qquad (1)$$

We now turn to the cleavage of DNA by thiols activated by Cu^{2+} ions, looking first at plasmid DNA cleavage for some of the overall properties of the Cu(II):RSH cleavage system. After that a more detailed look at cleavage of linear fragments of DNA of known sequence will be discussed.

Plasmid DNA Cleavage

In the presence of Cu(II) ions, supercoiled DNA is cleaved in neutral solution by very low concentrations (down to micromolar) of a range of thiols[23]. Supercoiled plasmid DNA is cleaved first to open circular DNA, which in turn produces linear DNA and eventually fragments. The cleavage is strongly dependent on temperature and is maximal at a sodium chloride concentration of 0.10-0.25M. In the presence of excess of either component of the Cu(II):RSH pair, the extent of cleavage is dependent on the concentration of the limiting partner and easily detectable down to micromolar concentrations of limiting GSH. The copper (II) concentration-dependence is complex.

The use of scavengers of oxygen-derived species such as hydrogen peroxide, superoxide ion and hydroxyl radical gave evidence that the hydroxyl radical may be involved in the cleavage mechanism but the results were not straightforward. DNA cleavage leads to some production of 2-thiobarbituric acid-reactive species and some of the cleavage sites, at least, had 5'-OH and/or 3'-OH groups detectable. Extensive base damage, prior to cleavage, was detectable. Studies with S1 nuclease indicated no gross or detailed sequence preference for Cu(II):glutathione cleavage of pSP64 plasmid DNA. The Cu(II):RSH system did not appear to target special structural features in the DNA such as Z-DNA inserts, cruciform structures or left-handed (but non-Z) DNA.

Linear DNA

Interestingly, and in apparent contrast to the plasmid case, thiol cleavage of linear fragments of linear DNA did show apparent sequence preference. Thus, the linear 160 base-pair tyr t fragment from pKMΔ98 plasmid DNA was cleaved with preference at five sites with the consensus (3'->5') sequence pur-pur-pyr-pur-pur (pur = purine base; pyr = pyrimidine base). The size of this consensus sequence (about a half-turn of a B-type helix) and the independence of the cleavage site of thiol size, structure and stereochemistry was interpreted as indicating that a common species (probably free radical) was produced by all the thiol systems and that this reactive species was then highlighting local conformation/dynamics in the tyr T chain[24]. For example, the reagent may be monitoring minor groove depth, weak sites of preferential Cu^{2+} binding etc.

We then studied end-labelled fragments of pBR322 DNA (517 bp and 167 bp) similarly. Again we detected thiol-independent sequence preferences in initial cutting sites. However, the sequences cut are characteristic of the piece of DNA used, presumably indicating some structural property of the DNA as the determinant of cleavage. Under the conditions used, we showed that Fe(II):edta cleavages of these DNA fragments were sequence neutral. We have not yet completely analysed the possible structural implications of the cleaved sites across all 3 sample of B-DNA linear fragments. However, cleavage might arise from a reagent generated by the Cu(II):RSH combination in free solution (probably HO•) or by attack involving Cu^{2+} (weakly) pre-bound to DNA.

We have synthesised a number of thiol-bearing intercalators and oligonucleotides and related molecules as controls. Studies of their interactions with DNA remain to be completed.

References

1. G.B. Dreyer and P.B. Dervan, Proc. Natl. Acad. Sci. (U.S.A.) 82: 968-72 (1985).

2. A.S. Boutorin, V.V. Vlassov, S.A. Kazakov, I.V. Kutiavin, and M.A. Podyminogin, FEBS Lett. 172: 43-46 (1984).

3. R.P. Hertzberg, and P.B. Dervan, Biochemistry, 23: 3934-45 (1984).

4. M. Boidet-Forget, M. Chassignol, M. Takasugi, N.T. Thuong, and C. Hélène, Gene 72: 361-71 (1988).

5. J.S. Taylor, P.G. Schultz, and P.B. Dervan, Tetrahedron, 40: 457-465 (1984).

6. S-B. Lin, K.R. Blake, P.S. Miller, and P.O.P. Ts'O, Biochemistry, 28: 1054-61 (1989).

7. A. Spassky, and D.S. Sigman, (1985) Biochemistry 24: 8050-56 (1985).

8. B.F. Chu, and L.E. Orgel, Proc. Natl. Acad. Sci. (U.S.A.) 82: 963-967 (1985).

9. J-C. Francois, T. Saison-Behmoaras, M. Chassignol, N.T. Thuong, and C. Hélène, J. Biol. Chem. 264: 5891-98 (1989).

10. J.W. Lown, and A.V. Joshua, J. Chem. Soc. Chem. Commun. 1298-1300 (1982).

11. J.W. Lown, S.M. Sondhi, C-W. Ong, A. Skorobogaty, H. Kishikawa, and J.C. Dabriowak, Biochemistry 25: 5111-17 (1986).

12. J. Bernadou, B. Lauretta, G. Pratviel, and B. Meunier, Compt. Rend. Acad. Sci. (Paris) 309: III, 409-414 (1989).

13. J.T. Groves, and T.P. Farrell, J. Amer. Chem. Soc. 111: 4998-5000 (1989).

14. K. Uchida, A.M. Pyle, T. Morii, and J.K. Barton, Nucl. Acids Res., 17: 10259-279 (1989).

15. I. Saito, T. Mori, T. Obayishi, T. Sera, H. Sugiyama, and T. Matsuura, J. Chem. Soc. Chem. Commun. 360-362 (1989).

16. C. Jeppesen, and P.E. Nielsen, Eur. J. Biochem. 182: 437-444 (1989).

17. C. Jeppesen, O. Buchardt, U. Henriksen, and P.E. Nielsen, Nucl. Acids Res. 16: 5755-5770 (1988).

18. L.Z. Benimetskaya, N.V. Bulychev, A.L. Kozionov, A.A. Koshkin, A.V. Lebedev, S. Novozhilov, M.I. Yu & Stockman, Biopolymers, 28: 1129-1147 (1989).

19. T.L. Doan, L. Perrouault, D. Praseuth, N. Habhoub, J.-L. Decout, N.T. Thuong, J. Lehomme, and C. Hélène, Nucl. Acids Res. 15: 7749-7760 (1987).

20. J-L. Sagripanti, and K.H. Kraemar, J. Biol. Chem. 264: 1729-1734 (1989).

21. K. Yamamoto, S. Inoue, A. Yamazaki, T. Yoshinaga, and S. Kawanishi, Chem. Res. Toxicol. 2: 234-239 (1989).

22. R.B. Meyer, Jr. J.C. Tabone, G-D. Hurst, T.M. Smith, and H. Gamper, J. Am. Chem. Soc. 111: 8517-8519 (1989).

23. C.J. Reed, and K.T. Douglas, Biochem. Biophys. Res. Commun. 162: 1111-1117 (1989).

24. D.C.A. John, and K.T. Douglas, Biochem. Biophys. Res. Commun. 165: 1235-1242 (1989).

INTERACTION OF REDUCTIVELY METHYLATED LYSYL-Fd GENE 5 PROTEIN WITH A NEGATIVELY CHARGED LANTHANIDE PARAMAGNETIC MACROCYCLIC CHELATE OR WITH OLIGONUCLEOTIDES USING 13-C NMR

C.F.G.C. Geraldes[&], A.D. Sherry[@], L.R. Dick[+], C.W. Gray[+] and D.M. Gray[+]

[&] Chemistry Department, University of Coimbra, Coimbra, Portugal
[@] Chemistry and [+] Molecular and Cell Biology Departments, University of Texas and Dallas, Richardson, Texas, U.S.A.

INTRODUCTION

The bacteriophage fd gene 5 protein (G5P) binds strongly to single stranded DNA. Electrostatic interactions, involving ion pairing of positively charged lysine and arginine protein side chains with the negatively charged phosphates of four nucleotidyl units of DNA, are very important in this interaction process[1–3]. However, implication of specific lysyl residues in the DNA-binding site is more difficult. X-ray crystal structure and modeling studies of G5P have been used[4,5] to obtain a structural binding model which only includes one (Lys-46) of its six lysyl residues. However, chemical modification studies[6] seem to implicate three lysyl residues (Lys-24, 46 and 69) in the nucleic acid binding site of G5P.

In the present work we describe ^{13}C NMR studies of the perturbations of the ^{13}C-enriched reductively methylated G5P upon binding of the deoxyribonucleotide $d(pA)_7$, the axially symmetric NMR shift probe $Tb(DOTP)^{5-}$ or relaxation probe $Gd(DOTP)^{5-}$, or competitive binding of oligonucleotide and probe[7].

MATERIALS AND METHODS

The isolation and concentration calculations of G5P were done by litterature methods[7]. Reductive methylation of the protein was carried out[7] to modify the lysyl residue ϵ-amino groups with [^{13}C] formaldehyde. Preparation of stock solutions of $d(pA)_7$, the macrocyclic ligand 1,4,7,10-tetraazacyclododecane-N,N',N'',N'''-tetrakis(methylenephosphonic acid) (DOTP) and the chelates $Tb(DOTP)^{5-}$, $Gd(DOTP)^{5-}$ or $La(DOTP)^{5-}$, was done according to published procedures[7]. ^{13}C NMR spectra were taken at 50.1 MHz on a Jeol FX-200 FT Spectrometer with WALTZ proton decoupling, at temperatures of 28 ± 2 °C. Computer analysis of the chelate binding position at the G5P surface was done via comparison of the experimental ^{13}C LIS (lanthanide induced shift) values with LIS's calculated using a computer program[7] and coordinates of the amino acid residues in the G5P crystallographic structure[4]. Molecular modeling of the $Ln(DOTP)^{5-}$.G5P complex was done on an Evans and Sutherland PS 300 Graphics System.

Bioorganic Chemistry in Healthcare and Technology, Edited by U.K. Pandit and F.C. Alderweireldt, Plenum Press, New York, 1991

RESULTS AND DISCUSSION

Electron microscopy has been used[7] to show that the methylated and non-methylated fd G5P complexes with single-stranded viral DNA are indistinguishable in their overall structure. Therefore, methylation of the lysines of fd G5P does not affect its structure or DNA-binding properties.

The 50.1 MHz ^{13}C NMR spectrum of the ^{13}C-methylated G5P at pH 8.8 consists of six resolved resonances (numbered 1-6) partially assigned to the six lysyl residues[6], and one resonance (numbered 7) corresponding to the α-amino group of the N-terminal Met-1 residue[6]. In the presence of increasing amounts of $Tb(DOTP)^{5-}$, these resonances are gradually and specifically shifted to lower or higher frequencies without much broadening (fast-exchange conditions)[7]. Plots of the chemical shift of each resonance as a function of the chelate/protein ratio indicated saturation of the binding of the probe to G5P. In fact, the maximum paramagnetic shift was obtained at a chelate to protein monomer ratio of 0.5. Therefore, a single high-affinity binding site (association constant $K_a \approx 10^5\ M^{-1}$) exists for the chelate on each protein dimer[7]. LIS values were obtained by subtraction of the ^{13}C shifts of a 1:1 mixture of the ^{13}C-G5P/$Tb(DOTP)^{5-}$ system (Table I). These data indicate that the largest LIS values occur for resonances 1, 2 and 5, previously assigned to Lys-69, Lys-24 and Lys-46[6]. Thus, these residues are nearest to the chelate binding site and also probably involved in the DNA-binding site.

Table I. Comparison of Experimental with Calculated Dipolar Shifts Induced by $Tb(DOTP)^{5-}$ for ^{13}C NMR Resonances of ^{13}C-Methylated G5P

Resonance	Exptl LIS (ppm)	Calcd		
		Residue	LIS (ppm)	Distance (Å)
1	0.91	Lys-69	0.90	11.5
2	2.38	Lys-24	-0.05	21.9
3	-0.03	Lys-7	-0.07	20.9
4	-0.10	Lys-87	-0.08	23.0
5	0.23	Lys-46	0.26	14.1
6	-0.10	Lys-3	-0.09	22.1
7	-0.05	Met-1	-0.06	26.1

^{13}C spectra of ^{13}C-G5P in the presence of differing amounts of $d(pA)_7$ and $Tb(DOTP)^{5-}$ showed that[7] increase of nucleotide concentrations and/or decrease of probe concentrations gradually decrease the paramagnetic ^{13}C LIS values of all protein lysyl residues. This indicates that both the chelate and the nucleotide binding to G5P are in fast exchange on the NMR time scale and that their binding to G5P are mutually exclusive, the chelate being a poor competitor of the oligonucleotide.

The ^{13}C spin-lattice relaxation (T_1) times were measured for the ^{13}C-G5P in the absence and the presence of various $Ln(DOTP)^{5-}$ chelates. In the presence of $Tb(DOTP)^{5-}$, only the T_1 values of the resonances 1 and 5 were significantly reduced (from 0.70 to 0.56 and from 0.60 to 0.53 s, respectively) by these paramagnetic ion chelates, relative to the effects of the diamagnetic ion chelate $La(DOTP)^{5-}$. However, resonance 2 was extensively broadened in the $Tb(DOTP)^{5-}$ G5P complex, with a large increase from observation at 50.1 MHz to 125.7 MHz. This indicates that there was a substantial exchange contribution to the line-width (and T_1) of this resonance.

The relative distances of the lysines corresponding to resonances 1, 2 and 5 and the chelate binding site were clarified by the use of $Gd(DOTP)^{5-}$, for which the enhancement of the relaxation rate is solely distance dependent (r^{-6}). The paramagnetic line-broadening effect (i.e., decrease of T_2) due to $Gd(DOTP)^{5-}$ was: 2 >> 1 > 5. Thus, the lysyl residue corresponding to resonance 2 is much closer to the chelate binding site than all the others.

A computer search for the $Tb(DOTP)^{5-}$ binding site on the surface of the G5P dimer was undertaken from the knowledge of the LIS values of the ^{13}C methyl residues. Firm assignments of resonance 4(Lys-87) and resonance 7(Met-1)[6] were used, and all possible assignments of the other resonances were considered. Initial searches made it clear that no solution was possible that included the crystallographic coordinates of Lys-24, which indicated that its position in the solution protein structure is quite different from the crystallographic structure. The 500.14 MHz ^1H NMR (General Electric GN-500) spectrum of nonmethylated G5P was also examined in the presence and absence $Tb(DOTP)^{5-}$ (Fig. 1). Resonances assigned to Phe-73 (3,4,5)H and Tyr-26(3,5)H completely disappear from the spectrum upon addition of only 20% of the probe-to-protein stoichiometry[7], indicating that these two resonances had large chemical shifts in the presence of the paramagnetic chelate and that the complexed protein was in intermediate exchange with the free protein. These and other[7] proton NMR LIS results, together with Gd-distance information and the ^{13}C NMR LIS data described above, omitting Lys-24, were used in further searches. These yielded the best solution structure of $Tb(DOTP)^{5-}$-G5P described in Table I by the calculated LIS values and metal-lysyl distances. Assignments are also presented in that Table.

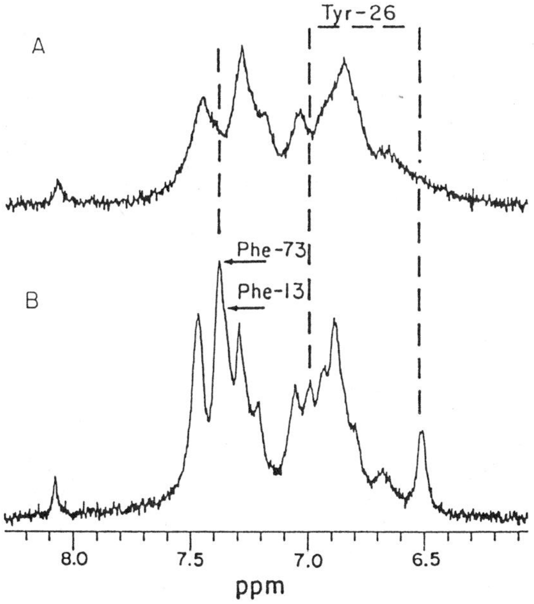

Fig. 1. The 500 MHz ^1H NMR of the aromatic region of nonmethylated fd G5P. (A) After addition of $Tb(DOTP)^{5-}$ to give a [chelate]/-[protein] molar ratio of 0.1. (B) Reference spectrum. The protein monomer concentration was 0.75 mM.

The binding site defined by this structure was in a region of the protein-dimer surface partially occupied by four hydrophobic residues: Leu-76, Leu-76 , Ile-78 and Ile-78 . Thus, $Tb(DOTP)^{5-}$ interacts with the G5P dimer by orienting its hydrophobic surface towards a hydrophobic pocket formed by the leucyl and isoleucyl residues. The negatively charged surface, formed by the phosphonate groups, faces the bulk solvent, and can interact electrostatically with Lys-46, Lys-69 and possibly Lys-24 and Arg-21.

CONCLUSIONS

Our competitive ^{13}C NMR experiments of ^{13}C-G5P with $Tb(DOTP)^{5-}$ and $d(pA)_7$ allowed to obtain the ^{13}C shifts of the three lysyl residues Lys-24, Lys-46 and Lys-69 in the presence of the DNA oligomers alone, and to support previous conclusions[6] that these three lysines are directly involved in the protein DNA-binding site.

The orientation of $Tb(DOTP)^{5-}$ on the protein surface was obtained from the experimental ^{13}C LIS values and relaxation data. All the ^{13}C resonances of the lysyl residues were assigned. Our data required that Lys-24 dramatically moves from a distance of 21.9 Å to about 6.8 Å of the chelate metal ion. This leads to the conclusion that the β-ribbon containing Lys-24 and Tyr-26 is flexible enough in the solution structure to change drastically its position relative to the crystal structure when the protein is complexed with $Tb(DOTP)^{5-}$, and perhaps DNA.[5] King and Coleman[8] have proposed a "dynamic clamp" model involving residues Tyr-26, Leu-28 and Phe-73', and fringed by a cluster of positively charged lysines and arginines. Our data support the notion of a clamp and provide evidence that the β-ribbon with Tyr-26, Leu-28 and Lys-24 can move to bring Lys-24 into the binding groove, in contact with a bound oligonucleotide.

ACKNOWLEDGEMENTS

We thank the Howard Medical Institute (University of Texas, Southwestern Medical Center at Dallas) for use of the Evans and Sutherland graphics system system and Dr. C.W. Hilbers (University of Nijmegen, The Netherlands) for useful discussions. Support from the R. Welsh Foundation, FLAD and INIC (Portugal) is also acknowledged.

REFERENCES

1. N.C.M. Alma, B.J.M. Harmsen, C.W. Hilbers, G. van der Marel and J.H. van Boom, FEBS Lett. 135:15 (1981).
2. N.C.M. Alma, B.J.M. Harmsen, E.A.M. de Jong, J. van der Ven and C.W. Hilbers, J. Mol. Biol. 163:47 (1983).
3. C. Otto, F.F.M. de Mul, B.J.M. Harmsen and J. Greve, Nucleic Acids Res. 15:6605 (1987).
4. G.D. Brayer and A. McPherson, J. Mol. Biol. 169:565 (1983).
5. G.D. Brayer and A. McPherson, Biochemistry 23:340 (1984).
6. L.R. Dick, A.D. Sherry, M.M. Newkirk and D.M. Gray, J. Biol. Chem. 263:18864 (1988)
7. L.R. Dick, C.F.G.C. Geraldes, A.D. Sherry, C.W. Gray and D.M. Gray, Biochemistry 28:7896 (1989).
8. G.C. King and J.E. Coleman, Biochemistry 27:6947 (1988).

STRUCTURAL AND KINETIC ANALYSIS OF FLAVINE ADENINE DINUCLEOTIDE MODIFICATION IN ALCOHOL OXIDASE FROM METHYLOTROPHIC YEASTS

Leonid V. Bystrykh, Richard M. Kellogg, Wim Kruizinga, Lubbert Dijkhuizen, Wim Harder

Institute of Biochemistry and Physiology of Microorganisms, USSR Acad. Sci., Pushinco, 142292, USSR Department of Organic Chemistry, University of Groningen, 9747 AG, Groningen, The Netherlands Department of Microbiology, University of Groningen, 9751 NN, Haren, The Netherlands

INTRODUCTION

Alcohol oxidase (MOX), a major peroxisomal protein of methanol- utilizing yeasts, has been shown to possess two different forms of flavine adenine dinucleotide, natural FAD and so called modified FAD (mFAD). A comparison of homogeneous preparations of MOX purified from different yeast strains revealed significant differences in the ratios of FAD and mFAD incorporated into the apoenzyme (1-3). Here we present results on the structure determination of mFAD by ^{1}H and ^{13}C NMR, HPLC and enzymatic analysis.

Results

^{1}H and ^{13}C NMR analysis of mFAD revealed that the isoalloxazine ring as well as the adenosine part of the molecule were intact. mFAD showed the same number of carbon atoms as FAD and an identical distribution of methyl, secondary, tertiary, and quaternary carbon atoms. Mild shifts were observed in the isoalloxazine and adenine moieties. However, significant differences were observed in the sugar chains. Two dimensional

Bioorganic Chemistry in Healthcare and Technology, Edited by U.K. Pandit and F.C. Alderweireldt, Plenum Press, New York, 1991

proton-proton spectra (COSY, NOESY, TOCSY) gave complicated pictures both for FAD and mFAD, especially in the sugar region where the overlap of absorptions (7 protons from ribitol and 5 protons from ribose) is severe. On the basis of literature partial assignments (4-6), it was possible to trace and assign both the ribose and ribitol chains in FAD entirely. In mFAD the ribose chain was also followed and assigned. However, the ribitol chain overlapped severely at H2′ with the ribose absorptions; moreover, coupling between H2′and H3′ virtually disappeared. In order to simplify the structural analysis of mFAD, it was cleaved by phosphodiesterase to mFMN and AMP. The latter compound was proven to be authentic both by enzymatic (with myokinase, pyruvate kinase and lactate dehydrogenase) and HPLC analysis. The ^{13}C spectra of FMN and mFMN (Fig.1) show small but important differences in the sugar regions. The connectivities are identical in both cases. Two interpretations seemed plausible at this point: a) the phosphate has migrated on the ribitol chain (7) or b) the sugar moiety in mFMN (and of course, also in mFAD) is not ribitol but a diastereoisomer thereof. Possibility a) was eliminated on the basis of ^{31}P NMR experiments and dephosphorylation to an <u>isomer</u> of riboflavin. This supports supposition b).

The COSY spectra of FMN and mFMN differed significantly (Fig. 2). A similar relationship was observed between riboflavin and m-ribofavine. In both FMN and mFMN the order of chemical shifts remained the same: (proceeding upfield) H1a′, H1b′ (diastereotopic protons) < H2′ < H4′. H5a′, H5b′ (diastereotopic protons) < H3′. However, proton-proton coupling strongly differs in FMN and mFMN. Most striking is the virtual absence of coupling between H2′ and H3′ in mFMN (resulting in a small J value), whereas this coupling is significant in FMN. Moreover, the coupling of H2′ to H1a′ and H1b′ is different in FMN and mFMN. All these observations are consistent with the hypothesis that in mFMN (and also in mFAD) the configuration of C2′ has changed from R to S absolute configuration. This suggests the presence of arabityl sugar chain in mFMN instead of ribityl in an FMN. A possible mechanism for this conversion is as shown in Fig. 3.

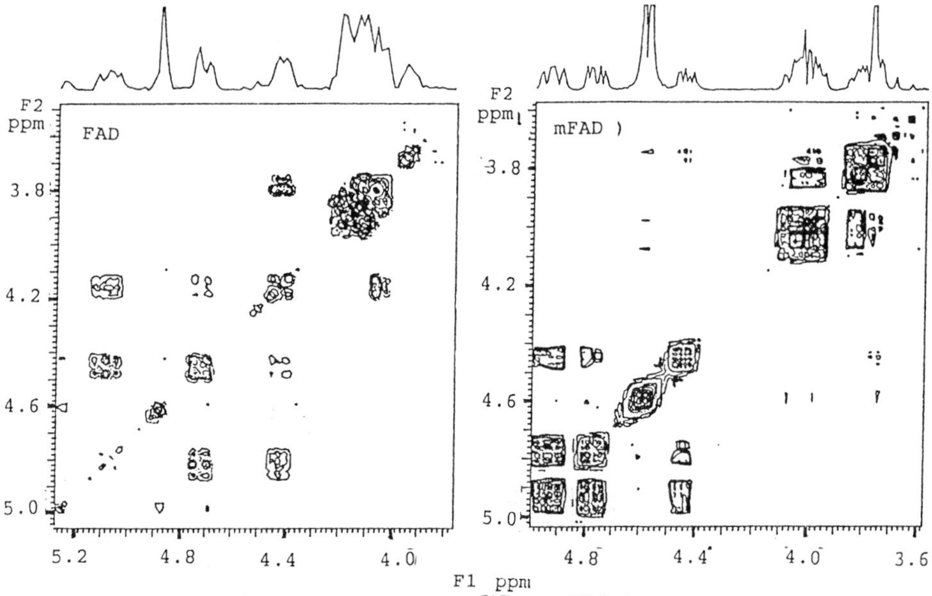

Fig. 1. Decoupled ^{13}C NMR spectra of FMN (left) and mFMN (right)

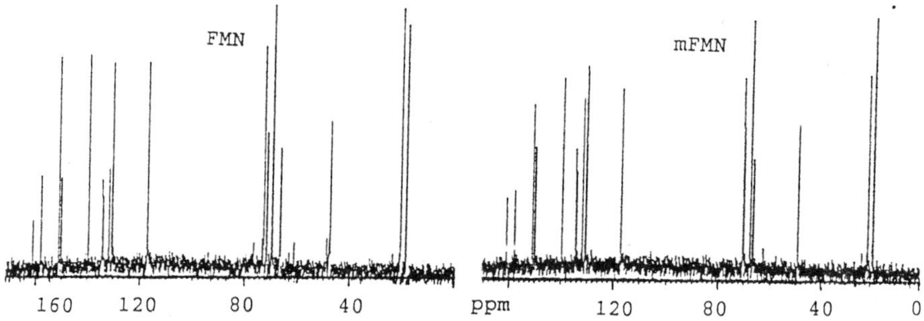

Fig. 2. COSY spectra of FMN (left) and mFMN (right). The diastereotopic protons H1a′, H1b′ are at lowest field in both spectra. The ordering of absorptions is as given in the text.

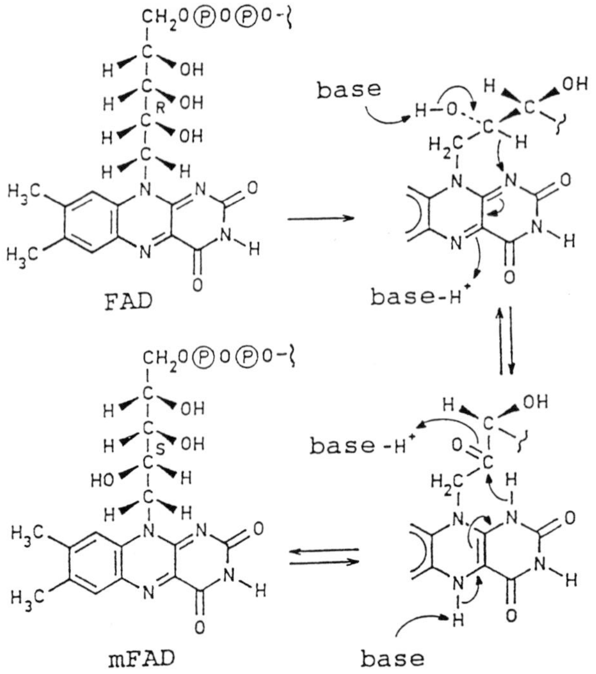

Fig. 3. A scheme for the inversion of configuration of C2′ of FAD.

References

1. Sherry, B., R.H. Abeles. 1985. Biochemistry, 24, 2594-2605.
2. Bystrykh, L.V., V.P. Romanov, J. Steczko, Y.A. Trotsenko. 1989. Biotechnol. Appl. Biochem. 11, 184.
3. van der Klei, I.J., L.V. Bystrykh, W. Harder. 1990. In: Methods in Enzymology, Academic Press Inc., 188, p. 420.
4. Ulrich, E.L., W.M. Westler, J.L. Markley. 1983. Tetrahedron Lett. 24, 473.
5. Breitmaier, E., W. Voelter. 1972. Eur. J. Biochem. 31, 234.
6. Kainosho, M., Y. Kyogoku. 1972. Biochemistry 11, 741.
7. Nielsen, P., P. Rauschenbach, A. Bacher. 1984. In: Flavins and Flavoproteins (R.C. Bray, P.C. Engel and S.G. Mayhew eds.). Walter de Gruyter, Berlin, p. 71.

ISOLATION, CHARACTERIZATION AND UTILIZATION OF PSYCHROPHILIC PROTEINASES FROM ATLANTIC COD

Bjarni Ásgeirsson, Einar Mäntylä,
and Jón Bragi Bjarnason

Science Institute
University of Iceland
Dunhagi 3, IS-107 Reykjavík
Iceland

INTRODUCTION

A mixture of psychrophilic proteolytic enzymes, called Cryotin, has been prepared from a neutral extract of pyloric caeca from Atlantic cod Gadus morhua. This proteinase mixture has many unique characteristics. The proteinases, studied so far, are more active at low temperatures, when compared to their mammalian counterparts. They are also unusually thermo-labile as well as acid labile. Cryotin has been shown to contain trypsin, chymototrypsin, elastase and a collogenolytic enzyme, as well as other proteolytic and peptidolytic activities, but it is practically devoid of lipase, amylase and nuclease activities.

TRYPSIN FROM ATLANTIC COD

Trypsin was purified from Cryotin on an p-aminobenzamidine Sepharose-4B affinity column. The bound trypsin was eluted from the column with 25 mM acetic acid pH 3.2, containing 10mM $CaCl_2$. Trypsin was further resolved into three differently charged species having pI values of 6.6, 6.2 and 5.5 on a chromatofocusing PBE-94 anion exchange column. All three trypsins were found to have similar molecular mass of 24.2 kDa. The amino terminal sequence of cod trypsin enzyme I, the predominant species from Atlantic cod, is similar to that of other known trypsins, in particular the porcine and rat trypsins, with 30 identical residues out of 37, bovine trypsin, with 29 residues identical out of 37, and dogfish trypsin having 26 residues identical out of 37.

Kinetic Properties

The catalytic efficiency of cod trypsin enzyme I at 25°C expressed as kcat/Km was 17 times greater than that for bovine trypsin when these enzymes were assayed as amidases using N-Benzoyl-L-arginine p-nitroanilide as substrate. This was

Bioorganic Chemistry in Healthcare and Technology, Edited by U.K. Pandit and F.C. Alderweireldt, Plenum Press, New York, 1991

revealed as differences in both apparent Km and Kcat values. The amidase activity of the cod trypsin displayed an apparent Km value of 0,077 mM, approximately eight times lower than that measured with the bovine enzyme of 0,650 mM. The maximum rate achieved at 25°C was also greater for the cod enzyme by a factor of two. The esterase activity of the two enzymes, using p-Tosyl-L-arginine methyl ester as substrate, also displayed dissimilar characteristics with the cod enzyme having a Kcat/Km value 2.5 times higher than bovine trypsin.

Thermal Stability

Atlantic cod trypsin demonstrated less resistance to thermal inactivation than bovine trypsin. The maximum temperature at which Atlantic cod trypsin remained fully active for at least 3 minutes was 55°C as compared to 65°C for bovine trypsin, suggestive of less structural stability in cod trypsin which possibly has evolved in response to the need of optimizing kinetic properties at low habitat temperatures. Such structural destabilization would not necessarily be brought about by fewer covalent links, notably disulfide bonds, but may perhaps be due to differences in the weak intramolecular interactions. The number of hydrophobic interactions expressed in terms of the average hydrophobicity were reduced in Atlantic cod trypsin as compared to bovine trypsin.

Acid Stability

Studies on the stability of cod trypsin at various pH values revealed that the enzyme is unstable in acidic solutions. Bovine trypsin is stable at pH 3.0 at low temperatures for weeks. The esterase activity of cod trypsin was quite stable in alkaline medium but displayed a marked acid lability. Esterase activity was lost when pH was lowered below pH 5.0, and this effect was apparent after 30 min (Ásgeirsson et al., 1989).

CHYMOTRYPSIN FROM ATLANTIC COD

Cod chymotrypsin was isolated on a Phenyl-Sepharose column following trypsin removal with the benzamidine affinity resin. Elastase was eluted of the phenyl-Sepharose column with a 25 mM tris buffer pH 7.5 containing 10 mM $CaCl_2$ and 20% (v/v) ethylene glycol, followed by chymotrypsin release from the column by washing with a 50% (v/v) ethylene glycol solution containing 20 mM $CaCl_2$. Chymotrypsin was further resolved into two differently charged species with isoelectric points of 6.2 (enzyme A) and 5.8 (enzyme B), but a similar molecular mass of 26 kDa. However, chymotrypsin B was clearly slightly larger than chymotrypsin A. The cod enzymes differed from bovine chymotrypsin (pI 8.5) in having more acidic isoelectric points and being unstable in weakly acidic solutions.Mammalian chymotrypsins have been found to be very stable in acidic solutions of pH 3.0, while the cod chymotrypsin displayed marked acid lability at pH values below 5.0.

Kinetic Properties

The kinetic properties of cod chymotrypsin were compared to those of the bovine enzyme. Initially when N-Benzoyl-L-tyrosine ethyl ester was used as substrate in the ester hydrolysis reaction the assay was performed in the presence of 25% methanol. The presence of alcohols and other nucleophiles in the assay mixture of serine proteinases is known to cause competition with water for the hydrolysis of the acyl-enzyme intermediate which affects apparent values obtained for kinetic constants. In view of this N-Benzoyl-L-tyrosine ethyl ester was dissolved in 10% dimethylsulfoxide instead of methanol, resulting in the reduction of the apparent Km values, whereas kcat values were increased five-fold for both cod and bovine chymotrypsin. Reducing the methanol concentration from 25% to 10% caused a doubling in kcat and a reduction in Km by a half for both enzymes. All measurements of chymotrypsin activity in the present study were performed under the latter conditions (i.e. 10% methanol), but clearly it would seem advantageous to replace methanol as solvent with a solvent such as dimethylsulfoxide which does not participate in esterase reactions.

The cod enzymes were found to be more active than bovine chymotrypsin towards both ester and amide substrates. The pseudo second order rate constant kcat/Km is about 3 to 4 fold higher for cod chymotrypsin than bovine chymotrypsin when ester hydrolysis is measured using N-Benzoyl-L-tyrosine ethyl ester as substrate at 25%C. When the amide substrate N-Benzoyl-L-arginine p-nitroanilide was employed, the cod chymotrypsin yielded two fold higher values of catalytis efficiency than the bovine enzyme. These values remained constantly higher for the cod enzymes at all temperatures measured, within the thermal stability of the enzymes. Under the experimental conditions employed the loss of half-maximal activity occurred at 52°C for the cod enzyme, compared to 57°C for the bovine enzyme.

Stability in Organic Solvents

The catalytic activity of cod chymotrypsin was studied in various organic solvents of differing concentrations. First the effects of ethylene glycol on the tolerance of cod chymotrypsin towards freezing was tested. The esterase activity of the enzyme was well preserved through three repetitive freezing trials (-26°C) in 25% ethylene glycol, with an activity loss of only about 20%. If 50% ethylene glycol was used no activity loss was detected. It should be noted that the 25% ethylene glycol solution is solid at -26°C, while the 50% solution is not frozen solid but is highly viscous.

The stability of cod chymotrypsin in organic solvents was measured as residual esterase activity at 25°C after incubation of the enzyme in organic solvents at ratios of 25% and 50% (v/v) in aqueous buffer for up to 30 days at 4°C. The organic solvents used in the experiment were dimethyl sulfoxide, dioxane, glycerol, methanol, ethanol, 1,3-propanediol, acetonitrile and dimethyl formamide. The cod enzyme retained constant activity for the total duration of the experiment of the 20 days tested in dimethyl sulfoxide, dioxane and glycerol solutions, and 30 days in the other organic solutions. The residual activity was approximately the same in all the organic solutions as in the aqueous buffer standard, except in dioxane, where the activity dropped immediately to about 20% of the value of the standard, but remained at that level for the duration of the experiment. The stability of cod chymotrypsin was thus measured in higher concentrations of dimethyl formamide. The enzyme remained stable in 60% dimethyl formamide for 30 days at 4°C, but rapidly lost activity in 70% and higher concentrations of the solvent.

ELASTASE AND COLLAGENASE FROM ATLANTIC COD

Elastase from cod has been purified to a high degree and is presently being investigated. Preliminary data suggests that it also is a psychrophilic proteinase. A collagenase, active on native interstitial collagen, Type I collagen, has also been isolated from the pyloric caeca of Atlantic cod and is now being purified and characterized. The cod collagenase appears to be a low molecular mass (30 kDa) serine proteinase.

CONCLUSION

Cryotin, a mixture of proteinases from Atlantic cod, has many unique characteristics for a pancreatic enzyme mixture. It contains practically no lipase, amylase or nuclease activities, which may be due to proteolytic breakdown of these enzymes in the initial homogenate. The proteinases are more active at low temperatures than their mammalian counterparts. They are thermo-labile and acid sensitive, and therefore easy to deactivate after use with mild heat treatment or slight acidification. Cryotin has many potential applications in industry and medicine, especially in food processing which require hydrolysis at low temperatures or inactivation under mild conditions. It has proven promising in various fish processing applications such as deskinning of fish, removal of membranes and ripening of herring. Cryotin also potential as a digestive aid, both for humans and animals, and could be used to assist in the natural healing of wounds.

REFERENCE

Ásgeirsson, B., Fox, J.W., and Bjarnason, J.B. 1989, Eur. J. Biochem., 180, 85-94.

CONFORMATIONAL ANALYSIS OF LEUKOTRIENES AND RELATED COMPOUNDS FOR MAPPING THE LEUKOTRIENE D_4 RECEPTOR: APPLICATION TO THE DESIGN OF NOVEL ANTI-ASTHMA DRUGS

R.N. Young, R. Zamboni, H. Williams, M. Bernstein and K. Metters

Merck Frosst Centre For Therapeutic Research

16711 Trans Canada, Kirkland, Quebec, Canada, H9R 4P8

INTRODUCTION

Leukotrienes D_4 (1a) and E_4 (1b) are potent contractile substances which have been implicated in the ethiology of human asthma. These leukotrienes exert their effects through interaction with specific receptors on pulmonary smooth muscle and other tissues and thus the design of specific antagonists of these receptors offers the potential for the discovery of novel anti-asthma drugs. We have undertaken to characterize and map the LTD_4 receptor via a variety of means. These include (a) analysis of structure-activity of the leukotrienes and of known LTD_4 antagonists; (b) physical characterization of the receptor using radioactive labelled agonists and antagonists and (c) NMR and molecular modelling conformational analysis studies on leukotriene agonists and antagonists. The initial studies allowed us to propose a hypothetical model of the LTD_4 receptor which was instrumental in the design of MK-571 (2), a LTD_4 receptor antagonist which has undergone clinical testing. We have subsequently attempted to further refine this model using MK-571 and novel synthetic agonists at the LTD_4 receptor as tools for such studies.

STRUCTURE-ACTIVITY STUDIES OF COMPOUNDS RELATED TO LTD_4

A large number of analogs of the leukotrienes C_4, D_4 and E_4 have been prepared over many years in order to define the structural requirements for potent binding at the LTD_4 receptor. These studies have been reviewed[1] and based on this information we were able to construct a hypothetical model of the LTD_4 receptor pharmacophore (Figure 1). The correlation of biological activity of MK-571 and its enantiomers suggests that the amide carbonyl in this antagonist and in LTD_4 itself is an important recognition site[2].

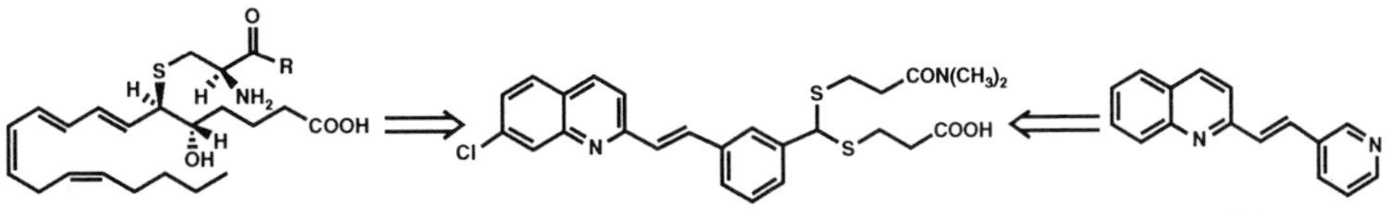

LTD_4 (1a) : R= $NHCH_2COOH$ MK-571 (2) (3)

LTE_4 (1b) : R= OH

Bioorganic Chemistry in Healthcare and Technology, Edited by U.K. Pandit and F.C. Alderweireldt, Plenum Press, New York, 1991

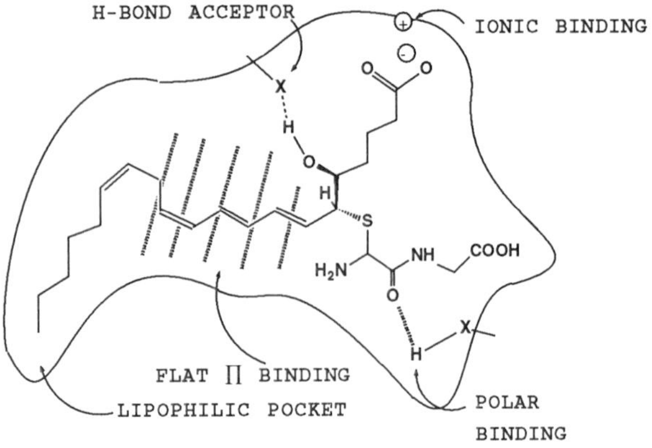

FIGURE 1. HYPOTHETICAL LTD_4 RECEPTOR

Literature examples show that at least one carboxy group is required for potent binding at the receptor and the removal of the hydroxy group of LTD_4 or analogs leads to up to 100-fold loss in biological activity[3,4].

APPLICATION OF THE MODEL TO THE DESIGN OF POTENT LTD_4 ANTAGONISTS

In spite of its simplicity, this receptor model has proved very useful in the design and discovery of MK-571, one of the most potent LTD_4 antagonists known to date. Thus simple structural lead (2-(2-(3-pyridinyl)ethenyl)quinoline(3)) discovered in large scale screening of the Merck sample collection, (IC_{50} = 3 μM versus [^{3}H]-LTD_4 binding) was hypothesized to occupy the flat central lipophilic binding region of the LTD_4 receptor. We proposed that enhanced activity could be obtained by introduction of two polar chains and by addition of lipophilic components. Extensive medicinal chemistry studies derived the final structure of MK-571 (2) which with IC_{50} = 1 nM versus [^{3}H]-LTD_4, represents a 3,000-fold increase in activity relative to the initial structural lead. MK-571 fits well with this model in that it contains a planar lipophilic component with extended conjugation coupled to two polar chains, one ionizable, the other not.

DEVELOPMENT OF A [^{35}S]MK-571 RECEPTOR BINDING ASSAY

In order to further characterize the LTD_4 receptor we have developed a receptor binding assay based on [^{35}S]MK-571 as the radioligand. This assay has allowed us to compare the ability of a variety of ligands to compete for the specific binding of both the antagonist, [^{35}S]MK-571, and the agonist, [^{3}H]LTD_4, to the LTD_4 receptor, and, therefore, to characterize more fully the properties of this receptor. The radiolabelled ligand was readily prepared from elemental sulfur $^{35}S_8$ using novel methodology, as previously described[5].

The radioreceptor binding assay was developed using [^{35}S]MK-571 (specific activity: 1,350 Ci/mmol) incubated with guinea-pig lung membranes under conditions previously reported for [^{3}H]LTD_4 receptor binding assays[6]. Specific binding was defined as the difference between total binding and non-specific binding determined in the presence of an excess (1000 fold) of unlabelled competing ligand, either LTD_4 or MK-571. Scatchard analysis of [^{35}S]MK-571

Table 1. Competition for $[^{35}S]$MK-571 and $[^{3}H]LTD_4$ binding by agonists and antagonists

	IC_{50} (nM) $[^{35}S]$MK-571	$[^{3}H]LTD_4$
Agonists		
LTD_4	1.0	0.32
LTC_4	53	54
LTE_4	6	3.5
5R,6S-LTD_4	40	19
YM-17690	-	0.06
Antagonists		
(+/-) MK-571	1.3	3.4
S-(+) L668018	2.5	0.77
R-(-) MK-679	5.5	3.2
ICI-198615	1.0	2.1
SKF-104353	10	20

binding, using either LTD_4 or MK-571 as competing ligand, revealed very similar binding characteristics, with K_D values of 12.5 and 18.3 pM and B_{max} values of 49 and 72 fmole/mg protein, respectively. Specific binding of both $[^{3}H]LTD_4$ and $[^{35}S]$MK-571 to the LTD_4 receptor was stimulated by divalent cations. However, although the specific binding of $[^{3}H]LTD_4$ was inhibited by sodiums ions, the specific binding of the antagonist was enhanced in the presence of this additive. Similarly, whilst the specific binding of $[^{3}H]LTD_4$ to guinea-pig lung membrane was inhibited by analogs of GTP (GTPγS and GMP-PNP) these compounds had no effect on the specific binding of $[^{35}S]$MK-571. When binding assays were performed to evaluate the ability of LTD_4 receptor antagonists and agonists to compete for $[^{3}H]LTD_4$ and $[^{35}S]$MK-571 specific binding comparable IC_{50} values were obtained, Table 1.

These and other data have allowed us to propose the following model for the specific binding of agonists and antagonists to the guinea-pig lung LTD_4 receptor. The LTD_4 receptor in this tissue behaves as a classical GTP binding (G)-protein linked receptor, in that high affinity binding of agonists is observed only when the receptor exists as a G-protein-receptor complex. In the presence of compounds, such as sodium ions and GTP analogs, which dissociate the G-protein from the receptor, the receptor moves to a low affinity state and high affinity agonist binding is abolished. In constrast the antagonist MK-571 exhibits high affinity binding in both the G-protein associated and uncoupled receptor states.

CONFORMATIONAL ANALYSIS AND MODELLING STUDIES

Conformational analysis on LTD_4 or LTE_4 has proved difficult due to the large number of freely rotatable bonds and high degree of conformational mobility of the molecule. NMR studies have been published which clearly demonstrate that the polyene lipophilic backbone exists in a generally transoid, planar extended conformation[7]. No physical evidence has been presented to define the preferred orientation of the polar chains in space. A greater understanding of the bioactive conformation of LTD_4 would be of great utility to the design of novel LTD_4 antagonists.

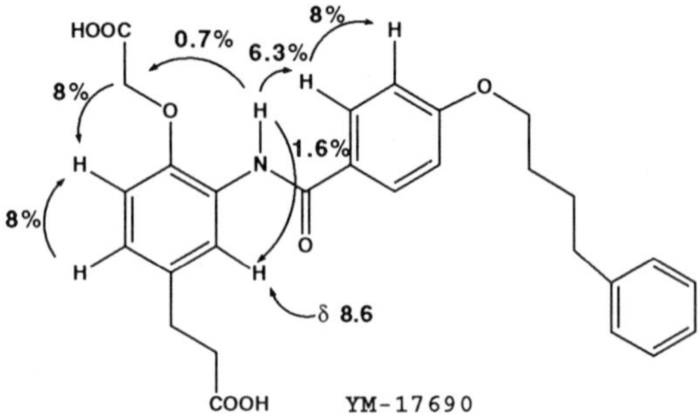

FIGURE 2. NOe STUDIES ON YM-17690

a) Studies on YM-17690

A potent agonist at the LTD_4 receptor[8], YM-17690, has recently been described (see Table I) and offers a unique opportunity to define the agonist conformation of leukotrienes at the receptor. YM-17690, unlike the natural ligand, is highly rigidized with relatively few degrees of freedom. We therefore undertook an NMR analysis to define by NOe experiment the solution conformation of YM-17690. The results of these experiments are depicted in Figure 2.

NOe experiments and MM2 calculations were used to quantitatively report on the conformational bias of the molecule. The following conclusions were made:

The data support the conclusion that the molecule is largely planar in the arylamide region with a trans amide bond.

The ether side chain is felt to prefer a planar conformation and NOe and steric arguments suggest that the conformation depicted here with the chain transoid from the amide group is preferred.

There is little reason to suggest significant conformational restraints on the rest of the molecule although one might reasonably expect extended transoid conformations to be preferable for the propionic acid side chain and the phenylbutoxy group. Using ^{13}C T_1s we examined the motional freedom of various parts of the molecule. These suppositions were largely confirmed.

b) Molecular Mechanics Calculations and Modelling of the Active Conformation of LTE_4

Based on the conformations derived from the NOe experimentation on YM-17690, truncated versions were modelled using MM2 calculations and families of low energy conformers were obtained. The lowest energy conformation (which agreed with results of the NMR studies above) was notable in that the two carboxyl groups are placed almost maximally distant from each other (9.2Å). We propose that in the agonist conformation of LTE_4 at the receptor site the distance between the two carbonyl groups of the polar chains should be similar (ie: ≈9Å) to that observed for the YM-17690.

CONCLUSIONS

A model of the LTD_4 receptor and of the active conformations of leukotrienes has been derived by study of structure activity data, the conformational analysis on rigidized leukotriene agonists and on leukotrienes themselves. These models have already proven useful in the design and development of a potent LTD_4 antagonist such as MK-

571. Further refinements to the understanding of the spacial arrangement of binding units in the receptor has been derived from molecular modelling and NMR studies on the leukotriene receptor agonist YM-17690. Biochemical probes derived from MK-571 have allowed us to characterize the receptor as a G-protein linked receptor which probably exists in two conformation states.

Further studies are ongoing in attempts to isolate and sequence and clone the receptor itself.

REFERENCES

1. J. Rokach, Y. Guindon, R.N. Young, J. Adams and J.G. Atkinson, "Synthesis of Leukotrienes" in "The Total Synthesis of Natural Products", Volume 7, E.J. Apsimon, Wiley-Interscience, New York, 144-273 (1988).

2. J.Y. Gauthier, T. Jones, E. Champion, L. Charette, R. Dehaven, A.W. Ford-Hutchinson, K.K. Hoogsteen, A. Lord, P. Masson, H. Piechuta, S.S. Pong, J.P. Springer, M. Thérien, R. Zamboni and R.N. Young, J. Med. Chem. 33, 2841-2845 (1990).

3. E.J. Corey and D.J. Hoover, Tetrahedron Lett., 23, 3463 (1982).

4. J.G. Gleason, R.F. Hull, C.D. Perchonock, K.F. Erhard, J.S. Fraszee, T.W. Ku, K. Kondrad, M.E. McCarthy, S. Mong, S.T. Crooke, G.G. Chi-Rosso, M.A. Wasserman, T.J. Torphy, R.M. Muccitelli, D.W. Hay, S.S. Tucker, L.J. Vickery-Clark, J. Med. Chem. 30, 959 (1987).

5. H.W.R. Williams, R.N. Young and R. Zamboni, J. Labelled Comp. Radiopharm., 28, 297 (1990).

6. K.M. Metters, E.A. Frey and A.W. Ford-Hutchinson, Eur. J. Pharmacol (in press)

7. M. Sugiura, H. Beierbeck, G. Kotovych and P.C. Belanger, Can. J. Chem. 62, 1640 (1984).

8. K. Tomioka, T. Yanada, K. Teramura, M. Terai, K. Hidaka, T. Mase, H. Hai, and K. Murase, J. Pharm. Pharmacol. 39, 819 (1987).

Closing Session

In the following account a summary is presented of the introductions given by the chairman, U. K. Pandit and the panel members H. J. Kooreman, A. Marquet, K. Müller and C. J. Suckling and of the general discussion which followed.

U. K. Pandit

The aim of the ARW, as visualized by the organizers, was to examine the translation of the molecular mechanisms of enzymes and biological processes into their implication in and impact upon the development of technological processes and the evolution of health-care agents. New developments in this area have in the first instance been stimulated to a large extent by our understanding of biological processes. However, in the context of further development, today interesting receptors and catalytic systems are being developed, which partly or completely depart from a biological heritage.
At the end of the scientific proceedings it is important to ask ourselves the following questions. In the light of the deliberations of the ARW, what are the identifiable gaps of knowledge in the field and which areas need to be emphasized in future research.

H. J. Kooreman

Kooreman indicated that his comments were presented as seen from his industrial R & D background.
In an ARW audience like this one there is enough synergy present to combine each other's knowledge and to integrate it into new concepts. This is how some fruitful concepts really grow and others slowly die. Looking at the title of the workshop, while all of the presentations certainly were in the field of bioorganic chemistry, sometimes the relationship with health-care and quite often with technology was rather remote. This is understandable because for the most part the topics discussed in this ARW are rather scattered pieces of science. It would be nice, to find a certain homogeneïty and to be able to clearly define a common denominator. This is certainly possible, since most of the lectures have focussed on (i) selectivity (enantioselectivity or regioselectivity) and (ii) the interaction of small molecules with macromolecules. On the frontiers of the field, some of the lectures have discussed the conceptually new idea on construction and properties of "programmed structures", exemplified by catalytic antibodies (Benkovic), supramolecular catalysts (Lehn) and the functional active site-like polymer cavities (G. Wulff).
The study of the macromolecular receptor-substrate interactions and the attempts to find the best fit of the small molecule, can be called a rational approach to drug design, although real drug design in the basic sense of the word is still far away. The best we can have at the moment is the possibility of avoiding random synthesis and steering the

synthesis in a specific direction. But a lot more best-fit screening and testing will be needed before we will be able to design active compounds : based upon concepts of a fit between a substrate and a receptor molecule. This still does not give us a drug, because there is much more to be done before a chemical can be considered as a drug. It will, however, be very worthwile to continue along these lines.

A. Marquet

As a University representative, I feel that, in spite of the title of the ARW, the main function of an audience like the one present is not the search for short term applications, but to gain a better understanding of biological systems. This does not mean, that one is not interested in applications. Research on the understanding of fundamental chemical and biological processes is indeed now entering into the area of applications, because there has been a big evolution in the last 10 years.

The topics of this conference covered two fields. The first one is biomimetic chemistry or the chemistry of model enzymes and - quoting J. M. Lehn - the problem is not to mimic enzymes or to make enzyme models, but just to use broad general concepts of enzymic catalysis to create completely different new systems. A lot of success has been reached in this area: asymmetric synthesis is now quite efficient. It has found its source in reflections upon enzymatic reactions. Supramolecular chemistry is now starting to have applications. These applications, however, are all based on old concepts.

The second topic is that of biological systems. Here again two aspects can be considered separately: enzymes and receptors. In the enzymic field a lot is known already, but although much work is still to be done, in this field also there is nothing really new. All the concepts are already present. In the receptor field, however, the situation is completely different and actually this is something that has not been sufficiently covered during this conference. A few receptors have been isolated. We can now have an idea about the receptor structure, just by taking the envelope of active and inactive molecules. One can hope that in the next years many more receptors will be cloned and isolated and that the interactions with small molecules will attain the same level of understanding as we now have of the enzymes. In this area much work needs to be done to catch up with the state of art achieved in the field of enzymes. This will require the close collaboration of organic chemists with biochemists and with biologists at a highly organized level because the big problems for the next 10 years make such an approach necessary. This opinion is not always shared by Industries. Universities should do something about this, especially university people who teach future managers of industry.

K. Müller

[The block diagram given below was constructed on the spot and commented by Dr. Müller. It is considered by him as an overview of the most important research and technology activities in the domain of health-care. The position of the activities or features in the diagram also gives their logical interdependence and the sequence to be followed in order to obtain new developments.]

The starting point of the diagram is the health-care issue. Much of the success of a study on a medicinal problem depends on how specifically it is identified. There is a diagnostic aspect, a prevention aspect, there is medication and there is cure. At some level folk-medicine comes into the picture. From the folk-medicine one recognizes opportunities in

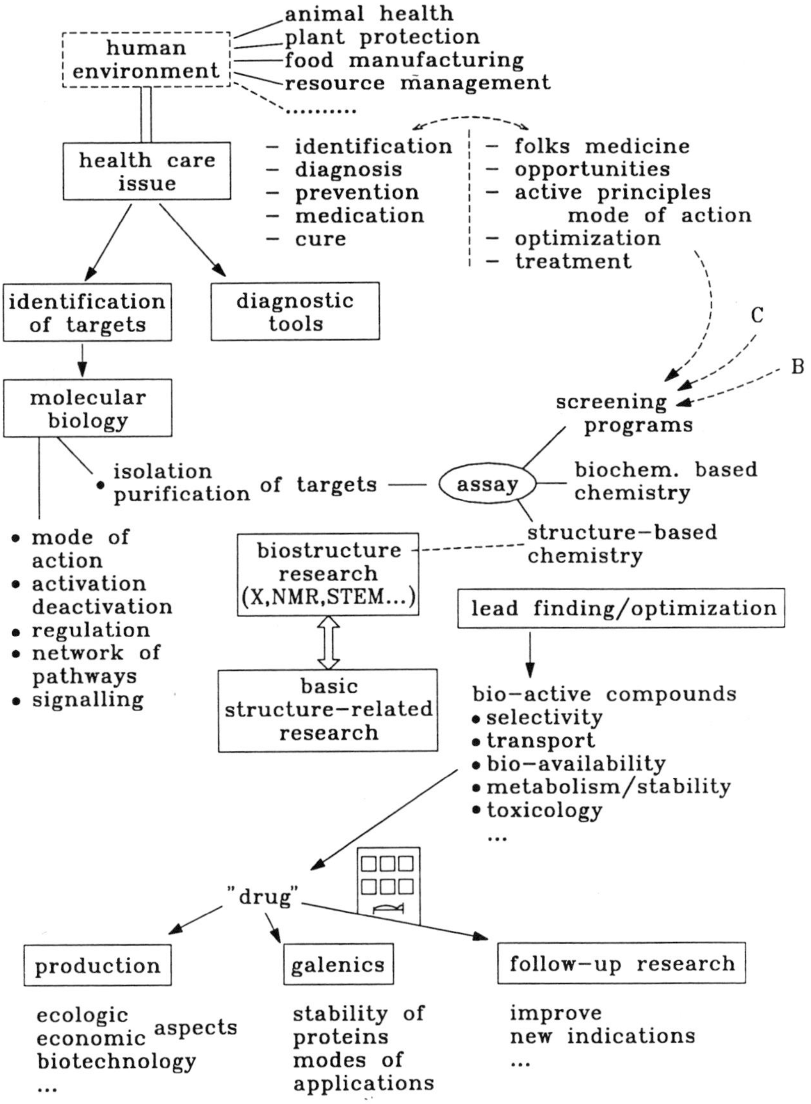

human environment
animal health
plant protection
food manufacturing
resource management
....
health care issue
– identification
– diagnosis
– prevention
– medication
– cure
– folks medicine
– opportunities
– active principles mode of action
– optimization
– treatment
C
B
identification of targets
diagnostic tools
molecular biology
screening programs
• isolation purification of targets
assay
biochem. based chemistry
structure-based chemistry
• mode of action
• activation deactivation
• regulation
• network of pathways
• signalling
biostructure research (X,NMR,STEM...)
lead finding/optimization
basic structure-related research
bio-active compounds
• selectivity
• transport
• bio-availability
• metabolism/stability
• toxicology
...
"drug"
production
galenics
follow-up research
ecologic economic aspects
biotechnology
...
stability of proteins
modes of applications
improve
new indications
...

health-care and from there one can try to isolate the active principles, the mode of action, optimize these to become more specific and optimize treatment. Another approach consists in identifying the targets : where is one able to interfere and which is the safest and the most efficient way to interfere. Parallel to this, all diagnostics related to a given issue are to be set up. Then there are a number of biochemical and biological events which need to be clarified, as much as possible on a structural level. The interplay of these pieces of information can give extremely important clues for the chemical programs to be set up. Next comes the isolation and purification of targets. These targets can be enzymes, receptors, DNA, RNA etc.. All instrumental methods useful for biostructural research have to be combined. Looking at elucidated structures, such as of inhibitor complexes or other macromolecular complexes, provides valuable insights and ideas for the design of new molecules. The availability of purified target systems is a prerequisite for setting up specific and well thought out assays.
Besides chemical programs also bacterial and fungal broths can be used for generating thousands of interesting metabolites. Furthermore, the hundreds of thousands of compounds already available in the industrial collections can be screened or rescreened in the light of new applications.

Once a bioactive compound has been identified, all the available information should be used to optimize the structure, taking into account subsequent problems such as selectivity, transport, bioavailability, metabolism, stability, toxicology, etc.. While most of these optimizations have still to be done empirically, some phenomena can already be understood at a molecular level and the situation will certainly improve in the decade to come. A new product which makes its way to a clinically tested drug raises questions of production, which are linked to aspects of ecology, economy and biotechnology. Galenics in the past has been somewhat relegated to the dark area, but has now matured to a highly sophisticated and "molecular" discipline. A particularly challenging problem is the safe administration of bioactive proteins, which implicates much more than classical galenics. It should not be forgotten, that health-care does not only include finding a drug, but also many technologies such as surgery, blood substitution, skin subtitution etc.. Finally, in health-care there is the issue "human environment", which not only means man itself, but the "whole human environment". There is animal health, plant protection, food manufacturing, resource management and many other such things which need to be considered.

It is evident that 'health-care and technology' defines an enormously vast area for bioorganic chemistry. So, it comes without surprise that only some topics could be touched during one single workshop.

C.J. Suckling

After making some general remarks on the investment of finances and time in the organization and attendance of scientific meetings, he continued with the following comments.
There is no doubt that the interaction between chemistry and biology has been one of the most exciting and powerful driving forces in the recent development of chemistry. The enormous advances in enantioselective synthetic chemistry would never have been brought forward as fast as they have if there had not been the challenge to the chemists to harness the potential of enzymes. Biological chemistry certainly must continue to provide a major drive forward for all areas of chemical science, the molecular aspects being the most important ones to focus on by teachers in universities.
One of the areas that should get more attention is molecular biology of the plant

kingdom, because of the enormous production possibilities of plants. It would be fascinating to get to the same level of knowledge about plant enzymes as the one reached for some mammalian and bacterial enzymes. The incisive molecular studies that one can do at the enzyme level should be extended much more into the cellular level and obviously technologies such as NMR would play a major part in that sort of development.
Perhaps, if one is focussing on biological chemistry as the main theme of meetings, then a series that mainly deals with biological chemistry in the context of health-care on one hand and environment on the other would be a reasonable way to progress. In a following meeting like this ARW, two strands should be followed : one is the technology with which one tries to make things and the other is the science of chemical and biological interactions. These two contiguous fields have many points of technical and conceptual contact and in meetings of this sort of length they seem to form a very happy balance because there are a lot that people involved in one area can learn from the other area.

General discussion

A number of participants took part in the open discussion which followed. The main points raised in this discussion are summarized below.

It was felt that there was a serious lack of basic knowledge of how molecules interact. This is important with reference to a further understanding of molecular recognition and transmission/amplification of signals. While some (empirical) notions of molecular recognition have been developed, this subject is still an art rather than a science. The new tools of molecular modelling and spectroscopy like NMR are of help in this field. A further black box relates to the transmission of the signal - following the substrate-receptor association - to the cellular level. Emphasis should also be placed on the study of dynamics and regulation in open systems, especially in connection with the function of drugs. Furthermore, it is necessary to vastly improve and increase our knowledge of receptors.

The basic molecular level information on the mechanism of enzymes is highly relevant to both health-care and the development of new methodologies for chemical transformation. However, for the latter, the judicious combination of chemical, biochemical and biological methods may constitute the optimum approach. It was also noted that despite our incomplete understanding of recognition, the empirical ideas on the phenomenon have led to its application in the development of molecular devices, in material science and in industrial use, as in the case of cyclodextrins.

The necessity for a broader vision on the part of chemists was stressed by a number of participants. Chemistry is the link between physics and biology, was one significant quotation. It is necessary to keep our eyes wide open about what we can bring in from other areas. To this end there is a big need for setting a stage for interdisciplinary research and communication. The importance of the role of public opinion towards science was also voiced. It could become a key issue in the next decennia. Consequently, it was considered an important task that chemists took it upon themselves to make clear to the public that their work had a positive contribution to the society.

PARTICIPANTS

ALDERWEIRELDT F.C.
R.U.C.A.
Organic Chemistry Laboratory
Groenenborgerlaan 171
2020 Antwerpen
Belgium

AZERAD R.
Université René Descartes
Laboratoire de Chimie et de
Biochimie Pharmacologique
45, rue des Saint-Pères
75270 Paris Cédex 06
France

BALDARO E.
Sclavo S.p.A.
Biochemical Division DE.BI
S.S. Padona Superiore km 160
I-20060 Cassina de' Pecchi (Mil)
Italy

BALLESTEROS A.
Spanish Research Council
Institute of Catalysis
Serrano 119
28006 Madrid
Spain

BENKOVIC S.J.
The Pennsylvania State University
Department of Chemistry
University Park, PA 16802
U.S.A.

BENVENISTE P.
Université Louis Pasteur
Laboratoire de Biochimie
Végétale
28, rue Goethe
67083 Strasbourg Cédex
France

BJARNASON Jon.B.
University of Iceland
Science Institut
Dept. of Chemistry
Reykjavik
Iceland

CASTEDO L.
Universidad De Santiago
Departamendo De Quimica Organica
Facultad De Quimica
Santiago
Spain

DOUGLAS T.D.
University of Manchester
Department of Pharmacy
Manchester, M13 9PL
United Kingdom

FEENEY J.
National Institute of
Medical Research
The Ridgeway Hill, Mill Hill
London, NW7 1AA
United Kingdom

FRÈRE J.M.
Université de Liège
Faculté des Sciences
Labo d'Enzymologie
Institut de Chimie - B6
Sart Tilman
B-4000 Liège
Belgium

FUGANTI C.
Politechnico di Milano
Dipartemento di Chimica
P.zza Leonardo da Vinci, 32
I-20133 MILANO
Italy

GAUDRY M.
Université P. et M. Curie
Laboratoire de Chimie
Organique Biologique
4, Place Jussieu
75252 Paris Cédex 05
France

GERALDES C.F.
Universidade De Coimbra
Departamendo De Quimica
3049 Coimbra
Portugal

GORRICHON L.
Université Paul Sabatier
Synthèse et Physicochimie
Organique
118, Route de Narbonne
31062 Toulouse Cédex
France

HACKSELL Ü.
Uppsala University
Department of Organic
Pharmaceutical Chemistry
Box 574
751 23 Uppsala
Sweden

JANSSEN M.
Janssen Pharmaceutica
Turnhoutseweg 30
2340 Beerse
Belgium

JOMMI G.
Universita' Degli studi di Milano
Dipartimento di Chimica Organica
e Industriale
Via G. Venezian 21
20133 Milano
Italy

KAASGAARD S.
NOVO Industri A/S
Novo Allé
2880 BAGSVAERD
Denmark

KELLOGG R.M.
University of Groningen
Organic Chemistry Laboratory
Nijenborgh 16
9747 AG Groningen
The Netherlands

KOOMEN G.-J.
University of Amsterdam
Organic Chemistry Laboratory
Nieuwe Achtergracht 129
1018 WS Amsterdam
The Netherlands

KOOREMAN H.J.
International Bio-Synthetics
Patentlaan 3
2288 EE Rijswijk
The Netherlands

KUTNEY J.P.
University of Britisch Columbia
Dept. of Chemistry
2036 Main Mall
Vancouver, B.C.
Canada V6T 1Y

LEHN J.-M.
Université Louis Pasteur
Institut le Bel
4, rue Blaise Pascal
67000 Strasbourg
France

LEMIÈRE G.L.
R.U.C.A.
Organic Chemistry Laboratory
Groenenborgerlaan 171
2020 Antwerpen
Belgium

MARQUET A.
Université P. et M. Curie
Laboratoire de Chimie
Organique Biologique
4, Place Jussieu
75252 Paris Cédex 05
France

MÜLLER K.
Hoffmann-La Roche AG
Central Research Units
B65/316
CH-4002 Basel
Switzerland

PANDIT U.K.
University of Amsterdam
Organic Chemistry Laboratory
Nieuwe Achtergracht 129
1018 WS Amsterdam
The Netherlands

PEGGION E.
University of Padua
Dept. of Organic Chemistry
Padua
Italy

POULSEN P.B.
NOVO Industri A/S
Novo Allé
2880 BAGSVAERD
Denmark

RAHIER A.
Université Louis Pasteur
Institut de Botanique
28, rue Goethe
67083 Strasbourg Cédex
France

RÉTEY J.
Universität Karlsruhe
Institut für Organische Chemie
Postfach 6380
7500 Karlsruhe 1
W. Germany

SANTANIELLO E.
Universita di Milano
Dipartimento di Chimica e
Biochimica Medica
Via Saldini, 50
20133 Milano
Italy

SCHMIDTCHEN F.P.
Techn. Universität München
Organ. Chem. Institut
Lichtenbergstrasse 4
8046 Garching
W. Germany

SCHNEIDER H.
Universität des Saarlandes
Fachrichtung Biochemie
6600 Saarbrücken
W. Germany

SCHNEIDER H.-J.
Universität des Saarlandes
Institut für Organische
Chemie
6600 Saarbrücken
W. Germany

SCHOEMAKER H.E.
DSM Research
Bio-organic Chemistry Section
P.O.Box 18
6160 MD Geleen
The Netherlands

SCOTT A.I.
Texas A & M University
Department of Chemistry
College Station
Texas 77843-3255
U.S.A.

ŞENER B.
Gazi University
Faculty of Pharmacy
PK. 143 06572
Maltepe, Ankara
Turkey

SUCKLING C.J.
University of Strathclyde
Dept. of Pure and Applied
Chemistry
295 Cathedral Street
Glasgow, G11XL
Scotland, United Kingdom

VAN DAMME E.
R.U.G.
Labo voor Algemene en Industriële
Microbiologie ; Landbouwfaculteit
Coupure links 653
Gent
Belgium

WULFF G.
Heinrich-Heine-Universität
Institut für Organische Chemie
und Makromolekulare Chemie
Universitätsstrasse 1
4000 Düsseldorf
W. Germany

YOUNG R.N.
Merck Frost Canada Inc.
P.O.Box 1005
Ponte Claire-Dorval
QUEBEC H9R 4P8
Canada

AUTHOR INDEX

SUBJECT INDEX*

*The page numbers indicated correspond to the title-page of the articles.